SEABIRD ENERGETICS

A white tern (*Gygis alba*) returns to its chick with a beakful of fish
(Photograph by George H. Balazs)

SEABIRD ENERGETICS

Edited by

G. Causey Whittow

University of Hawaii
Honolulu, Hawaii

and

Hermann Rahn

State University of New York
Buffalo, New York

Plenum Press • New York and London

Library of Congress Cataloging in Publication Data

Main entry under title:

Seabird energetics.

"Based on the proceedings of a symposium sponsored by the Comparative
Physiology Section of the American Physiological Society, held August 23-24, 1983, in
Honolulu, Hawaii" — T.p. verso.
   Bibliography: p.
   Includes index.
   1. Seabirds — Physiology — Congresses. 2. Bioenergetics — Congresses. 3. Birds —
Physiology — Congresses. I. Whittow, G. Causey, 1930-   . II. Rahn, Hermann,
1912-   . III. American Physiological Society (1887-   ). Comparative Physiology
Section.
QL698.S364   1984   598.29′24   84-15998
   ISBN-13: 978-1-4684-4861-0   e-ISBN-13: 978-1-4684-4859-7
   DOI: 10.1007/978-1-4684-4859-7

Based on the proceedings of a symposium sponsored by the Comparative Physiology
Section of the American Physiological Society, held August 23-24, 1983,
in Honolulu, Hawaii

© 1984 Plenum Press, New York
Softcover reprint of the hardcover 1st edition 1984
A Division of Plenum Publishing Corporation
233 Spring Street, New York, N.Y. 10013

"Seabird Energetics" is a composite volume with a coherent theme. It makes a valuable contribution to our understanding of the costs of breeding, the significance of which goes far beyond physiology as a brief historical perspective may illustrate.

After decades of mainly anecdotal observations by naturalists with an interest in seabirds, there was still so little information that in 1954, David Lack in his book "The Natural Regulation of Animal Numbers" was forced to ignore seabirds in a way that would be unthinkable today. The late fifties, however, produced a seminal contribution to seabird ecology and behaviour in the series of papers which stemmed from the Centenary Expedition of the British Ornithologists' Union to Ascension Island. Not only had quantitative ecology become the norm but the interest aroused by the European Ethological approach to bird behaviour had led to properly descriptive and analytical studies of seabird behaviour. The complex interactions between social behaviour and ecology then received more attention and the sixties and seventies brought a flood of papers on ecology and on some social aspects of breeding ecology. V.C. Wynne-Edwards linked these two as part of his attempt to understand the mechanism of the regulation of animal populations in his book "Animal Dispersion in Relation to Social Behaviour" (1962). He paid considerable attention to seabirds and the phenomena of clutch and brood-size, deferred breeding, "rest" years, etc., although, unfortunately, the most relevant studies were yet to come.

This wholly inadequate sketch introduces my principal comment with regard to the present volume. Just as Mendelian Genetics vitalised Darwinism, the current interest in "strategies" lends new point to the detailed study of social behaviour. As an integral part of this new approach, we now urgently need to know much more about the energetics of seabird ecology and behaviour so that the cost/benefit analyses of the various options which have been adopted by breeding seabirds can be evaluated. By 'options' I mean the age of first breeding, the

energy invested in egg, clutch and brood, the frequency of
breeding, the nature of the adult's activities before laying and
after the young have fledged, etc.

In addition to assessing the energy-cost of moult, body-
maintenance, egg-production and the care of the young (including
the cost of foraging) it would be of inestimable value to be able
to assess the fitness of an individual bird. With this potent
tool it might become possible, for example, to relate the age of
first-breeding or the adoption of a rest-year to some physiological variable. Furthermore, and importantly, the effect of
breeding, on the condition of the adult, could be gauged. It
might, for instance, be predicted that selection would ensure
that seabirds which lie towards the K end of the r_{max}-K spectrum
would adhere strictly to the strategy of maintaining adult
fitness at all costs. (The defining characteristics of r_{max}-
selection include higher population growth and productivity
linked to larger fluctuations in population whilst K-selected
species tend to reproduce more slowly, live longer and have more
stable populations; K-selected species typically live in relatively impoverished or 'difficult' habitats). They would thus
allow their young of any age, regardless of the energy already
invested in them, to die, rather than themselves accept stress by
increasing their labours in an effort to feed young under adverse
conditions. This in fact appears to happen in such extremely
K-selected seabirds as frigates and Abbott's booby. Species
tending more towards r_{max} selection, on the other hand, might
accept adult stress in a trade-off between reduced longevity and
increased annual productivity.

Of course the ecological correlates of breeding success and
recruitment rate, such as the habit of foraging inshore as
against far-distant from the colony, can be established independently of energetics. But still the two alternative strategies,
to stress or not to stress, remain in theory open to all seabirds
except the few which hardly feed their young at all. Whilst it
is not necessary to demonstrate adult stress in order to establish a correlation between recruitment rate and annual mortality
(supposing one exists) it is necessary to establish the
cost/benefit relationships, of which adult stress is one, if the
mechanism is to be understood.

So it seems timely and important to investigate, as this
volume does for some aspects, the energetics of seabird breeding
biology. It is a step towards a more complete synthesis which
the Darwinists among us believe will confirm that natural
selection at the individual (or kinship) level shapes even the
small details of seabird breeding biology, via its effect on
gene-propagation. These details, in their turn, are equations in
energetics.

Concerning feeding ecology, usually the big unknown in seabird studies, the combination of energetics and radio-telemetric tracking of individuals, via satellites, may yield important discoveries. And the current interest in computer-modelling of the competition for fish between seabirds and man, is well served by the study of the energy requirements of seabird populations, to which this volume contributes.

The physiological approach cannot supplant ecology or behaviour but will always supplement them. "Seabird energetics" is relevant to all seabird biologists.

J. Bryan Nelson
University of Aberdeen
Scotland

REFERENCE

MacArthur, R. H., 1962, Some generalized theorems of natural selection, Proc. Nat. Acad. Sci., 48: 1893.

PREFACE

Seabirds are linked to their oceanic world by energy; energy considerations pervade all aspects of the life history of seabirds and also, in the sense that they may consume large quantities of fish, squid and other marine organisms, determine the impact of the birds themselves on the oceans. This volume traces the energetics of seabirds in an orderly sequence from the formation of the egg to the energy flow through the entire population. This has not been attempted before and it is a unique feature of the present volume. The colonial nesting habits of seabirds have greatly facilitated the study of their eggs, chicks and of the adult birds themselves. As a result, new insights into the physiology of avian incubation and the energetics of avian growth have emerged from the study of seabirds and they receive prominence in this book. One particular group of birds has been singled out for special treatment not because it represents the "typical" seabird, but rather because it is the most specialized of all seabirds. The penguins have given up flight altogether and they display profound adaptations to the oceans and to a harsh terrestrial environment.

This book is based on a symposium sponsored by the Comparative Physiology Section of the American Physiological Society at its Fall Meeting held in Honolulu on August 23-24, 1983. One of the purposes of the symposium was to draw attention to the unevenness in the present state of our knowledge of seabird energetics, and to highlight current strengths and weaknesses. The published proceedings reflect this imbalance and, far from being a weakness, this is seen as one of the strengths of symposia volumes—they represent the state of affairs at a given time and, moreover, stimulate research to fill the gaps.

We are grateful to the National Science Foundation for generously providing travel funds, without which many participants would not have been able to attend the symposium. The editing of this book was supported by a grant (NI/R-14) from the Sea Grant College Program (NOAA).

G. Causey Whittow
Hermann Rahn

CONTENTS

INTRODUCTION

H. Rahn[1] and G. C. Whittow[2]

[1]Department of Physiology
Schools of Medicine and Dentistry
State University of New York
Buffalo, N.Y.
[2]Department of Physiology
John A. Burns School of Medicine and
P.B.R.C. Kewalo Marine Laboratory
University of Hawaii
Honolulu, Hawaii

FOOD, ENERGY AND THE NUMBER OF SEABIRDS

The productivity of the oceans ultimately determines the numbers of seabirds, and regional variations in the abundance of food have a profound effect on their distribution. It has been argued that food is the resource that is most commonly limited in supply, as far as seabirds are concerned (Evans, 1980), and restricted food niches may be partly responsible for the relatively small number of species of seabirds. Thus, out of a total of 8,600 species of birds, approximately 285 may be classified as seabirds. They belong to fifteen families representing four orders (Fig. 1). The exact number of species depends on the definition of a seabird on the one hand, and a species or race on the other. It is apparent from Fig. 2 that approximately 97% of all species of birds are confined to one third of the world's surface, while only 3% utilize the remaining surface, the oceans. This distribution suggests a general scarcity of available food for seabirds, and indeed vast central areas of the oceans are relatively unproductive, particularly in the tropics and subtropics. In general, the areas of highest productivity are the seas overlying the continental shelves, and in the best areas, the sea can grow as much food as the most fertile land. The convergences, where water masses of different properties meet and extensive turbulence occurs, support large populations of seabirds, and so,

1

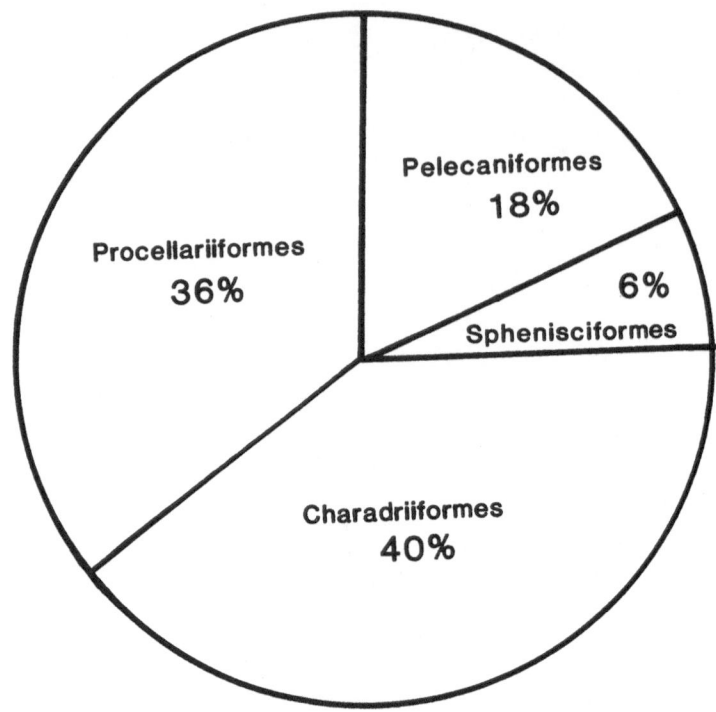

Fig. 1. Orders of seabirds. Percentages are the precentage of the total number of species of seabirds in each order.

also, do areas of upwelling in which cold water rich in nutrients, from the depths of the ocean, replaces surface waters. Superimposed on this broad canvas of ocean characteristics are numerous smaller areas of local richness both in terms of the productivity of the sea and the numbers of seabirds. In addition, there are diurnal and seasonal variations in productivity (Ashmole, 1971). Nevertheless, it has to be remembered that birds are mainly adapted to life in the air and that they are tied, by their breeding characteristics, to land. Even the penguins, which have lost their ancestral ability to fly, have to come ashore to lay their eggs. Although the availability of food and energy may have played an important part in the speciation of seabirds and in the determination of their numbers and distribution, other factors, unrelated to food and energy, must not be forgotten.

El Niño

Four major upwelling areas in the world lie in the eastern boundary currents of four anticyclones: off California, Peru, Northwest Africa, and Southwest Africa (see Cushing, 1982). The

NUMBER OF AVIAN SPECIES   AREA OF WORLD SURFACE

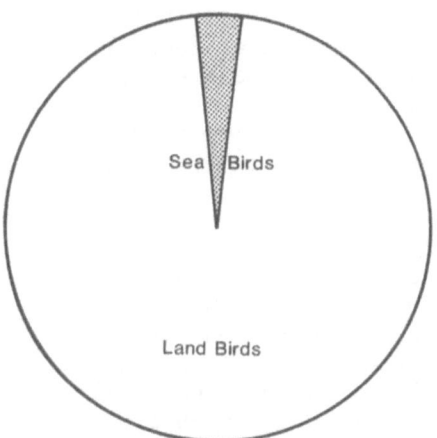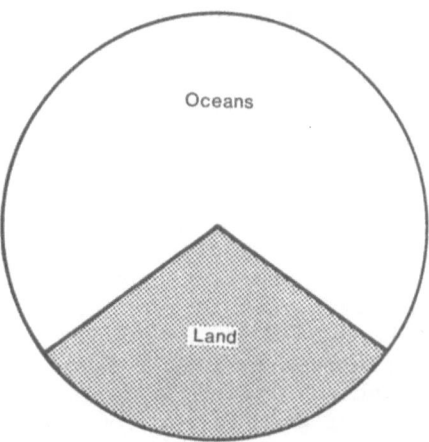

Fig. 2.   Number of species of seabirds and proportion of world's
          surface represented by land and oceans.

Peruvian Coast with its cold upwelling currents has one of the
richest waters in the world, and beginning in 1960 was the site of
a huge single-species fishery, reaching 12 million tons in 1970,
then collapsing by 1972. This area also has one of the world's
larger seabird populations. Changes in the yearly seabird population (Guanay Cormorant, Peruvian Booby, and Chilean Pelican),
nesting off the Peruvian coastline have been estimated by Jordan
and Fuentes (1966) on the basis of the guano deposits. From 1910
to 1964 they showed periodic and massive declines following the
failure of the upwelling and the arrival of warm waters from the
north around Christmas time. The latter phenomenon has been
referred to as the El Niño (in reference to the Christ Child).
This phenomenon is followed by the abandonment of eggs and young,
by the birds, as a result of the disruption of their food supply.
From 1909 to 1956 there was an overall increase in the seabird
population from 4 million to 20 million. In 1957, after an El Niño,
the numbers fell to 7 million, and recovered to 18 million by 1962.
After the 1965-66 El Niño the number fell to 4 million and did not
recover (Cushing, 1982). After the 1972-73 El Niño the anchoveta
catch fell to 2 million tons per year and the bird population to 1
million (Jordan, 1976). Alternative figures for the effect on the
bird and fish populations are presented by Nelson (1978); they lead
to similar conclusions viz. that huge declines in both fish and
birds have occurred.

     These observations provide an interesting example of a link

between physical phenomena and biological consequences and the competition between seabirds which feed on anchoveta and the commercial fisheries. Nelson (1978) and Cushing (1982) believe that the failure of the present seabird population to recover is in large part attributable to the overfishing of anchoveta. According to Table 1, the daily food consumption of the birds is 6,500 metric tons, a value which agrees with the daily food requirement of 430 g of anchoveta per bird (Jordan, 1967; Nelson, 1978).

The periodic nature of El Niño and its devastating effect upon the bird and fish populations of Peru have stimulated cooperation between meterologists and physical oceanographers to a point where the phenomenon can now be forecast, even though forecasting the difference between a major and a minor event is still not possible (Cushing, 1982). A feature of the El Niño is a trans-equatorial flow of warm water from the west to the east, which upon arrival in the Gulf of Panama is guided down the west coast of South America where it displaces the cold upwelling from the Peruvian Current and brings rain to the usually arid coastline. Between December 1982 and June 1983 2.5 m of rain was deposited on the Peruvian coast, where 12 cm is the usual amount (Schumacher, 1983). El Niño is preceded by a year or two of differences in atmospheric pressure between the South Pacific High and the Indonesian Low, augmented by strong Easterlies which stack up the water in the Western Pacific. When winds abate, a Kelvin wave is generated in the equatorial thermocline which travels relatively fast in the eastward direction along the equatorial countercurrent to flood the eastern tropical ocean with warm waters. This may have happened recently at Christmas Island (Pacific Ocean) which lies close to the equatorial countercurrent. During the southern spring-summer of 1982-1983, an estimated population of 13 million birds most of which were Sooty Terns (Schreiber and Ashmole, 1970), abandoned their eggs and fledglings (Schreiber and Schreiber, 1983).

FEEDING HABITS

In order to take advantage of the great variety of marine food available in the surface layers of the oceans, seabirds evince an astonishing array of feeding methods (Ashmole, 1971; Nelson, 1979). These methods range from deep diving in the ocean to outright piracy in the air, and each technique is associated with particular modifications of the bill, the wings, and other features of the birds' structure. Feeding has, in fact, played a large part in shaping the morphology of seabirds and, in many instances, reproductive activity has been modified to adapt to feeding habits (Nelson, 1979). Little is known about the energy costs of different modes of feeding but, in the search for food, many birds conserve energy by virtue of the character of their

4

Table 1. Summary of the total numbers, biomass, food and energy requirements of six seabird populations. mt = metric ton = 1000 kg

Region	Numbers n	Mass Range kg	Biomass mt	Biomass/Number kg	Existence Metabolism# kcal·day⁻¹ x 10⁶	Food Requirement mt·day⁻¹
Alaska[1]	55,270,000	0.05 - 2.5	30,847	0.56	6,177	9,141
South Georgia[2]	31,150,232	0.04 - 9.8	53,758	1.72	5,612	8,305
California[3]	679,609	0.04 - 3.5	641	0.94	103	152
Hawaiian Archipelago[4]	3,518,832	0.05 - 2.6	2,464	0.70	351	520
Christmas Island (Pacific Ocean)[5]	14,060,500	0.12 - 1.4	2,556	0.18	734	1,086
Peru[6]	18,500,000	1.52 - 6.0	39,080	2.11	4,453	6,590
Total	122,000,000		129,346			25,794

[1]based on 1050 colonies along the entire coast (Sowls et al., 1978)
[2]based on Croxall and Prince (1981)
[3]based on Sowls et al. (1980)
[4]based on Harrison et al. (1983)
[5]based on Schreiber and Ashmole (1970). Eight species with populations of less than 3,000 were omitted from the calculations
[6]based on Nelson (1979)

#calculated according to Kendeigh et al. (1977); $kcal \cdot day^{-1} = a - bt$, a = metabolic rate at $0°C$ = $4.142 \ W^{0.544}$, b = temperature coefficient = $0.2761 \ W^{0.2818}$, t = ambient temperature $(°C)$; W = body mass (g).

flight. Thus, the largest and most pelagic of seabirds, the albatrosses, are, for the most part, birds of southern high latitudes where strong westerly winds blow continuously. The albatrosses have taken advantage of these conditions by using a form of gliding known as "dynamic soaring". Their long, narrow wings with high aspect ratios (length:breadth) are adaptations to this type of flight. The advantage of gliding across the wind is that considerable distances can be covered without flapping the wings, with a substantial saving of energy. Frigatebirds soar on columns of warm air rising above sun-heated islands or calm lagoons. Their light weight and large wing area enable them to do this with facility.

Seabirds have featured in the few direct measurements that have been made of the metabolic rate of birds flying in a wind tunnel. Thus, the metabolic rate of Laughing Gulls (Larus atricilla) during level flight was 6-8 times that of birds at rest (Tucker, 1972). On the other hand, during the gliding flight of the Herring Gull (Larus argentatus) the metabolic rate was only 1.7 times the basal metabolic rate (Baudinette and Schmidt-Nielsen, 1974). In Chapter 10 of the present volume additional information on the energy expenditure of free-flying seabirds, measured by the doubly-labeled water technique, in their natural environment, is presented.

A great deal is known about the types of food that seabirds eat, as a result of studies such as those of Ashmole and Ashmole (1967) at Christmas Island in the Pacific Ocean, Croxall and Prince (1980) at South Georgia and Harrison et al. (1983) in the Hawaiian Islands. The energy content of many, but by no means all, prey species is known. The energy content is especially important during the breeding season because the parent birds have to transport food from the feeding grounds back to the nesting colony. Clearly, the greater the amount of energy per unit mass of food, the more energy can be delivered to the nestlings per unit time. Procellariiformes have substantially increased their capacity to deliver energy to the chicks by manufacturing stomach oils, which are fed to the chicks (Clarke and Prince, 1976; Warham et al., 1976). Oils have the highest energy content of any food material; consequently, the birds are able to minimize the mass of food that they have to transport, often over considerable distances.

Seabirds have been classified according to their feeding sites. Some species feed in waters within sight of land - the inshore feeders. Others feed out of sight of land, and they have been broken down into separate categories: the offshore feeders that feed out of sight of land, as far as the continental shelf, but return to land at night, and the pelagic species which feed over deep waters and return to land only to breed (Ashmole, 1971; Diamond, 1978; Buckley and Buckley, 1980). This classification has merit because the numbers and variety of prey species are vastly different in inshore waters and the deep oceans. In addition, it separates seabirds into categories according to the distances that they travel in order to feed. However, there are some pitfalls attendant upon the use of these distinctions. For example, the terms "inshore" and "offshore", implying feeding grounds within sight of, and out of sight of, land, respectively, provide a rough index of the distances that seabirds travel in order to feed. However, inshore and offshore carry different connotations with regard to the availability of food when they are used in connection with seabirds of continents on the one hand, and of oceanic islands on the other. An additional consideration is that pelagic species, although feeding over the open ocean, may travel relatively short distances in order to feed, because they roost at sea. A further complication arises in the instance of pelagic species, in that during the nesting season they do have to travel considerable distances between the feeding grounds and the nesting colony ashore. The distances that birds have to travel to feed are an important consideration from an energetic point of view because flight is expensive in terms of energy and because the greater the distance a seabird has to travel in order to feed, the lower the amount of food, and energy, that it is able to deliver to its chicks, in a given time.

EGGS

The seabird's life may be taken to begin with the laying of the fertilized egg. The formation of the egg and the events leading up to the laying of the egg are described in the following chapter. In general, seabirds are larger than land birds and consequently they lay larger eggs (Fig. 3). However, the mass of the egg relative to that of the female adult bird is less in large than in small birds. Therefore the eggs of seabirds are not proportionately larger than those of terrestrial birds. In fact, egg mass is proportional to the basal energy expenditure (in W or kcal/hr) of the adult, which increases with the 3/4 power of adult body mass. This notwithstanding, the ratio egg mass/adult body mass varies as much as five fold in different seabirds, and for eggs of comparable mass there are large differences in shell mass and thickness (Table 2). Thus, all four orders of seabirds have representatives with a body mass of 1,500 g. Using this as a basis for comparison it is evident that the relative egg mass varies from 2% in cormorants to 10% in tube-nosed birds (Fig. 4). Procellarii-form seabirds, which lay only one egg, have the largest eggs relative to their body mass and as they do not have particularly thick shells, their larger eggs are due to a greater content, especially of the yolk. This implies a greater energy content. There are large discrepancies in egg size between families of the order Pelecaniformes (Fig. 4). The smallest eggs relative to the

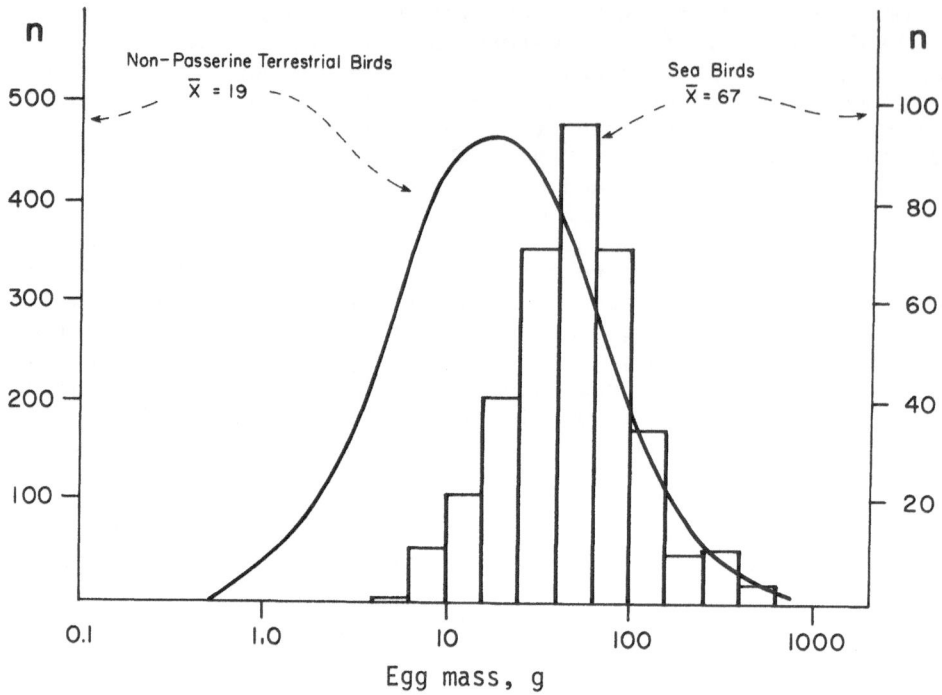

Fig. 3. Egg mass of terrestrial, non-passerine birds (curve) and
         seabirds (columns). x = mean egg mass in g; n = number of
         species and subspecies.

size of the adults belong to the cormorants, that lay several eggs
in a clutch, and the largest to the uniparous tropicbirds. The
egg shells of the Sphenisciformes and Pelecaniformes are heavier
and thicker than those of the Procellariiformes and Charadriiformes
(Table 2).

        There is a tendency in seabirds, particularly among tropical
species and in the order Procellariiformes, toward extended
incubation times. Whittow (1980) defined prolonged incubation in
seabirds as incubation times greater than the upper 95% confidence
limit of the regression of incubation time on fresh-egg mass
presented by Rahn and Ar (1974). While this definition ensured
that all eggs so designated did indeed have long incubation
times, its arbitrary nature excluded many species in which incubation was moderately prolonged. A more meaningful treatment of
incubation time would designate the degree to which the incubation
time of a particular species exceeded the value predicted, from
the mass of the freshly-laid egg, by Rahn and Ar's (1974) equation.
This permits a comparison between families of seabirds and even
between species.

Table 2. Egg mass (W), shell mass (W_{shell}) and shell thickness (L) in four orders of seabirds, calculated from the data of Schönwetter (1960). r^2 = coefficient of determination.

Order (no. of species)	W (range) g	$W_{shell} = aW^b$ (g)			$L = aW^b$ (mm)			Shell dimensions (100 g egg)		
		a	b	r^2	a	b	r^2	W_{shell} g	L mm	density g·cm⁻³
Sphenisciformes (19)	52 – 425	0.052	1.16	0.98	0.042	0.53	0.92	10.9	0.49	2.20
Procellariiformes (88)	5.5 – 455	0.047	1.09	0.99	0.055	0.40	0.95	7.1	0.34	2.04
Pelecaniformes (87)	19 – 185	0.073	1.07	0.93	0.080	0.39	0.62	10.1	0.48	2.02
Charadriiformes (175)	8 – 115	0.038	1.15	0.99	0.045	0.45	0.88	7.6	0.36	1.95

9

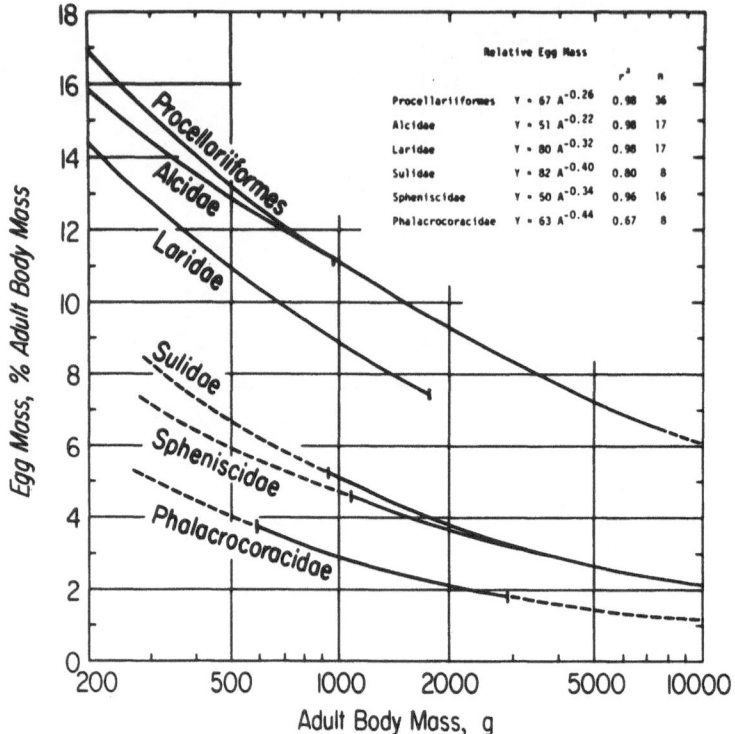

Fig. 4. The relationship between egg mass as a percentage of adult
body mass and body mass in 4 orders of seabirds (data from
Schönwetter, 1960; Rahn et al., 1975).

Such a comparison confirms the very long incubation periods
of the Procellariiformes, regardless of geographical distribution.
However, incubation periods are not uniformly long within the
Procellariiformes: they are relatively longer among the smaller
than among the larger species (Fig. 5). Within the Pelecaniformes,
the tropical families have conspicuously longer incubation times,
and in the Charadriiformes also, prolonged incubation is a feature
of the tropical larids but not the larids from higher latitudes.

The significance of prolonged incubation as far as energetics
are concerned is simply this: the longer an energy-consuming embryo
is present in the egg, the greater the total amount of energy consumed by the embryo, for maintenance, over the entire incubation
period. In order to provide for this greater energy requirement,
the freshly-laid eggs of birds with prolonged incubation have a
greater energy content, and as most of an egg's energy is contained in the yolk, this implies a greater yolk content also.
There is no doubt that seabirds with prolonged incubation have

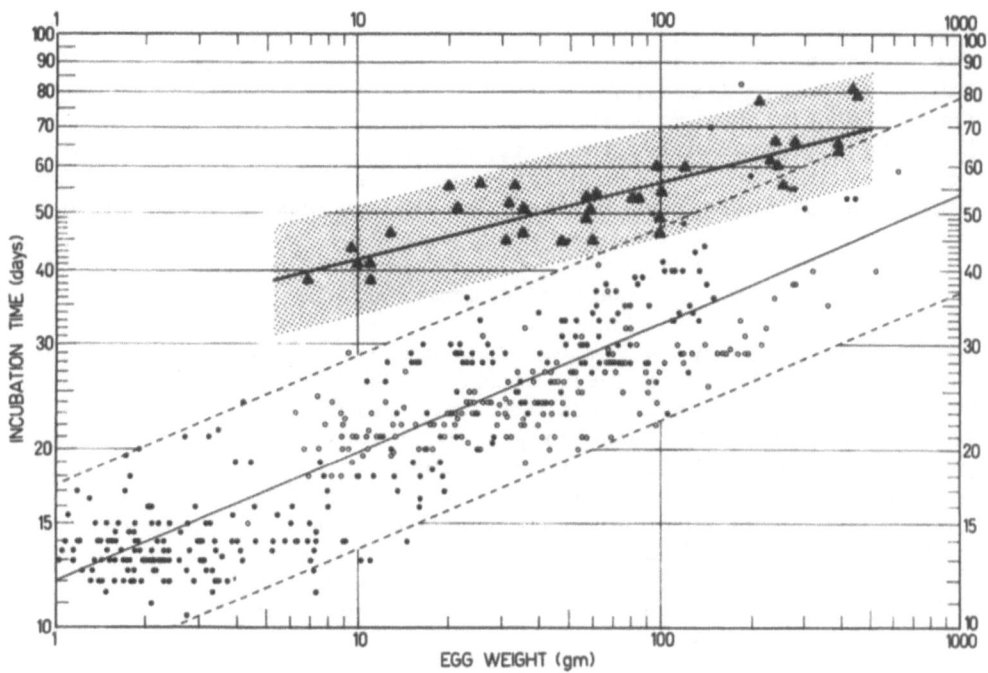

Fig. 5. Relationship between incubation time and fresh-egg mass in the Procellariiformes (solid triangles) and in other birds. The area enclosed by the 95% confidence limits for the Procellariiformes is stippled. Reproduced, with permission, from Grant et al., 1982.

relatively high yolk contents. Interestingly enough, in the Procellariiformes, Warham (1983) has presented evidence that the relative yolk content increases as the size of the egg diminishes. This phenomenon parallels the increase in incubation time, relatively speaking, in the smaller eggs of the Procellariiformes (see above). Consequently, in the small eggs of petrels, the energy and yolk content, as well as the total energy cost of incubation, are relatively greater, and the incubation time longer, than in the large eggs of the albatross (Pettit et al., 1982b). On the other hand, and as Nelson (1971;1978) has pointed out, prolonged incubation may also provide a species with flexibility in the timing of its egg-laying in adaptive correlation with the availability of food. Thus, Abbott's Booby (Sula abbotti) is a tropical sulid that lays a large egg with a long incubation period. Its laying period is flexible and its hatchling is relatively mature. Presumably it forms its egg when adequate food is available and the large egg and extended incubation permit the maximal utilization of the energy and nutrients in the egg for growth. In contrast, the Atlantic Gannet (Sula bassana) lays a small egg, its laying period is fixed and, unlike its tropical counterpart, this temperate sulid has an assured food supply for its chick after hatching. Consequently, there is not the pressure for this species to achieve the maximal amount of growth during the embryonic stage, making use of energy incorporated in the egg when food is abundant.

However, the energy content (yolk content) of the egg is determined also by the state of maturity of the newly-hatched chick (Ricklefs, 1977; Carey et al., 1980). Clearly a greater investment of energy in the egg is necessary if the end product of incubation is a relatively mature, independent precocial chick rather than a helpless, naked altricial chick. Accordingly, in the eggs of precocial and semi-precocial seabirds the percentage yolk content is much higher than in those of altricial seabirds. Williams et al. (1982) concluded, with reference to nineteen avian taxa, that "there is no consistent decrease in the proportion of yolk from the most precocial to the most altricial types", although a definite relationship albeit not a consistent one is evident in their Fig. 5. The yolk and energy content of seabird eggs and the energetics of embryonic growth are discussed in Chapters 5 and 6. The energy content of an egg is only one dimension of the energetics of incubation. The other is the number of eggs per clutch. In general, seabirds have small clutches and this is particularly true of tropical species and species, including many Procellariiformes, which have both long incubation times and high yolk (energy) contents.

In many seabirds, and again this is particularly true of species with prolonged incubation, the rate of embryonic growth is extremely low. There are clear energetic implications to this because a reduced rate of synthesis of tissue, and storage of energy in the synthesized tissue, means that the daily energy requirement of the embryo is correspondingly low. This is offset by the fact that in species with prolonged incubation, the total amount of energy expended over the entire incubation period is large because the maintenance energy costs of a living embryo are sustained for a longer time (see above). There are differences in embryonic growth rates and in embryonic oxygen consumption between precocial and altricial embryos. Thus, in altricial species the oxygen consumption increases continuously during incubation but in precocial embryos the rate of oxygen consumption may slow down prior to pipping of the eggs (Vleck et al., 1980). However the interval between pipping and hatching is an extremely important phase in the life history of the egg in many seabirds. Thus, in the Bonin Petrel (Pterodroma hypoleuca) no less than 50% of the total amount of oxygen consumed by the egg during the pip-to-hatch interval (Pettit et al., 1982b; Whittow, 1983). Embryonic growth rates are discussed in Chapters 5 and 6.

CHICKS

The end result of incubation, the newly-hatched chick, differs in its degree of maturity in different species (see above). However, in a small sample of tropical Procellariiformes and Charadriiformes, the basal metabolic rates of newly-hatched chicks are related to their mass in a manner which appears to be similar in neonates of different size, regardless of their degree of maturity, or the duration of incubation (Fig. 6). This is surprising in view of the obvious disparity between an altricial and a precocial chick although it must be pointed out that the species represented in Fig. 6 vary from semi-altricial to semi-precocial and they do not therefore cover the entire range from fully altricial to precocial. Nevertheless, it means that each embryo has to attain a certain level of metabolic activity, which is related to its mass, by the time that it hatches. In this respect, the metabolic rate of the neonate is a constraint on the events that precede it ie during incubation. The data that provided the regression line in Fig. 6 were all obtained by the same group of investigators. Also shown in Fig. 6, but not included in the computation of the regression line, are data obtained by other investigators from species breeding outside the tropics. At first sight, the metabolic rates of the latter species are greater, in relation to their body mass, than in the tropical neonates. However, in three of the five non-tropical species, it is uncertain whether the metabolic rates were measured within the thermoneutral zone of the neonate. In the other two species, the metabolic rates were truly basal because they were measured under thermoneutral conditions.

Subsequent to hatching, the rate of growth of the chick, and its metabolic rate, vary enormously in different species and in a way that does seem to be related to the degree of maturity of the freshly-hatched chick. In precocial and altricial chicks the basal metabolic rate per unit body mass increases to a peak and then declines as the chick increases in size (Ricklefs, 1974). In both cases the maximal value exceeds that for an adult bird of the same size. However, the peak in mass-specific metabolic rate occurs earlier in the life of the chick in precocial than in altricial species. Furthermore, the energetics of growth are closely related to the development of thermoregulation (Evans, 1980). Thus, in a comparison of the altricial Double-crested Cormorant (Phalacrocorax auritus) with the semi-precocial Herring Gull (Larus argentatus), Dunn (1976) estimated that a much smaller percentage of the total energy expenditure was attributed to thermoregulation in the former species, which depended on its parents for protection from the climatic environment. The allocation of energy is also affected by the rate of growth: Ricklefs and White (1981) compared the energy budgets of two closely related terns with markedly different rates of growth. In the more rapidly growing Common Tern (Sterna hirundo), the allocation of energy to the synthesis of both fat and other

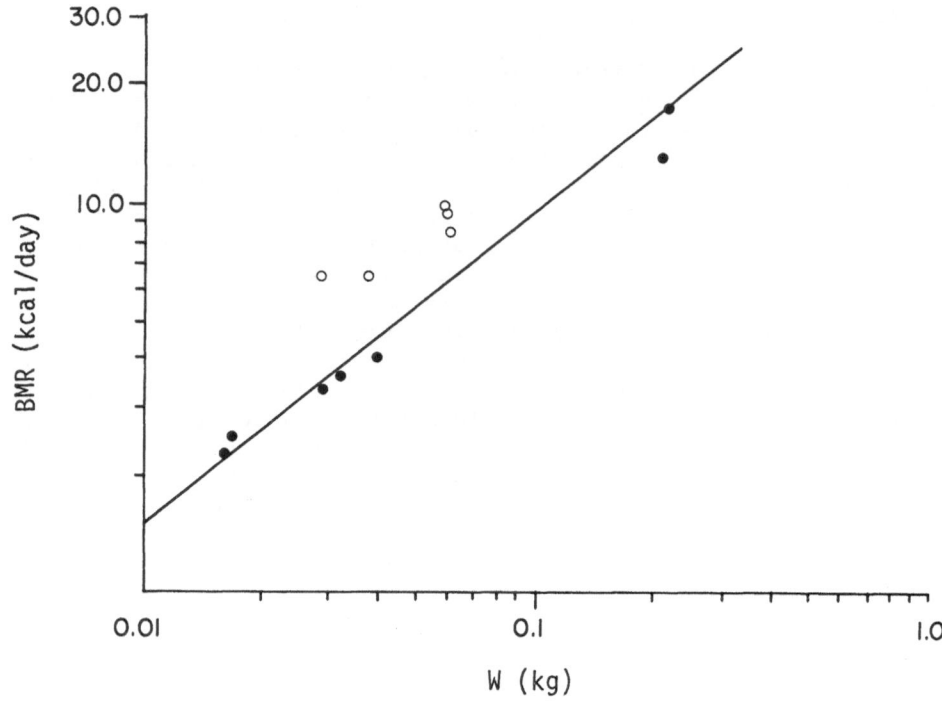

Fig. 6. Basal metabolic rates (BMR) of newly-hatched chicks in re-
lation to their body mass (W). ● = tropical species;
O = species breeding outside the tropics. The fitted re-
gression line (BMR = 61.37 $W^{0.808}$; n=7) is for the tropi-
cal species. Data from following sources: Drent, 1970;
Dawson et al., 1972; Ackerman et al., 1980; Dawson and
Bennett, 1981; Pettit et al., 1981, 1982a,b; Pettit and
Whittow, 1983.

tissues increased during the first half of the nestling period and
then declined. In the more slowly growing Sooty Tern (Sterna
fuscata), the allocation of energy to the formation of body tissues
declined throughout the nestling period while the energy expended
in maintenance increased.

Precocity of development is relatively rare in seabirds, only
four species of alcids being fully precocial (Evans, 1980;
Montevecchi and Porter, 1980). This probably reflects the fact
that the self-feeding precocial bird is largely restricted to areas
where food is readily available and little skill or energy is required for its acquisition. In general, these conditions are met
only on land. Although the precocial mode of development absolves
the parent birds from a considerable investment of time and energy

in the care and nurture of the young, this is offset to an undetermined extent by an increased investment of energy in the egg. The precocial mode of development may well be less efficient in terms of energy utilization than is the altricial pattern (Dawson and Hudson, 1970). Thus, in the altricial condition, there is a trend towards a more direct transfer of energy (either in the form of food or heat) from the parent bird to the embryo and chick, rather than through the intermediary of a greater amount of yolk. The development of thermoregulation in the embryo, its maturation in the chick and the energetics of post-natal growth are the subject matter of Chapters 7 and 8.

DELAYED MATURITY

While a long life span may increase the energetic demands that seabirds make on their marine environment (see below), their relatively low intrinsic reproductive rates have the opposite effects. Their low rate of reproduction derives from the combined effects of small clutches, to which allusion has already been made, and delayed reproductive maturity (see Evans, 1980). The Wandering and Royal Albatrosses do not breed until they are ten years old, for example (see Nelson, 1979). Of the various reasons advanced for delayed breeding in seabirds, the most convincing is that it is related to the time taken to acquire feeding skills, and experience variations in the availability of food (Nelson, 1979). This is highlighted by the Greater Frigatebird, which never alights on the ocean. It's feeding repertoire includes cannibalism, aerial food piracy and the preying on chicks of other species of seabirds, in addition to the more conventional seizure of prey from the ocean. It takes longer to begin breeding than do most other species of seabird.

BASAL METABOLIC RATE

The basal energy expenditure is an important part of the total energy budget of a bird (see Whittow, 1984b) and multiples of the basal metabolic rate (BMR) are often used to arrive at estimates of the total energy expenditure of individual birds and also of populations. In the latter instance, errors in the BMR are magnified enormously and so, therefore, are estimates of the food consumption of a community of birds, derived from their energy expenditure (see Whittow, 1984b). The available information on the BMR's of seabirds is summarized in Fig. 7. The slope of the relationship between basal metabolic rate and body mass in the Procellariiformes was somewhat lower than that of both other orders of seabirds (Fig. 7) and birds in general (see Whittow, 1984a). On the other hand, the slope of the regression of basal metabolic rate on body mass in Charadriiformes was greater than that for

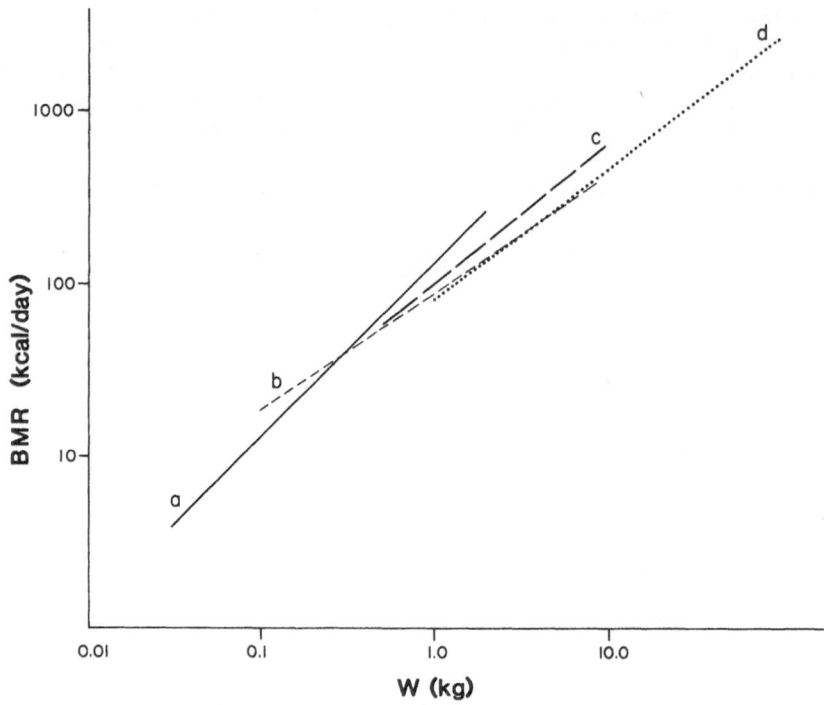

Fig. 7. Relationship between the basal metabolic rate (BMR) and body mass (W) in: (a) Charadriiformes, BMR = 131.6 $W^{1.014}$, n=15; (b) Procellariiformes, BMR = 88.8 $W^{0.68}$, n=21; (c) Pelecaniformes, BMR = 99.72 $W^{0.804}$, n=4; (d) Sphenisciformes, BMR = 81.98 $W^{0.749}$, n=7. BMR in kcal/day; W in kg. The regression lines were derived from data in the following sources: Benedict and Fox, 1927; Scholander et al., 1950; Enger, 1957; Drent and Stonehouse, 1971; Iverson and Krog, 1972; Baudinette and Schmidt-Nielsen, 1974; Johnson and West, 1975; Kooyman et al., 1976; LeMaho et al., 1976; Pinshow et al., 1976; Kendeigh et al., 1977; MacMillen et al., 1977; Vleck and Kenagy, 1980; Ricklefs et al., 1981; Ellis et al., 1982a,b; Stahel and Nicol, 1982; Adams and Brown, 1983; Grant and Whittow, 1983; Simons, 1983; Brown and Adams, personal communication; Pettit, Ellis and Whittow, unpublished data.

other birds (Fig. 7). This latter phenomenon reflects the fact that, in general, the larger members of the order had relatively high basal metabolic rates, and they were outwith the tropics, while many of the smaller, tropical species had relatively low basal metabolic rates. In the case of species resident in temperate or polar climates there is some question as to whether the birds were tested under thermoneutral conditions and whether, therefore, the metabolic rates were truly basal. In future compilations of allometric relationships it would seem important to include only data for species in which the thermoneutral zone has been determined by experiment. Combining the data from 39 seabirds belonging to different orders yielded a regression equation of:

$$kcal \cdot day^{-1} = 95 \ W^{0.744} \quad (W = body \ mass \ in \ kg)$$

The slope of the relationship between basal metabolic rate and body mass in seabirds was similar to that of nonpasserine birds in general (Lasiewski and Dawson, 1967; Dawson and Hudson, 1970) but the intercept was 22% greater in seabirds. The basal metabolic rates of seabirds are discussed further in Chapter 10.

THERMOREGULATION

Thermoregulatory demands make up a significant part of the total energy expenditure of many seabirds, both of adults and chicks. Most seabirds have the navigational abilities and power to undertake long migrations or dispersions and many do, outside the breeding season. This enables them to avoid exposure to the extremes of climate, the energy cost of migration being more than offset, presumably, by a less harsh environment and greater opportunities for feeding. During the breeding season, however, they are constrained to remain in the vicinity of their nesting colonies. As far as tropical seabirds are concerned, this means exposure to very hot conditions. The burden that this places on the birds, both adults and chicks, is more in terms of replacement of the water lost during evaporative cooling than of meeting greatly increased energy demands. Most seabirds have effective salt glands so they are able to maintain the osmolarity of their body fluids, in the face of excessive evaporative water loss, by excreting the excess salt (Schmidt-Nielsen, 1975). The foraging parent is able to replace the lost water by drinking seawater as well as feeding. The chicks, on the other hand, are entirely dependent on the water contained in their food; their water balance is correspondingly more precarious than is that of the adult birds.

Polar species, such as the penguins, incubate their eggs under extremely cold conditions and some species fast during incubation. Thus, when their energy requirements are greatest they are not able to meet these requirements from their food, relying, instead, on the breakdown of body reserves. Because of their harsh environment, their flightlessness, and their adaptations to an aquatic life, penguins represent a special case of seabird energetics, covered in Chapters 11 and 12.

The geographical distribution of birds, and of their breeding sites, is determined by many factors (Buckley and Buckley, 1980) including the availability of food and nesting sites, and in the physiological and behavioral thermoregulatory capacities of the birds themselves. Thermoregulation in adult seabirds is dealt with in Chapter 9.

MOLTING

The replacement of feathers consumes a great deal of energy in a variety of different ways (see Whittow, 1984b). In addition to the cost of synthesizing new tissue, there is the energy locked up in the tissues themselves. Furthermore, the loss of plumage entails a reduction of thermal insulation, requiring, in many instances, a compensatory increase in heat production. Some species (penguins) compound their energetic problems by fasting during the molt, so that a period of increased energy requirements coincides with a time when their only source of energy is their own body tissue. Although the total energy cost of the molt is substantial, the daily cost to the bird depends on how quickly the molt is completed. Thus, penguins replace their feathers within a period of 3-4 weeks, losing 3% of their body weight per day in order to supply the energy. They prepare for the molt by feeding intensively after breeding has been completed. Dr. Davis and Dr. Kooyman enlarge upon the situation in penguins, in their chapter. In general, sea-birds molt after the breeding season is over and when long migrations have been completed, so that different energy - demanding activities do not coincide. However, there are many exceptions to this gener-alization. Some tropical species, eg. tropical sulids, molt while they are breeding (Nelson, 1978) and the molt of the Ashy Storm-petrel takes no less than 257 days. The energetic implications of molting are discussed further in Chapter 13.

LIFE-SPAN ENERGETICS

The life span of seabirds is germane to their energetics be-cause it affects the total amount of food removed from the ocean, in a bird's lifetime, and, ipso facto, the impact of the birds themselves on the oceans. Other things being equal, the longer a seabird lives, the more energy it requires. In addition, a bird that lives longer produces more offspring, although this is offset, to some extent, by delayed maturity (see above). On the whole, seabirds are long lived. Thus, albatross may live for forty-six years or more, gulls for thirty-one years and terns for twenty-six years (Nelson, 1979). These are probably conservative estimates. Male longevity often differs from that of the female bird and mortality rates vary in different age groups. The mortality of young seabirds can be very high. For example, the mortality of Atlantic Gannets (Sula bassana) during the first year of life is

70% (Nelson, 1966). Consequently, at any given instant, there are chicks and adults of different ages and with different life expectancies.

That energetics may be viewed from the perspective of life span was first conceived by Max Rubner. Rubner (1908) showed that the daily metabolic rate times the life span, when expressed per kilogram body mass, was essentially similar for animals from guinea pigs to the horse and cow, namely, ca. 200,000 kcal.kg^{-1}. His data base was possibly too small for such a generalization but the concept survived and the evidence was most recently reviewed by Boddington (1978) who showed that if the mass-specific metabolic rate regressed against body mass has an exponent of -0.25, and life span regressed against body mass has an exponent of +0.25, the product of these two relations must yield a constant which he called the "absolute metabolic scope". In Fig. 8 the mass-specific metabolic rate of passerine birds is plotted against body mass. The

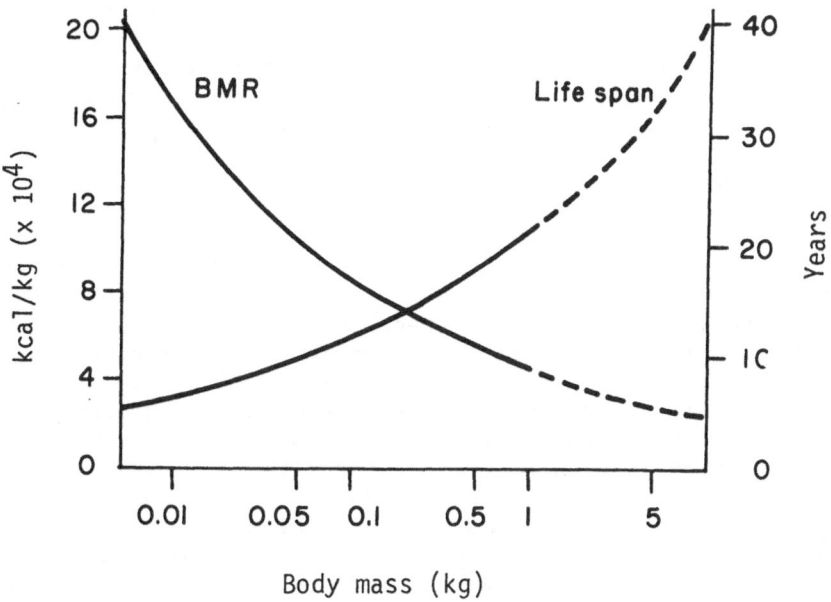

Fig. 8. Basal metabolic rate and life span of passerine birds plotted against body mass. The broken lines are extrapolations beyond the maximal body mass of Passeriformes.

exponent for the regression, taken from Lasiewski and Dawson (1967), is -0.28. Also plotted in Fig. 8 is the regression of life span against body mass in which the exponent is 0.26 (Lindstedt and Calder, 1976). The product of the regressions, the absolute metabolic scope, is approximately 1,000 kcal/g. The absolute metabolic scope is based upon basal metabolic rate. A refinement would take into account the energy cost of daily activities over and above the basal metabolic rate.

From the basal metabolic rates of seabirds (see Fig. 7 for sources of data) and calculations of their life span from the allometric equation of Lindstedt and Calder (1976), it is possible to arrive at an absolute metabolic scope for each species from the following equation:

$$AMS = (MR \times 365 \times LS)/Kg$$

where AMS = absolute metabolic scope, $kcal \cdot kg^{-1}$
     BMR = basal metabolic rate, $kcal \cdot day^{-1}$
     365 = $days \cdot year^{-1}$
     LS = life span, years = $16.6 \ M^{0.18}$, where M = body mass, kg (Lindstedt and Calder, 1976) based on 81 wild, non-passerine species.

The average absolute metabolic scope, calculated for 36 species of seabirds was 570,000 $kcal \cdot kg^{-1}$ with a coefficient of variation of 26%. This is approximately half the value for passerine birds. In view of the assumptions that were made, particularly regarding the life span of seabirds, the figure for seabirds is probably an underestimate of their absolute metabolic scope.

More recently, it was suggested that the total life span-metabolic concept in birds might be conveniently divided into three separate periods, the embryonic life span, the juvenile life span, and the adult life span, and that each period might be identified with its own absolute metabolic scope (Rahn and Ar, 1980). This would be realized if the mass-specific metabolic rate for each period, when regressed against embryonic or juvenile mass had an exponent of ca. -0.25, and the embryonic and juvenile life span regressed against embryonic or juvenile mass had an exponent of ca. +0.25. The latter relationships are shown in Fig. 9, where the incubation time is plotted against hatchling mass (from Rahn, 1982), fledging time for altricial birds (days to grow from 10% to 90% of asymptotic fledgling mass) is plotted against fledgling mass (Dunn, 1979) and adult life span in passerine birds is plotted against adult mass (Lindstedt and Calder, 1976). In each case the slopes are close to +0.25 and fit into the general pattern of periodic phenomena as recently reviewed by Lindstedt and Calder (1982).

An absolute metabolic scope for embryonic life span is suggested

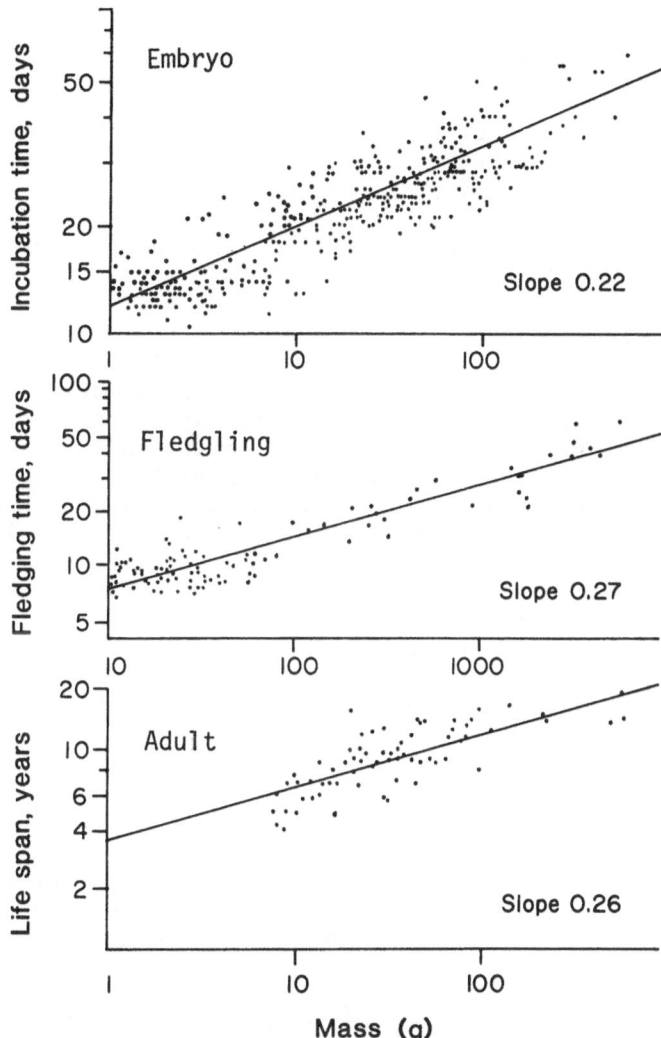

Fig. 9. Regression of incubation time, fledging time and life span
on hatchling, asymptotic nestling and adult mass, respectively, for passerines (Rahn, 1982; Dunn, 1979; Lindstedt
and Calder, 1976).

by the fact that the mass-specific metabolic rate of embryos at the
pre-internal pipping stage has a slope of -0.29 (Hoyt and Rahn, 1980).
As the metabolic rate changes during development, a better criterion
would be to assess the total O_2 uptake. In 34 cases the total O_2
uptake divided by egg mass yielded a mean value of 102 $\pm$ 20 ml $O_2 \cdot g^{-1}$
of egg. As the average hatchling mass is 0.67 of egg mass (Heinroth,
1922), the absolute metabolic scope per gram of embryo is 152 ml O_2

or ca. 0.73 kcal.g^{-1}. While such generalizations are attractive, there are exceptions, particularly among the Procellariiformes (Pettit et al., 1982a,b).

It will be of interest to assess the caloric requirements throughout the fledging period of birds and to see whether or not these would conform to a similar concept. If an absolute metabolic scope can be established for this stage, then it becomes eventually possible to describe the whole energetic history of birds in terms of these three quantities.

Life Span Food Requirements

The average <u>basal</u> metabolic rate of seabirds during a life span was shown to be 570,000 kcal.kg^{-1} body mass (see above). Assuming an <u>active</u> metabolic rate throughout this period which is 2.5 times larger, this would increase to 1.42 x 10^6 kcal·kg^{-1}. Converting this to food intake using a conversion factor of 1.5 g. kcal^{-1} (see below), yields a figure of 2.1 x 10^6 g·kg^{-1} or ca. 2 metric tons of food consumed per kilogram body mass during the life span. Thus a 100 g tern would require 200 kg, a 10 kg penguin, 20,000 kg. These are obviously approximations at best, based upon the various assumptions along each step of the calculations. Nevertheless, they can eventually be refined and possibly begin to shed some light on a common denominator linking body size, life span, metabolic rate, and food consumption.

POPULATION ENERGETICS

Knowledge of the energy requirements of entire populations of seabirds provides information of both ecological and practical value. The energy requirements of a seabird community together with knowledge of the energy content of the prey species provides the only practical means of estimating the bio-mass of organisms removed from the oceans by the birds. Methods of making such assessments are described in Dr. Wiens' contribution to this volume. A specific example of a seabird community, that of South Georgia, is presented by the final contributors in Chapter 14. Information is available for the numbers of seabirds in other areas and, if the mean body mass of each species is known, an estimate of the energy expenditure of the population can be derived using the equation for existence metabolism presented by Kendeigh et al. (1977). This equation provides a conservative estimate of energy expenditure based on the body mass of the species. It's usefulness lies mainly in the comparison that it permits between different populations rather than in the absolute values. Recent observations using a doubly-labeled water technique indicated that the average daily metabolic rate of two species of free-ranging penguins was 67 and 51% higher than those calculated from existence metabolism. Direct measurements of the energy expenditure of a few species of sea birds are discussed in Chapters 10 and 12 of the present volume.

Energy Expenditure

In the absence of more extensive direct measurements of the energy expenditure of free-flying birds, it is possible to estimate the biomass of bird populations, their energy requirements and food consumption from knowledge of the numbers of birds, their body mass, and Kendeigh et al.'s (1977) equation. The results of such estimates for six seabird populations are summarized in Table 1. The average bird mass (biomass/number) varies greatly from region to region. Thus, in South Georgia it is 1.72 kg because of the prevalence of penguins in the population, whereas on tropical Christmas Island (Pacific Ocean) it is only 0.18 kg because the predominant species, the Sooty Tern, is a small bird. There is no simple relationship between the total biomass of the birds and their energy requirements. For example, in the two high-latitude populations, the total biomass of the South Georgian seabird population was 42.6% greater than that of Alaska, but the energy expenditure of the South Georgian birds was only 90.9% of that of the Alaskan population. This apparent anomaly is explained by the lower mass-specific metabolic rate of the larger birds at South Georgia.

Food Requirements

Assuming an assimilation efficiency of 75% (Kendeigh et al., 1977) and a mean calorific value of crustacea, squid and fish of 0.9 kcal.g^{-1} (Croxall and Prince, 1981), then (1/.75) x (0.9/1.0), or 1.48 g of food, will yield 1 kcal of energy. Thus the daily food requirement (g.day-1) = energy requirement (kcal.day^{-1}) x 1.48 (g.kcal^{-1}). A direct approach to estimating the daily food consumption on the basis of body mass can also be obtained from Fig. 10 where the mass-specific food consumption (g.day^{-1}.g^{-1}) has been computed from the above equations and plotted against body mass for an ambient temperature of 0°, 10°, and 20°C. The general shape of these curves not only emphasizes the large decline in mass-specific metabolism, and therefore food consumption, as body mass increases, but also the reduced requirements as ambient temperature increases, particularly at the smaller body masses. For example, the daily food consumption of a 100 g bird at Ta = 0°C is 75 g.day^{-1}, but it is 45 g at 20°C. A 1 kg bird, on the other hand, would require only a daily food intake equal to 26% of its body mass at 0°C and 20% at 20°C. Fig. 10 may be used to estimate the daily food requirements of all antarctic penguins using data supplied by Bengston (1978). The daily food requirement of 50,160 metric tons was twice that of the total food consumption for all the seabird colonies listed in Table 1.

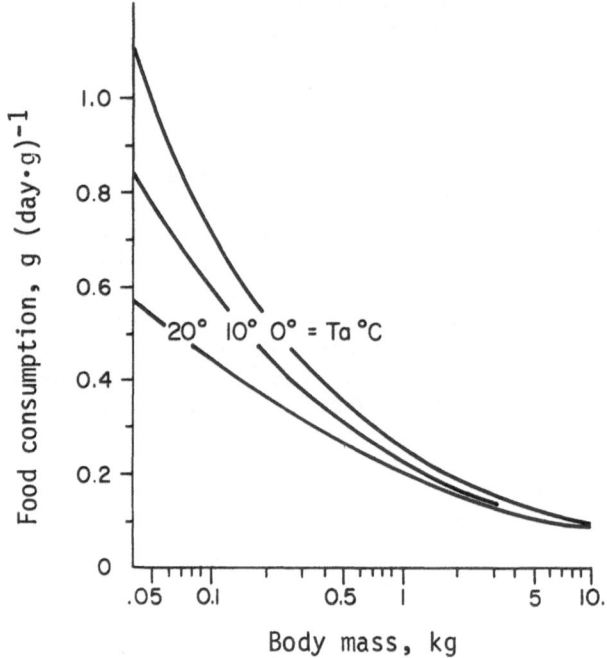

Fig. 10.   Relationship between mass-specific food consumption and
body mass in birds, at three air temperatures (Ta),
based on the equations of Kendeigh et al. (1977).

Taxonomic Representation

In Table 3 is shown the taxonomic representation of seabirds in different regions under the heading of numbers, biomass,
and metabolism or food consumption. For example, in the Hawaiian
Archipelago the population is evenly divided between Procellariiformes and Laridae, yet 70% of the food requirement goes to the
former which includes many large birds such as albatross. In
South Georgia the penguins constitute 33% of the total population
but account for 76% of the food consumption. The order Pelecaniformes is not strongly represented in this table, yet it is
probably the dominant species in each category in Peru (where a
population census for other species was not available).

Yearly Food Consumption

When the residence time for the Alaskan birds is taken into
consideration, the estimated food consumption is 2.02 million tons
a year for Alaskan waters, which can be compared (1) with the
estimates of the National Marine Fisheries Service for the total
Alaskan fish catch (Bering Sea and Gulf of Alaska) which was 2.05

Table 3. Taxonomic representation of seabirds (% of total population) in terms of their abundance, biomass, and metabolic requirements or food consumption.

	Order or Family	Alaska	South Georgia	California	Hawaiian Islands	Christmas Island (Pacific Ocean)
Numbers	Spheniscidae	–	33	–	–	–
	Procellariiformes	47	67	4	50	<1
	Pelecaniformes	<1	–	12	2	<1
	Laridae	6	–	8	48	>99
	Alcidae	47	–	76	–	–
Biomass	Spheniscidae	–	93	–	–	–
	Procellariiformes	41	7	<1	83	<1
	Pelecaniformes	<2	–	31	3	1
	Laridae	6	–	7	14	99
	Alcidae	51	–	62	–	–
Existence Metabolism or Food Consumption	Spheniscidae	–	76	–	–	–
	Procellariiformes	43	24	<1	70	<1
	Pelecaniformes	1	–	22	3	<1
	Laridae	6	–	8	27	>99
	Alcidae	50	–	70	–	–

and 1.93 million metric tons for 1981 and 1982, respectively; and (2) with the estimates of Hunt et al. (1981) who estimated a yearly food consumption of seabirds in the Eastern Bering Sea of 1.09 million tons. The latter estimate was derived by assuming a daily food consumption of 20% (not 40% as indicated) of body mass for all species. The food consumption during the breeding season of the South Georgia bird colony was estimated by Croxall and Prince (1981) to be 1.7 million metric tons; and if the daily food consumption of the Antarctic Penguins (see above) is multiplied by 365 days, a yearly consumption of ca. 18 million tons is obtained, which is about halfway between the range of 12 to 26 million tons estimated by Bengston (1978). One can finally multiply by 365 days the total food consumption of all the areas listed in Table 1, including the antarctic penguins, and arrive at an annual rate of ca. 30 million tons. This figure represents only part of the global food consumption of seabirds, yet it represents a volume equal to one-half the global commercial fish catch of ca. 60 million tons a year (Kanwisher and Ridgway, 1983). It has to be borne in mind, however, that, at best, these are estimates and, moreover, that a great deal of the food consumption of seabirds comprizes species (eg. flying fish) that are not commercially valuable.

GUANO

It might be appropriate to conclude this introduction with a brief reference to guano - the most visible impact of seabird energetics on the human economy. The word guano is derived from the Quechua language of the Peruvian Indians - huanu, meaning dung. It was first applied to the enormous deposits of bird droppings on the Peruvian coast, which have been mined and exported since 1810. In fertilizer terminology, they contain 11-16% nitrogen, 8-12% phosphoric acid, and 2-3% potash. While many seabirds deposit guano on land, the most productive are members of the Pelecaniformes, certain cormorants, boobies, and pelicans. It requires highly social birds, breeding in large colonies, using guano in constructing their nests, and depositing an appreciable fraction of their excreta on land. It also requires a very dry climate so that the rains do not wash it away. Such ideal conditions are met along the coast of Peru where the layering of each annual guano deposit can be used to estimate the number of breeding birds. On the basis of such an analysis Jordan and Fuentes (1966) have reconstructed figures for the numbers of Peruvian guano birds since 1909 - an increase from 4 million in 1909 to 20 million in 1956, with large fluctuations depending upon the El Niño discussed above.

The amount of nitrogen in guano deposits varies in different parts of the world, and it is possible to distinguish between nitrogenous guano as found in Peru and phosphatic guano in which the organic nitrogenous fraction derived from uric acid is lost, leaving a material that usually consists of calcium phosphate minerals. The yearly world-wide deposition of guano based on mining statistics was estimated by Hutchinson (1950) at 185,000 tons, 92% of which is deposited in South America. It is of interest to inquire how large a guano-bird population is required to achieve such an annual deposit. According to Jordan (1967) the daily anchoveta consumption by each bird is 0.43 kg, of which 6 to 10% is converted to dry guano. Thus, the average guano deposit per bird is 0.027 to 0.043 $kg \cdot day^{-1}$ or 10 to 16 $kg \cdot year^{-1}$. Dividing this into the total world-wide yearly guano deposit gives an estimate of ca. 19 to 11 million birds. Since 92% of the guano is produced in Peru, this population estimate is not far from the values estimated by Nelson (see Table 1) or Jordan and Fuentes (1966).

ACKNOWLEDGEMENTS

The preparation of this chapter was supported, in part, by a grant, to one of us (GCW) from the Sea Grant College Program (Nl/R-14). The authors are indebted to Dr. J. Bryan Nelson for his most perceptive comments on a draft of the manuscript.

REFERENCES

Ackerman, R. A., Whittow, G. C., Paganelli, C. V., and Pettit, T. N., 1980, Oxygen consumption, gas exchange, and growth of embryonic Wedge-tailed Shearwaters (Puffinus pacificus chlororhynchus), Physiol. Zool., 53:210.

Adams, N. J., and Brown, C. R., 1983, Metabolic rates of subantarctic Procellariiformes: a comparative study, Comp. Biochem. Physiol., in the press.

Ashmole, N. P., 1971, Seabird ecology and the marine environment, in: "Avian Biology", Volume 1, D. S. Farner, J. R. King, and K. C. Parkes, eds., Academic Press, New York.

Ashmole, N. P., and Ashmole, M. J., 1967, Comparative Feeding Ecology of Seabirds of a Tropical Oceanic Island, Peabody Museum of Natural History, Yale University, Bull. 24.

Baudinette, R. V., and Schmidt-Nielsen, K., 1974, Energy cost of gliding flight in Herring Gulls, Nature, 248:83.

Benedict, F. G., and Fox, E. L., 1927, The gaseous metabolism of large wild birds under aviary life, Proc. Amer. Philos. Soc., 66:511.

Bengtson, J. L., 1978, Review of information regarding the conservation of living resources of the Antarctic marine ecosystem, Marine Mammal Commission, Nat. Tech. Inform. Service PB-289-496, U. S. Dept. Commerce, Washington, D. C.

Boddington, M. J., 1978, An absolute metabolic scope for activity, J. Theor. Biol., 75:443.

Buckley, F. G., and Buckley, P. A., 1980, Habitat selection and marine birds, in: "Behavior of marine animals, Vol. 4: Marine Birds", J. Burger, B. L. Olla, and H. E. Winn, eds., Plenum Press, New York.

Carey, C., Rahn, H., and Parisi, P., 1980, Calories, water, lipid and yolk in avian eggs, Condor, 82:335.

Clarke, A., and Prince, P. A., 1976, The origin of stomach oil in marine birds: analysis of the stomach oil from six species of subantarctic procellariiform birds, J. exp. mar. Biol. Ecol., 23:15.

Croxall, J. P., and Prince, P. A., 1981, A preliminary assessment of the impact of sea birds on marine resources at South Georgia, Colloque sur les Ecosystemes Subantarctiques. Palmpont, C.N.F.R.A., Olla, No. 51, 501.

Croxall, J. P., and Prince, P. A., 1980, Food, feeding ecology and ecological segregation of seabirds at South Georgia, Biol. J. Linn. Soc., 14:103.

Cushing, D. H., 1982, "Climate and Fisheries", Academic Press, London.

Dawson, W. R., and Bennett, A. F., 1981, Field and laboratory studies of the thermal relations of hatchling Western Gulls, Physiol. Zool., 54:155.

Dawson, W. R., and Hudson, J. W., 1970, Birds, in: "Comparative physiology of thermoregulation", G. C. Whittow, ed., Academic Press, New York.

Dawson, W. R., Hudson, J. W., and Hill, R. W., 1972, Temperature regulation in newly hatched Laughing Gulls (Larus atricilla), Condor, 74:177.

Diamond, A. W., 1978, Feeding strategies and population size in tropical seabirds, Amer. Nat., 112:215.

Drent, R. H., 1970, Functional aspects of incubation in the Herring Gull, in: "The Herring Gull and its egg", G. P. Baerends, and R. H. Drent, eds., Brill, Leiden.

Drent, R. J., and Stonehouse, B., 1971, Thermoregulatory responses of the Peruvian Penguin, Spheniscus humboldti, Comp. Biochem. Physiol., 40A:689.

Dunn, E. H., 1976, The development of endothermy and existence energy expenditure in Herring Gull chicks, Condor, 78:493.

Dunn, E. H., 1979, Time-energy use and life history. Strategies of northern seabirds, in: "Conservation of Marine Birds in Northern North America," J. C. Bartonek, and D. N. Nettleship, eds., U. S. Fish and Wildlife Service, Wildl. Res. Rept., No. 11.

Ellis, H. I., Maskrey, M., Pettit, T. N., and Whittow, G. C., 1982a, Temperature regulation in Hawaiian Brown Noddies (Anous stolidus pileatus), Physiologist, 25:279.

Ellis, H. I., Maskrey, M., Pettit, T. N., and Whittow, G. C., 1982b, Temperature regulation in Hawaiian Red-footed Boobies, Amer. Zool., 22:394.

Enger, P. S., 1957, Heat regulation and metabolism in some tropical

28

mammals and birds, <u>Acta Physiol. Scand.</u>, 40:161.

Evans, R. M., 1980, Development of behavior in seabirds: an ecological perspective, in: "Behavior of marine animals, Vol. 4: marine birds", J. Burger, B. L. Olla, and H. E. Winn, eds., Plenum Press, New York.

Grant, G. S., and Whittow, G. C., 1983, Metabolic cost of incubation in the Laysan Albatross and Bonin Petrel, <u>Comp. Biochem. Physiol.</u>, 74A:77.

Grant, G. S., Pettit, T. N., Rahn, H., Whittow, G. C., and Paganelli, C. V., 1982, Water loss from Laysan and Black-footed Albatross eggs, <u>Physiol. Zool.</u>, 55:405.

Harrison, C. S., Hida, T. S., and Seki, M. P., 1983, Hawaiian seabird feeding ecology, <u>Wildlife Monographs</u>, No. 85.

Harrison, C. S., Naughton, M. B., and Fefer, S. I., 1984, The status and conservation of seabirds in the Hawaiian Archipelago and Johnston Atoll. ICBP Technical Bulletin. In the press.

Heinroth, O., 1922, Die beziehungen zwischen vogelfewicht, eigewicht, gelegegewicht und brutdauer, <u>J. Ornithol.</u>, 70:172.

Hoyt, D., and Rahn, H., 1980, Respiration of avian embryos - a comparative analysis, <u>Resp. Physiol.</u>, 39:255.

Hunt, G. L., Burgeson, B., and Sanger, G. A., 1981, Feeding ecology of seabirds of the Eastern Bering Sea, in: "The Eastern Bering Sea Shelf," D. W. Hood and J. A. Calder, eds., Oceanography and Resources, Vol. 2, NOAA.

Hutchinson, G. E., 1950, Survey of Contemporary Knowledge of Biogeochemistry. 3. The Biogeochemistry of Vertebrate Excretion, <u>Bull. Amer. Mus. Nat. Hist.</u>, 96:1.

Iverson, J. A., and Krog, J., 1972, Body temperatures and resting metabolic rates in small petrels, <u>Norw. J. Zool.</u>, 20:141.

Johnson, S. R., and West, G. C., 1975, Growth and development of heat regulation in nestlings, and metabolism of adult Common and Thick-billed Murres, <u>Ornis. Scand.</u>, 6:109.

Jordan, R., 1967, The predation of guano birds on the Peruvian anchovy (<u>Engraulin ringens Jenyns</u>), <u>Calcofi. Rep.</u>, XI:105.

Jordan, R., 1976, Biologia de la anchoveta. Parte I. Resumen del Conocimiento actual. Actas de la Reunion de Trabajo Sobre el Fenomeno Conocido como El Nino, <u>FAO informes de Pesca</u>, 185:359.

Jordan, R., and Fuentes, H., 1966, La poblaciones de aves guarneras y su situacion actual, <u>Informe del Inst. del Mar del Peru</u>, 10:30 pp.

Kendeigh, S. C., Kanwisher, J. W., and Ridgway, S. H., 1983, The physiological ecology of whales and porpoises, <u>Sci. Amer.</u>, 248:111.

Kendeigh, S. C., Dol'nik, V. R., and Gavrilov, V. M., 1977, Avian energetics, in: "Granivorous Birds in Ecosystems", J. Pinowski and S. C. Kendeigh, eds., International Biological Programme, Vol. 12, Cambridge Univ. Press, London.

Kooyman, G. L., Gentry, R. L., Bergman, W. P., and Hammel, H. T., 1976, Heat loss in penguins during immersion and compression, <u>Comp. Biochem. Physiol.</u>, 54A:75.

Lasiewski, R. C., and Dawson, W. R., 1967, A re-examination of the relation between standard metabolic rate and body weight in birds, Condor, 69:13.

LeMaho, Y., Delclitte, P., and Chatonnet, J., 1976, Thermoregulation in fasting Emperor Penguins under natural conditions, Am. J. Physiol., 231:913.

Linstedt, S. L., and Calder, W. A., 1976, Body size and longevity in birds, Condor, 78:91.

MacMillen, R. E., Whittow, G. C., Christopher, E. A., and Ebisu, R. J., 1977, Oxygen consumption, evaporative water loss and body temperature in the Sooty Tern, Auk, 94:72.

Montevecchi, W. A., and Porter, J. M., 1980, Parental investments by seabirds at breeding area with emphasis on Northern Gannets, Morus bassanus, in: Behavior of marine animals, Volume 4: marine birds" J. Burger, B. L. Olla, and H. E. Winn, eds., Plenum Press, New York.

Nelson, J. B., 1971, The biology of Abbott's Booby Sula abbotti, Ibis, 113:429.

Nelson, J. B., 1978, The Sulidae, Oxford Univ. Press, Oxford.

Nelson, J. B., 1979, "Seabirds: their biology and ecology", A & W Publishers Inc., New York.

Pettit, T. N., and Whittow, G. C., 1983, Embryonic respiration and growth in two species of noddy terns, Physiol. Zool., in the press.

Pettit, T. N., Grant, G. S., Whittow, G. C., Rahn, H., and Paganelli, C. V., 1981, Respiratory gas exchange and growth of White Tern embryos, Condor, 83:355.

Pettit, T. N., Grant, G. S., Whittow, G. C., Rahn, H., and Paganelli, C. V., 1982a, Embryonic oxygen consumption and growth of Laysan and Black-footed albatross, Am. J. Physiol., 242:R121.

Pettit, T. N., Grant, G. S., Whittow, G. C., Rahn, H., and Paganelli, C. V., 1982b, Respiratory gas exchange and growth of Bonin Petrel embryos, Physiol. Zool., 55:162.

Pinshow, B., Fedak, M. A., Battles, D. R., and Schmidt-Nielsen, K., 1976, Energy expenditure for thermoregulation and locomotion in Emperor Penguins, Am. J. Physiol., 231:903.

Rahn, H., 1982, Comparison of embryonic development in birds and mammals: birth weight, tissue and cost, in: "A comparison to animal physiology", C. R. Taylor, K. Johansen, and L. Bolis, eds., Cambridge Univ. Press, Cambridge.

Rahn, H., and Ar, A., 1974, The avian egg: incubation time and water loss, Condor, 76:147.

Rahn, H., and Ar, A., 1980, Gas exchange of the avian egg: time, structure and function, Amer. Zool., 20:477.

Ricklefs, R. E., 1974, Energetics of reproduction in birds, in: "Avian Energetics", ed. R. A. Paynter, Publications of the Nuttall Ornithological Club, No. 15, Cambridge.

Ricklefs, R. E., 1977, Composition of eggs of several bird species, Auk, 94:350.

Ricklefs, R. E., and White, S. C., 1981, Growth and energetics of

chicks of the Sooty Tern (<u>Sterna fuscata</u>) and Common Tern
(<u>S. hirundo</u>), <u>Auk</u>, 98:361.

Ricklefs, R. E., White, S. C., and Cullen, J., 1981, Energetics
of postnatal growth in Leach's Storm-Petrel, <u>Auk</u>, 97:566.

Rubner, M., 1908, "Das problem der lebensdauer und seine beziehungen
zu wachstum und ernahrung", Oldenburg, Munich.

Schmidt-Nielsen, K., 1975, "Animal Physiology: adaptation and
environment", Cambridge Univ. Press, London.

Scholander, P. F., Hock, R., Walters, V., and Irving, L., 1950,
Adaptation to cold in arctic and tropical mammals and birds
in relation to body temperature, insulation and basal metabolic rate, <u>Biol. Bull.</u>, 99:259.

Schreiber, R. W., and Ashmole, N. P., 1970, Seabird breeding seasons
on Christmas Island, Pacific Ocean, <u>Ibis</u> 112:363.

Schönwetter, A., 1960, Handbuch der oologie, Lief. 1., ed. W. Meise,
Akademie Verlag, Berlin.

Schreiber, R. W., and Schreiber, E. A., 1983, Reproductive failure
of marine birds on Christmas Island, Fall 1982, <u>Trop. Ocean.
Atmos. Newsletter</u>, February, 10.

Simons, T. R., 1983, The breeding biology and conservation of the
endangered Dark-rumped Petrel (<u>Pterodroma phaeopygia</u>) in the
Hawaiian Islands, Ph.D. Thesis, University of Washington.

Sowls, A. L., Hatch, S. A., and Lensink, C. J., 1978, Catalog of
Alaskan seabird colonies, U. S. Dept. of the Interior, Fish
and Wildlife Service, <u>FSW/OBS</u>-78/78.

Sowls, A. L., DeGrange, A. R., Nelson, J. W., and Lester, G. S.,
1980, Catalog of California Seabird Colonies, U. S. Dept.
Interior, Fish and Wildl. Serv., Biol. Services Program <u>FWS/
OBS</u>, 37/80.

Stahel, C. D., and Nicol, S. C., 1982, Temperature regulation in
the Little Penguin, <u>Eudyptula minor</u>, in air and water, <u>J.
Comp. Physiol.</u>, B, 148:93.

Tucker, V. A., 1972, Metabolism during flight in the Laughing
Gull, <u>Larus atricilla</u>, <u>Am. J. Physiol.</u>, 222:237.

Vleck, C. M., and Kenagy, G. J., 1980, Embryonic metabolism of the
Fork-tailed Storm Petrel: physiological patterns during prolonged and interrupted incubation, <u>Physiol. Zool.</u>, 53:32.

Vleck, C. M., Vleck, D., and Hoyt, D. F., 1980, Patterns of metabolism and growth in avian embryos, <u>Amer. Zool.</u>, 20:405.

Warham, J., 1983, The composition of petrel eggs, <u>Condor</u>, 85:194.

Warham, J., Watts, R., and Dainty, R. J., 1976, The composition,
energy content and function of the stomach oils of petrels
(order Procellariiformes), <u>J. exp. Mar. Biol. Ecol.</u>, 23:1.

Whittow, G. C., 1980, Physiological and ecological correlates of
prolonged incubation in seabirds, <u>Amer. Zool.</u>, 20:427.

Whittow, G. C., 1983, Physiological ecology of incubation in tropical seabirds, <u>Studies in Avian Biol.</u>, in the press.

Whittow, G. C., 1984a, Regulation of body temperature, <u>in</u>: "Avian
Physiology", 4th ed., P. D. Sturkie, ed., Springer-Verlag,
New York.

Whittow, G. C., 1984b, Energy metabolism, <u>in</u>: "Avian Physiology",
    4th ed., P. D. Sturkie, ed., Springer-Verlag, New York.
Williams, A. J., Siegfried, W. R., and Cooper, J., 1982, Egg com-
    position and hatchling precocity in seabirds, <u>Ibis</u>, 124:456.

EGG FORMATION

C. R. Grau

Department of Avian Sciences
University of California
Davis, California 95616

The avian egg is the bridge between generations, the link that permits the mother to provide the food to its developing offspring at a time and place separate from her food supply. The unique combination of energy-rich lipids, utilizable proteins, minerals, vitamins, and water provides the embryo with an efficient source of all necessary nutrients except oxygen, packed into the egg over a short time.

Eggs first become evident at the primitive streak stage of embryo development when the primordial germ cells become differentiated from somatic cells. After the female chick hatches, the germ cells in the ovary are surrounded by granulosa, interstitial, and supporting cells (Romanoff, 1960). Typically only the left ovary develops, but some females have two functional ovaries; examples are kiwis, the Little Gull (Larus minutus), and several Falconiformes (Kinsky, 1971). The oviduct, the other half of the reproductive apparatus, develops from the Müllerian duct, derived from the urogenital ridge (Romanoff, 1960). It remains small and inactive until the hypothalamus and pituitary respond to genetic and environmental influences, providing the essential endocrine stimulation to reproduction. Birds with two oviducts are uncommon, but have been reported (Kinsky, 1971).

Descriptions of structures and events that lead to egg formation rest solidly on information derived from such well studied birds as chickens (Gallus gallus domesticus), Japanese quail (Coturnix coturnix japonica), ducks (Anas platyrhynchos), and White-crowned Sparrows (Zonotricia leucophyrs). From studies of the comparative reproductive anatomy and from gross, microscopic, and chemical examinations of eggs of diverse size and origin, it appears that the basic pattern of egg production is similar in all members of the Class Aves. The ovum contains a large germinal disc closely adhered to the perivitelline layer (yolk membrane); its abundant yolk material is primarily lipoprotein and is similar among species; protein-rich albumen surrounds the yolk; and a hard shell of calcium carbonate covers the enclosing shell membranes. No species exceptions to these generalizations have been reported, but variations occur in proportions of components and chemical compositions. For example, eggs of the Northern Gannet (Sula bassana) contain 15.5% yolk, 72.5% albumen, and 9.6% shell (Ricklefs and Montevecchi, 1979), whereas Bonin Petrel (Pterodroma hypoleuca) eggs contain 37.6% yolk, 56.1% albumen, and 6.3% shell (Warham, 1983). In addition to proportions of yolk and albumen, comparisons may be made of solids, protein, lipid, and energy composition of yolk, albumen, and total contents (Ricklefs, 1974). Information on seabird egg composition is included in Romanoff and Romanoff (1949), Kuroda (1963), Reid (1965), Lack (1968), Collins and LeCroy (1972) Lawrence and Schreiber (1974), Ricklefs (1977), Jones (1978), Carey et al. (1980), Siegfried et al. (1978), Williams et al. (1982), and Pettit et al. (this volume).

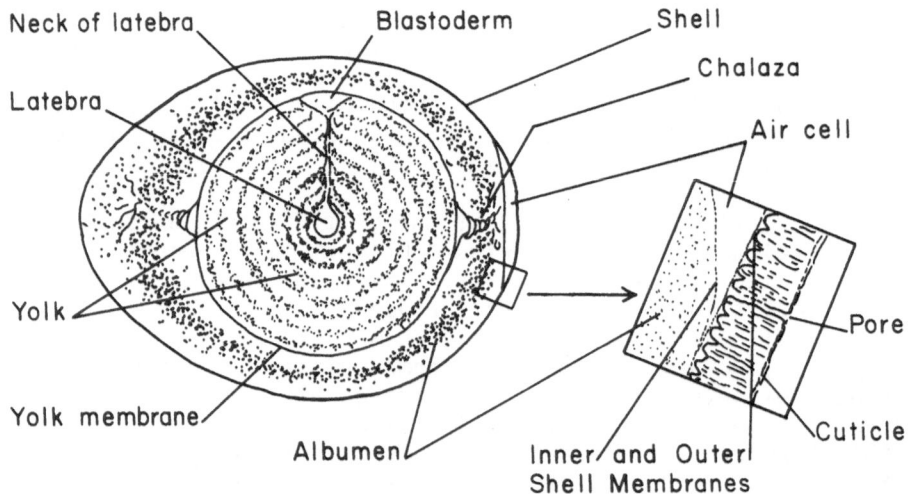

Figure 1.   Structure of Cassin's Auklet egg. Mass 25.5 g; dimensions 44.2 x 32.2 mm; yolk volume 7.8 ml; yolk diameter 24.6 mm. Enlargement shows air-cell area of the shell.

ORIGINS OF EGG COMPONENTS

The organs that produce the egg are the ovary, which contributes the egg cell and its accompanying nutrient-rich yolk, and the oviduct, which provides conditions for fertilization, protein and water in the form of albumen, and a hard shell. The times of egg-formation events vary widely among species, as shown in Figure 2.

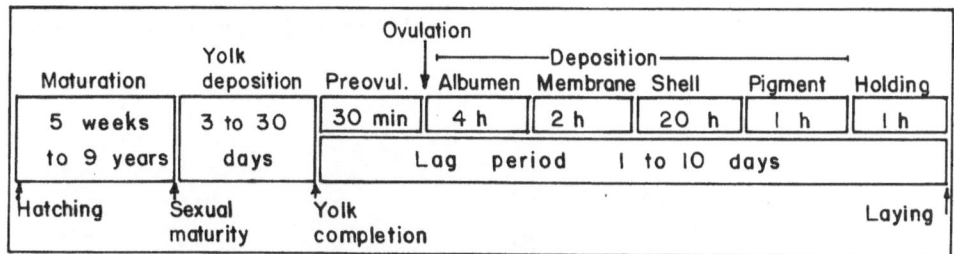

Figure 2. General time sequence of egg formation events among selected bird species. Examples of maturity ranges are Japanese quail, 5 weeks; Royal Albatross (Diomedea epomophora), 9 years. Yolk deposition: Brewer's Blackbird (Euphagus cyanocephalus), 3 days; Royal Albatross, 30 days. Albumen and shell deposition have been studied only in domestic species such as the chicken. Holding in the uterus is normally less than an hour in chickens. Times in other species are not known. The lag period is one day in chickens, ten days in the Royal Albatross.

The Ovary

There is remarkable uniformity in the development of the nucleus and cytoplasm of egg cells in the avian embryo. Several stages of prophase are completed by the time the chick hatches (Gilbert, 1979), but meiosis is not completed until the final stages of yolk deposition after the chick becomes reproductively active. In such slow maturing birds as albatrosses or large penguins the period of gonadal inactivity may continue for 10 or more years.

The premature ovary and its oocytes remain small; the nucleus is displaced from the center but remains spherical and follicle cells surround each oocyte. Over a period of months, some oocytes grow slowly to 1 mm or more, after which rapid yolk formation is possible (Gilbert, 1971). The nucleus, now called the germinal disc, has become flattened and closely apposes the surrounding follicle cells (Romanoff, 1960).

The stimulus for rapid yolk deposition to begin is unknown, but once initiated in a single follicle yolk deposition continues until yolk growth is completed. This process may be as short as 3 days in some passerines or as long as 30 days in an albatross.

Until the approach of the breeding season, the ovary remains small, approximately 0.2 % of body mass. Under the influence of follicle-stimulating hormone (FSH), the ovary grows enormously, increasing its mass to more than 2% of body mass, most of which is yolk itself (King, 1973; Ricklefs, 1974). Essentially all ovarian needs for energy and other nutrients are those required to form the ova, with a minimum necessary for follicle and supporting tissue maintenance.

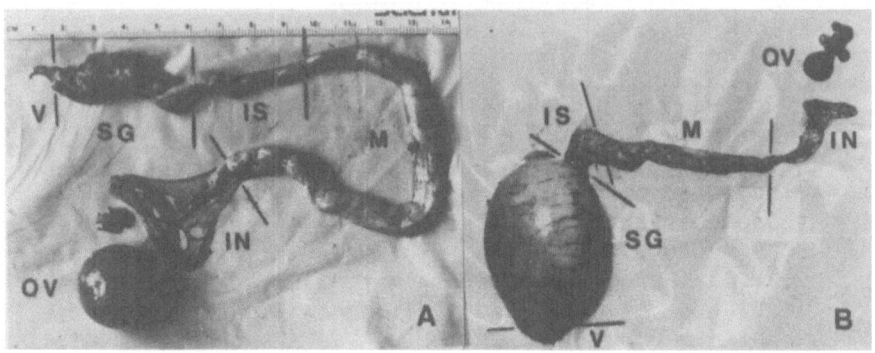

Figure 3.   Reproductive organs of Cassin's Auklet.   A. Ovum ready for ovulation.   B. Hard-shelled egg in the shell gland. OV, ovary; IN, infundibulum; M, magnum; IS, isthmus; SG, shell gland (uterus); V, vagina.

The Oviduct

In preparation for egg formation, the oviduct undergoes rapid, massive growth. In the immature or regressed state, the oviduct has a mass of approximately 0.25% of body mass, which increases to approximately 4% during egg formation when it is active in albumen synthesis (Ricklefs, 1974). The additional nutrients needed by a 3 kg penguin, for example, to grow its 120 g oviduct are approximately 24 g protein and 6 g lipid, a total of 183 kcal energy, amounts easily furnished by a few extra meals.

Following ovulation, the infundibulum, the funnel-like upper end of the oviduct, surrounds and engulfs the ovum. Sperm are released from short-term storage sites in the folds of the lower infundibulum and penetrate the perivitelline layer which surrounds the newly ovulated yolk, thus fertilizing the ovum. The ovum is enveloped by a secretion of the magnum which becomes the outer layer of the yolk membrane, and by the albumen layer, thus effectively blocking further contact of the egg cell by sperm. The yolk and albumen move down the oviduct into the isthmus, a thin-walled section of the oviduct which produces fibrous secretions that enclose the albumen-covered yolk in a membranous sac. Water and salts are added and the shell gland (uterus) deposits calcium carbonate in a protein matrix to produce the characteristic rigid shell.

Pigments are incorporated into the shell as it is deposited, and shortly before oviposition surface pigments are added as spots or streaks. Finally a thin cuticle of complex protein and lipid covers the entire shell, and the egg is laid.

YOLK DEPOSITION

As domestic fowl enter a reproductive cycle, the hypothalamus acts on the pituitary via various releasing factors, and it, in turn, secretes gonadotrophins into the blood stream. In response, the ovary grows, produces progesterone in the granulosa cells of the follicle, converts progesterone to testosterone in the thecal cells, and from testosterone latter, estradiol and other estrogens are synthesized in the theca and the granulosa cells (Gilbert, 1979). In response to the increased blood estrogen levels, the liver produces large amounts of lipoproteins, raising the blood concentration of these yolk precursors as much as 30-fold. Simultaneously, the estrogens stimulate the growth of the oviduct, setting the stage for egg formation (Gilbert, 1979).

Initiation of Rapid Yolk Deposition

The combination of factors which results in the initiation of yolk deposition remain obscure. The female must, of course, have a mature ovary which can support the development of one or more ova, and the oviduct must be ready to receive the released ovum at the appropriate time. Interactions of light, rain, and other environmental variables with the genetic information inherent in the birds stimulate the final chain of events. Courtship display, nest or burrow preparation, territory defense, feeding, copulation, and other behavioral factors may also influence initiation of yolk growth. From birds that lay large clutches it appears that the selection of the first follicle to begin its growth is the most important control feature of all the processes of egg formation: when one follicle is marked to proceed with depositing yolk, all other candidates for development are in some way inhibited. This takes place each day for birds that lay large clutches and only once for each single egg clutch.

In some seabird populations, initiation must occur over a relatively short time, perhaps one to two weeks, because all birds lay within a period of only a few days, and egg formation times are constant. For example, eggs of most Short-tailed Shearwaters (Puffinus tenuirostris) are laid within a 5-day period (Serventy, 1963).

Follicle Structure

Superficially, avian follicles show little variation between species except size. However, there have been no comparative studies of their gross or microscopic anatomy, and for such information we must presently rely on observations of domestic fowl. It is only recently that improved fixing and staining techniques for electron microscopy have revealed the fine structure of the follicle wall during yolk formation (Perry, et al., 1978).

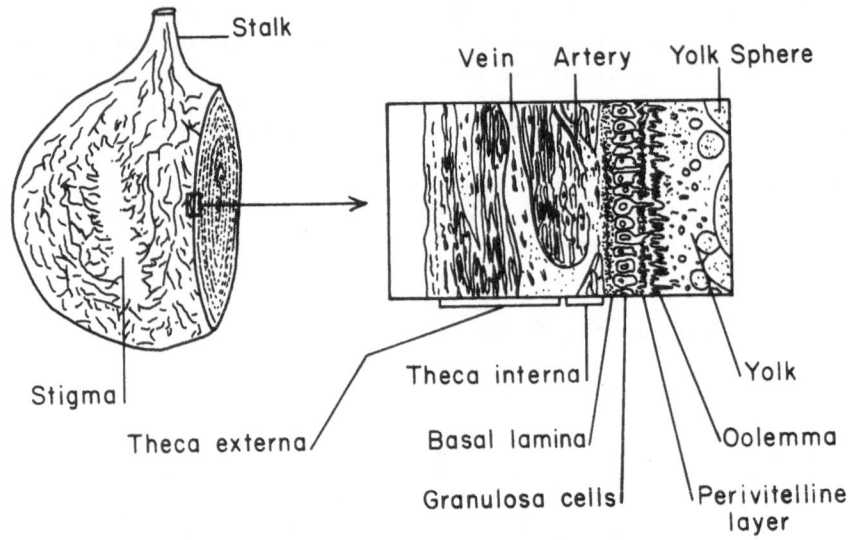

Figure 4. Follicle of ovary cut to show wall. Enlarged structure at right.

The complex structures by which plasma components are selectively concentrated and deposited in the ovum develop rapidly, function predictably, and when their work is completed after ovulation, regress to a quiescent state. The structural details are shown in Figure 4. The follicular stalk provides a strong ovarian connection and carries arteries, veins, and nerves. The outer follicular capsule and the theca externa are stretched tightly over the expanding ovum. The arterial supply, inconspicuous on the surface compared with the venous drainage (Nalbandov and James, 1949), penetrates the theca interna and becomes a capillary network with a fenestrated endothelium that permits direct contact of the blood with the basal lamina. This membrane completely surrounds the granulosa cells and the enclosed ovum, acting as a sieve of plasma components (Perry and Gilbert, 1980). The perivitelline layer, which is probably produced by the granulosa layer (Gilbert, 1979), is a network of protein fibers approximately 0.2-0.6 μm thick surrounding the yolk mass and the germinal disc. After ovulation, sperm may move through its 2 μm spaces between fibers or possibly they may dissolve part of it in order to fertilize the ovum (Bakst and Howarth, 1977). But the yolk spheres which constitute the yolk mass are 80-120 μm in diameter and cannot escape through the fibers. The perivitelline layer separates from the granulosa layer at ovulation and is the final contribution of the ovary to the laid egg.

The follicle is generously supplied with nerves, but the possible functions of the nerves in yolk initiation, formation, or ovulation are not known (Gilbert, 1979).

Yolk Formation and Structure

Yolk is a mixture of concentrated lipoproteins and other plasma components. The plasma lipoproteins exist as particles of 15-260 nm diameter, with most being about 25 nm, a size capable of passing through the basal lamina, between the granulosa cells, and through the perivitelline layer to the ovum, where the oolemma incorporates the concentrated material into the ovum (Evans et al., 1979). The yolk, as compared with the plasma from which it is derived, is low in water, sodium, and chloride, and is high in lipoproteins, in potassium, calcium. phosphorus, and some free amino acids (see Table 1). It is acidic compared with plasma.

Recently, Perry and Gilbert (1981) found that lipoprotein particles become attached to the oolemma at pH 7.5 and are released at pH 7.2. Thus the pH shift may help to move lipoproteins from plasma to ovum.

The proportions of proteins, lipids including cholesterol, and other major yolk components do not fluctuate with environmental changes such as food supply, but the fatty acids present are influenced by the female's diet (Shenstone, 1968).

Because the yolk is formed directly from plasma, any substance that has reached the circulation may be present in the yolk. Minerals such as iron and calcium are carried in tightly bound forms; iron by transferrin and phosvitin; calcium by lipovitellin and phosvitin. Vitamins are also bound to proteins. Fat-soluble pigments such as lutein and astaxanthin are effectively transported to the yolk, where they may serve as convenient markers of feeding activity. Lipophilic pollutants such as chlorinated hydrocarbons are concentrated in yolk, which can serve as an effective route of excretion of toxic substances for the female but may be potentially hazardous to the embryo.

Yolk appears to be a viscous liquid, but in reality, whether
domestic fowl or seabird, it is a collection of individual yolk
spheres pressed together inside the yolk membrane with almost no
free fluid between the spheres (Fujii et al., 1973); within the
yolk, the spheres are stable for months under conditions of
refrigeration, but when diluted the spheres are easily ruptured,
and under the microscope the contents can be seen to pour out of
the spheres.

Table 1. Yolk and plasma compositions. Concentrations of selected
components of domestic fowl yolk and hen plasma, and
ratios indicating the concentrating actions of the ovary.

	Plasma g per 100 ml	Yolk g per 100 g	yolk ÷ plasma
Proteins	5.2	16.0	3.1
Phosvitin	0.027	0.75	28.0
Lipids	3.0	32.0	10.7
Cholesterol	0.2	1.15	5.8
Water	94.0	45.0	0.48
Solids	6.0	55.0	9.2
Glucose	0.25	0.2	0.8
Uric acid	0.025	0.045	1.8
Sodium	0.39	0.06	0.15
Potassium	0.026	0.11	4.2
Calcium	0.038	0.16	4.2
Phosphorus			
Total	0.04-0.12	0.52	13-4.3
Inorganic	0.005-0.01	0.02	4-2
Chlorine	0.42	0.13	0.31
Hydrogen ion	4×10^{-7}	1×10^{-5} to 5×10^{-6}	25.0
pH	7.4	6.0-6.3	
Free amino acids			
Aspartic acid	0.0007	0.027	39.0
Isoleucine	0.0026	0.026	10.0
Glycine	0.0053	0.011	2.1
Cystine	0.0053	0.00046	0.9

Daily Yolk Rings

Until Riddle (1908) conceived the idea of feeding a
fat-soluble dye to a hen in order to trace the history of yolk
formation, little thought had been given to this part of avian
reproduction. It had been recognized by Thompson in 1859 (quoted
by Gilbert, 1979) that yolk was laid down in layers by adding
material to the outer margin. It was Riddle, however, who first
fed the lipophilic red dye Sudan III to a hen and observed that an
egg laid several days later contained the dye. He hardcooked and

cut open yolks and observed a red ring that marked the day the dye was fed. He used this technique to estimate the growth of the yolk to be about 2 mm in diameter each day (Riddle, 1910). He observed that even without dye being fed, light and dark yellow rings could often be seen, and he attempted to determine the composition of "yellow" and "white" yolk rings as well as that of the central yolk, which was always light. He thought that during times of feeding, yellow yolk was laid down rapidly and efficiently, whereas white yolk was deposited during non-feeding periods, particularly at night (Riddle, 1911).

Warren and Conrad (1939) extended Riddle's basic idea to quantify yolk growth in the hen, and later determined rates of growth during the day and the night (Conrad and Warren, 1939). They concluded that there was only a 5% difference in rate of yolk deposition between day and night periods, and that the hen used her tissue reserves during the night, consequently depositing yolk that was less deeply pigmented than during the day when the hen was eating a diet rich in natural yellow or orange xanthophyll pigments. They also concluded that yolk formation rate was increased only slightly by feeding a diet enriched with yolk, and in general held the view that perceived differences between yellow and white yolk were functions of the intake of foods containing more or less yellow pigment, not differences in the kind of yolk being deposited.

It now appears that there are indeed differences between "day" and "night" yolk that can be distinguished by freezing, fixing in formalin, and staining with a potassium dichromate solution (Grau, 1976), and that these differences are unrelated to pigments present in the diet. The differences in appearance are due to variation in transparency, with clearer, dark-staining yolk revealing dichromate-marked lipids, whereas light-staining yolk is more opaque, and reflects more incident light. This staining technique can be applied to wild as well as to captive bird eggs, permitting counting of dark- and light-staining rings that represent day (dark) or night (light) periods and thus the total days of yolk formation. The rings are not simply related to feeding activity, however, because faint rings can be seen in yolks of birds that do not feed during egg formation such as the Fiordland Crested Penguin (Warham, 1974; Grau, 1982). These rings are poorly differentiated as compared with those of other penguins that feed during yolk formation such as the White-flippered Penguin (Eudyptula minor albosignata), as shown in Figure 5. Adelie Penguins feeding on pigment-rich crustacea during the early part of yolk formation have distinct, pink inner rings. After they come ashore and are no longer eating, the yolk rings they deposit are pale and less well defined (Astheimer and Grau, 1983).

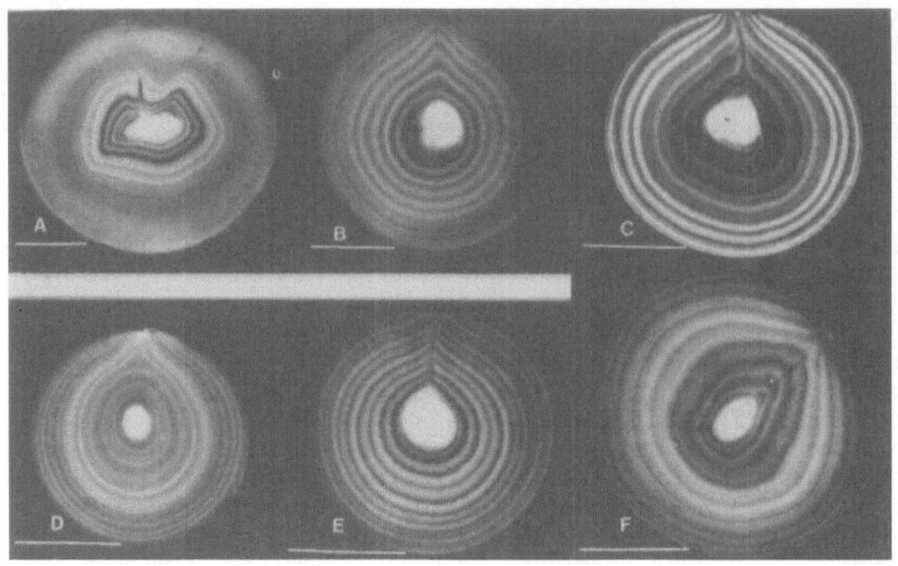

Figure 5. Slices of yolks after dichromate staining. A. Adelie
Penguin. B. White-flippered Penguin. C. Red-billed
Gull. D. White-faced Storm petrel. E. Spotted Shag.
F. Cassin's Auklet.   Scale marker is 10 mm.

The yolk in a particular ring is a partial sample of plasma
that was circulating on the day the yolk was deposited. Nutrients,
pollutants, steroids, and other blood-carried substances are among
the materials that could be measured in yolk and related to the
female's status days or weeks before the egg was laid.

Yolk Growth in Seabirds

Because the dye-feeding technique of determining yolk growth
depends on the accessibility of the female for repeated dye dosing
during the period of rapid yolk deposition, as well as on
convenience of collecting eggs laid subsequently, the method has
not been adaptable to free-living birds. An indirect method of
estimating yolk growth by measuring follicle diameters at necropsy
of birds was used by Stieve (1918) to construct growth curves for
the yolks of chickens. Romanoff (1943), using this technique for
chickens, ducks, pheasants and quail, concluded that yolk growth
rates were similar when corrected for final yolk weights. Seabirds
have not been studied in this way.

An indirect means of determining the period of yolk growth is
to record the time required to lay a replacement egg after the
clutch has been completed and the ovary has regressed. Natural
disasters during incubation have provided material for such

studies: for example, Paludan (1951) observed synchronous egg laying in Herring Gulls (Larus argentatus) 12 days after completed clutches being incubated were destroyed by a snow storm. The time between removal of fresh eggs as they are laid and relaying gives a maximum estimate based on the assumption that a new yolk starts to grow soon after the clutch is removed, and that the time between yolk completion and laying is not variable. Examples of the regularity of relaying are Cassin's Auklet (Ptychoramphus aleuticus) and Western Gull (Larus occidentalis). In the former, removal of the single egg before incubation was followed by laying a second egg 15 days later (Astheimer, unpublished), while in the gull the normal clutch of three was extended to four, followed by a gap of 6 days, when the first of three more eggs was laid (Table 2) (Roudybush and Grau, unpublished). Unfortunately, it is not possible from these data alone to draw firm conclusions about yolk growth because the time of initiation of yolk formation cannot be determined. From ring counts after dichromate staining of the gull yolks, however, it appears that the first yolk of the second clutch began to grow the day after the first egg of the first clutch was removed, giving a total yolk formation time of 10 days. Other examples of relaying include Thick-billed Murre (Uria lomvia) 14 (range 11-17) (Gaston and Nettleship, 1981), Common Murre (Uria aalge) 14 (R. Boekelheide, pers.com.), and Gannet (Sula bassana) 18 (range 6-39) (Nelson, 1978).

Table 2. Egg laying by Western Gulls (Larus occidentalis) in nests from which eggs were removed daily.

	2	3	4	5	6	7	8	9	10	11	12	13	14	15	16	17	18	19	20	21
Nest 1	X		X		X			X							X			X		X
Nest 2	X		X		X			X							X		X		X	
Nest 3	X		X		X										X		X		X	
May	2	3	4	5	6	7	8	9	10	11	12	13	14	15	16	17	18	19	20	21

The natural diurnal variation in yolk structure seen in dichromate-stained yolk permitted estimation of days for yolk formation in numerous charadriiform birds (Roudybush et al., 1979) and led to the general conclusion that yolk formation time increased with egg size. Application to birds of other orders revealed that there is considerable variation in yolk deposition time and that it is not necessarily related to yolk size. The techniques used to quantify yolk formation were essentially those used by Warren and Conrad (1939) with hens, except that they were able to inject a dye marker intravenously each day and subsequently measure the radius of each daily ring to calculate the volume, assuming the yolk was spherical. They plotted cumulative volume and daily increment against time. Because such daily feeding or injecting of dye is virtually impossible in wild seabirds, Grau and

Astheimer (1982) obtain data from dichromate-stained yolks by measuring the radius of each daily ring on enlarged black and white photographs or color photocopy prints made from color transparencies of slices of fixed and stained yolk. From the radius measured on the photo and the diameter measured on the yolk slice, the actual yolk volume included in each ring was calculated for yolks of 24 seabird species representing four orders. Plots of cumulative volume against time revealed striking differences in growth curves. Common Murres (Uria aalge) and Tufted Puffins (Lunda cirrhata), for example, deposit yolk at high rates almost comparable to those for domestic fowl, whereas some Procellariiformes deposit yolk slowly. In many species the first third of yolk deposition is relatively slow, an example being the Grey-faced Petel (Pterodroma macroptera) (Figure 6). Data for albatrosses show that their yolk growth rates are higher than in other Procellariiformes studied. Unique among the birds studied are the cormorants and shags. Their yolks are small, with beautiful, regular rings (Figure 5). The growth curve of these yolks is flat as compared with other seabirds (Figure 6).

Because of the diversity of yolk sizes and times for formation, it is difficult to express comparative yolk formation rates. In curve-fitting tests, cumulative yolk growth generally conforms best to a second-order polynomial expression; coefficients of species curves can be compared via cluster analysis (Astheimer and Grau, ms). In addition to the plots of volume vs. time, proportions of final weight were plotted against time, or proportions of total time. Such manipulations did not prove to be as useful in understanding the processes involved as simple plots of volume against time.

Another method of expressing growth rates, relating volume deposited per day against time, is subject to considerable variation because a small difference in the radius results in a large difference in the daily increment.

OVULATION

The cessation of yolk deposition within a follicle is as mysterious an event as was the initiation of yolk growth some days earlier. After the yolk is completed, the next landmark in egg formation is ovulation, the release of the ovum into the body cavity. In domestic birds that lay almost every day, the time relationships between oviposition and the next ovulation are well known and highly predictable: 15-45 minutes after an egg is laid the next ovum is released (Warren and Scott, 1935; Woodard and Mather, 1964; Gilbert, 1971; Simmons and Hetzel, 1983). This conclusion was drawn by internal examination of chickens, turkeys, quail, and ducks that had been anesthesized or killed at intervals after oviposition. Non-invasive time markers are not available to trace these events in free-living birds, hence accurate prediction of ovulation times are not yet possible for seabirds. However, because of the regularity of relaying following egg removal in some species, it may in the future be possible to make such predictions.

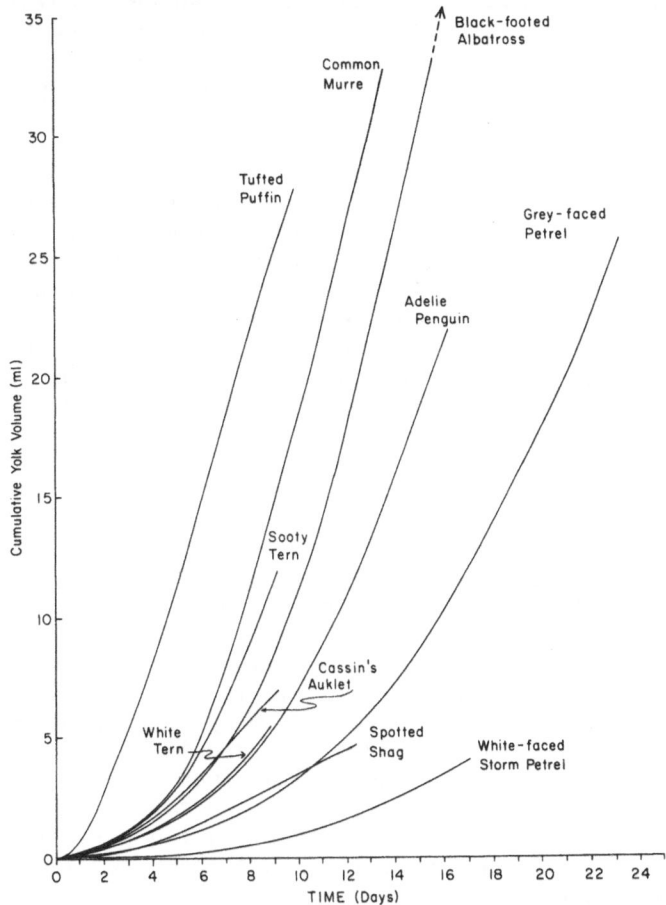

Figure 6. Yolk growth in some representative seabirds.

In chickens, yolk deposition usually continues until a few hours before ovulation (Smith, 1959; Gilbert, 1972) and the egg is laid a day later; in turkeys the timing appears to be less clear, possibly with a delay of 1-3 days before laying (Bacon and Cherms, 1968). In Emus (Dromaius novaehollandiae), Hirsch and Grau (1981) found a 10 day lag between yolk completion and laying.

The unexpectedly long lag in Emus, which have irregular laying intervals of several days, led us to investigate seabirds, many of which lay clutches of one or two eggs, and do not operate under the same time constraints as birds that lay one egg daily for several days. By feeding a dye capsule once and counting the dichromate-stained yolk rings of the egg laid several days later, it was found that Cassin's Auklets have a lag period of 4-6 days (Asthimer et al., 1980), gulls have a 1-3 day lag, and penguins 4-7 days, depending on the species. The egg of one Northern Royal Albatross showed a lag of 10 days; other Procellariiformes studied have not yet yielded definitive results, primarily because the prelaying exodus does not permit the administration of a dye capsule during the 10-15 days before laying.

It is not presently possible to offer an explanation for the lag between yolk completion and laying, but the evidence favors a delay in ovulation rather than prolonged stay of the egg in the oviduct. Observations of the reproductive tracts of Cassin's Auklets at various stages of egg formation after having been induced to lay replacement eggs were consistent with ovulation delay (Astheimer, pers. com.), but other species have not been similarly studied. The lag appears to be fixed within a population, with little variation among eggs. One possible advantage to a lag period of several days is that more time would be provided to synthesize and store albumen and shell membrane materials in anticipation of completion of the egg after ovulation.

The constancy of the lag period and physical and spatial constraints that would be imposed by large clutches and long laying intervals suggest that long lag periods may have been a feature of reproduction in ancient birds. Present-day reptiles exhibit a range of developmental options from oviparity to pseudoplacental viviparity (Packard et al., 1977). It is possible that as birds with their hard-shelled eggs evolved, many species produced more eggs per clutch, laying intervals were shortened, and lag times necessarily were reduced to the minimum seen today in galliforms and anseriforms. Single-egg clutches, and small clutches laid over several days, could easily accomodate long lag periods, however, and in the absence of pressure to reduce them, lag times may have remained unchanged.

The actual mechanism by which the ovum is released from the follicle is not known (Yoshimura and Koga, 1983), but the hormonal events that lead to ovulation have been investigated in domestic fowl and may point the way to a general understanding (Sharp, 1980). For example, progesterone is produced in the follicle to be ovulated, and there is a surge in the plasma LH, progesterone, and other steroids approximately 6 hours before ovulation, but the relation of these changes to the actual ovulation is not clear (Etches and Cunningham, 1976; Bluhm et al., 1983). Rupture of the follicle wall occurs along the stigma, a wide, scar-like line on the surface of the follicle where there are few collecting veins (Figure 4). The ovum, enclosed only by the perivitelline layer, is separated from the granulosa cells, the follicle wall contracts, and the ovum oozes out of the follicle, soon to be engulfed by the finger-like projections of the infundibulum.

Table 3. Egg formation in selected seabirds. Lag times are days between yolk completion and laying. A, first egg; B, second.

	Clutch size	Egg mass	Laying interval	Yolk formation time	Lag time	Total egg formation time
	n	g	d	d	d	d
Sphenisciformes						
Pygoscelis adeliae	2	120	3	15	6	21
Eudyptes pachyrhynchus	2	A 97 B120	4	16 17	7	23 24
Eudyptula minor albosignata	2	61	2-3	14	4	18
Procellariiformes						
Diomedea epomophora sanfordi	1	440	-	30	10	40
D. nigripes	1	300	-	25		
Pterodroma macroptera gouldi	1	86	-	23		
Pelagrodroma marina maoriana	1	13	-	14		
Pelicaniformes						
Stictocarbo punctatus punctatus	3	40	3	12		
Phalacrocorax penicillatus	3	50	3-4	16		
Charadriiformes						
Larus occidentalis	3	85	2	10	3	3
L. novahollandiae scopulinus	3	38	2	12	2	14
Sterna fuscata	1	35	-	9		
Gygis alba	1	22	-	9		
Lunda cirrhata	1	28	-	10		
Ptychoramphus aleuticus	1	26	-	9	4	13

Prostaglandins are apparently not involved in ovulation as they are in mammalian ovulation. However, prostaglandins produced by the postovulatory follicle affect the shell gland and ovipostion (Day and Nalbandov, 1977) explaining in part the observation that excision or ligation of the postovulatory follicle inhibits laying of the egg which had been ovulated by that follicle (Rothchild and Fraps, 1946).

FERTILIZATION

In contrast to voluminous observations on copulation and other breeding behavior, the information on fertilization and embryonic development in birds other than domestic species is sparse.

Fertility does not appear to be a problem in wild birds, in
contrast to captive birds; field observations reported as poor
fertility (i.e., not hatched) are usually early embryo mortality.
The relations of timing of single or multiple copulations, sperm
longevity in the female reproductive tract, sperm transport,
motility, and fertilizing capacity have not been studied in
relation to formation of seabird eggs as they have in domestic fowl
(Lake, 1975), where the electron microscope has permitted a better
understanding of the penetration of the perivitelline layer by
sperm (Bakst and Howarth, 1977), and formation of the female and
male pronuclei preparatory to fertilization.

In chickens, turkeys, and quail, mature sperm are found in
sperm-storage glands of the uterovaginal junction, from which they
are transferred to the infundibulum-magnum IM) storage sites at the
upper end of the oviduct before the release of the ovum from the
follicle (Lake, 1975; Ogasawara, unpublished data). Because
fertilization occurs soon after the ovum leaves the infundibulum,
active sperm must be available at the time of ovulation. Hatch
(1983) examined the oviducts of Northern Fulmars (Fulmarus
glacialus), Horned Puffins (Fratercula corniculata), and Leach's
Storm-petrels (Oceanodroma leucorhoa) before or shortly after
egg-laying, and found sperm-storage glands in all species,
essentially similar to domestic fowl. Such sites were also found
in Cassin's Auklet oviducts (Astheimer and Grau, unpub.). Hatch
concluded that, for petrels, the presence of these glands permits
the male and female to be separate for the several weeks after
copulation that constitute the prelaying exodus. Puffin and
Cassin's Auklet pairs remain together before laying and have no
such need, but possess the glands, possibly as a general feature of
avian reproduction. Previously, Imber (1976) had reported the
possibility of long-term sperm storage in Grey-faced Petrels
(Pterodroma macroptera) which have a prelaying exodus of 6 weeks,
not an impossible length of time as turkeys are known to store
sperm up to 42 days (Lake, 1975).

The ovum moves rather quickly through the magnum and isthmus of
the domestic fowl oviduct, and cleavage usually does not occur
before the shell gland is reached. By the time of oviposition, the
embryo of a chicken egg is made up of some 20 thousand cells. No
data are available for seabirds.

FORMATION AND SECRETION OF ALBUMEN

Unlike the ovarian follicle, which concentrates plasma
lipoproteins to form yolk, the oviduct tissue utilizes free amino
acids and other small molecules to synthesize the more than 10
proteins that constitute the albumen solids (Feeney and Allison,
1969). The major products are ovalbumin, ovotransferrin,

ovomucoid, ovomucin, lysozyme, avidin, and ovoflavoprotein. Their
water solubility has made them relatively easy to characterize
physically and chemically (as compared with yolk), and comparative
studies have included some seabirds, particularly two penguins,
Pygoscelis adeliae (Feeney et al., 1966), and Eudyptula minor
(Baker and Manwell, 1975). No evolutionary advantage can be
assigned to one kind of protein or another but they have proved
useful in taxonomic associations (Ho et al.,1976; Sibley, 1960).

In birds that lay an egg each day, albumen is synthesized and
stored in the tubular gland cells of the magnum. The secretion of
the albumen into the lumen completely surrounds the ovum with a
thick protein gel over a short period of time, 3-4 hours in
chickens. As the ovum and the surrounding albumen are moved by
peristaltic action down the oviduct, the shape of the egg becomes
apparent, with the small end first.

After the egg enters the uterus and fluid is added to the
level found in the laid egg, the albumen contains approximately 11%
dry weight, almost entirely protein, and sodium 150 mg per 100g,
potassium 140, and chloride 130, compared with plasma values of
390, 26, and 420, respectively. Other elements such as calcium,
magnesium, phosphorus, and iron are present in minor amounts
(Everson and Souders, 1957; Shenstone, 1968). Fat-soluble vitamins
are absent while water-soluble vitamin composition reflects amounts
present in the diet. The protein avidin tightly binds any biotin
present.

SHELL MEMBRANES, SHELL, AND PIGMENTS

The ovum, enclosed in albumen, moves into the isthmus, where
in chickens the fibrous shell membranes are secreted over several
hours. The membrane fibers are composed of a mucoprotein core
surrounded by a unique protein (Leach and Rucker, 1978), the outer
surface of which is well supplied with sites where calcification
can begin. The egg moves on into the so-called "red" isthmus, or
perhaps more properly, the red-appearing part of the shell gland.
Crystalline deposits of a calcium salt accumulate at points on the
fibers (Wyburn et al., 1973), while a fluid secretion of
approximately equal volume as the original albumen moves into the
egg to cause plumping. The loosely fitting membranes are stretched
to make a turgid base for calcification. After this significant
addition of fluid, which is approximately 50% of the total albumen,
the proportions of solids in albumen remain constant, and do not
vary markedly among birds of different species. The variations
that occur among eggs are in the amounts of albumen and their
proportions in the egg (Carey et al., 1980), not in the proportions
of water that they contain.

Fluid addition and calcification of the shell membrane proceed simultaneously for a time, the shell becomes rigid and the calcium carbonate which constitutes the shell continues to be deposited in a protein matrix (Simkiss, 1975). In chickens, turkeys, quail, and ducks the process of shell deposition continues for 18-20 hours.

The general shape of the egg can be observed in eggs still in the magnum, but to what extent modification occurs in the shell gland is not known. Speculation about the possible adaptive advantages of the pyriform shape of murre eggs to prevent their rolling off cliff ledges, was expressed as early as 1858 (Laishley), but contrary views were stated (Dixon, 1888), and real evidence is lacking.

Peristaltic rotation of the egg within the shell gland spins out fibers from the ovomucin fraction, leaving a less viscous albumen close around the ovum and forming the chalazae. This results from the constant upward position of the blastoderm and the movement of the albumen around the yolk.

Seabird egg shells vary widely in color, design, and texture. Procellariiform eggs are primarily white, but some (e.g., Laysan Albatross) have red-brown spots on the blunt end. Penguin and cormorant shells are pale blue with parts covered by a chalky coat, which is noticeably rough in cormorant eggs. Charadriiform eggs vary widely from white (auklets), to greenish brown with heavy spotting (gulls), to the extremes of murres, which may be pale or dark, blue or brown, with light or heavy streaking (Harrison, 1978). A recent study of the infrared reflectances of eggs indicates that eggs such as those of gulls that are dark in visible light are protected against heat by their properties of reflecting infrared light (Bakken et al., 1978). Heavy spots and streaks of porphyrin pigments are deposited shortly before oviposition.

OVIPOSITION

It has long been recognized that oviposition is a closely regulated event in egg formation related to the ovulation which occurred hours earlier (Rothchild and Fraps, 1946). Because an injection of oxytocin was known to induce oviposition of eggs with hard shells, attempts were made to link shell gland contractions with secretions of the postovulatory follicle and the posterior pituitary. It now appears that production of prostaglandins by the follicle and their release into the blood causes oviposition (Day and Nalbandov, 1977).

The egg proceeds down the oviduct small-end first, and presumably is usually laid the same way, but in chickens many eggs are reversed before laying (Bradfield, 1950).

50

ENERGY AND PROTEIN NEEDS

The nutrient needs of birds for egg production have been estimated by King (1973) and Ricklefs (1974), principally in galliforms, anseriforms, larids, and passeriforms from data on egg size and composition, laying interval, clutch size, and ovary and oviduct growth. They concluded that these requirements may place significant strains on birds with restricted food resources.

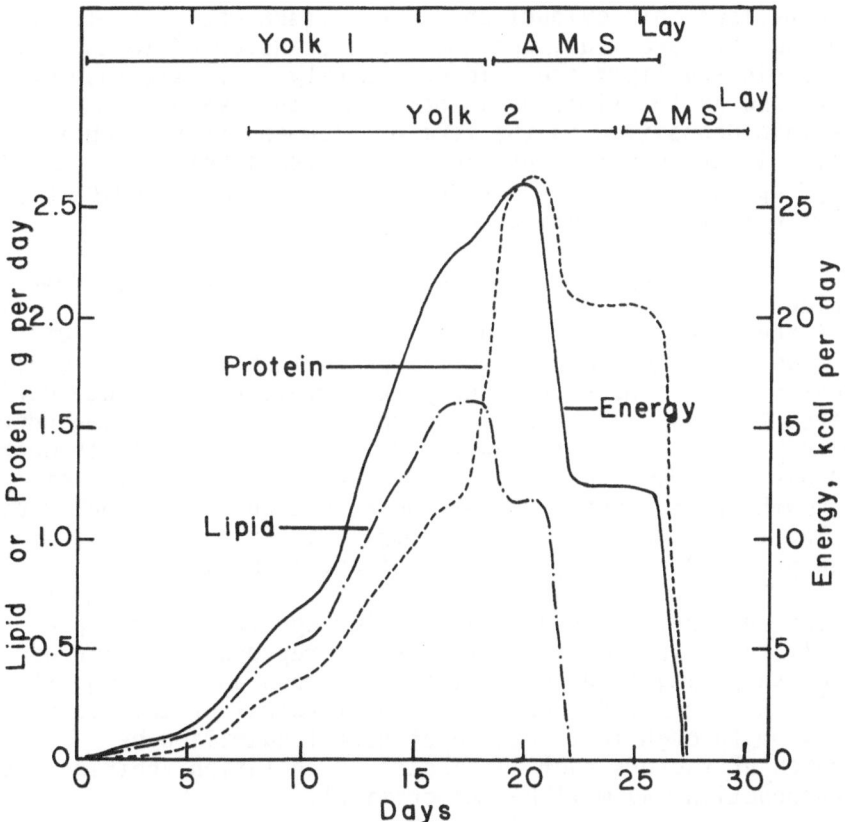

Figure 7. Amounts of protein, lipid, and energy deposted daily in the two eggs of a Fiordland Crested Penguin. AMS is period of deposition of albumen, membranes and shell.

It is difficult to determine the significance of nutrition in the production of eggs of most seabirds because of uncertainties in the time needed to synthesize and deposit the egg components. From data such as those presented in Figure 6 and in Table 3, it is possible to calculate the daily increments of yolk, albumen, and shell that are deposited, and from composition data for these components the total nutrient amounts deposited each day. An example of plots relating protein, lipid, and total energy to time is shown in Figure 7 for the Fiordland Crested Penguin (Grau, 1982). Crested penguins lay two eggs 4 days apart, with the second being significantly larger than the first (Warham, 1975). In these birds, which do not eat during the entire process of egg formation, nutrients are mobilized from the adipose and organ tissue stores and synthesized into oviduct and liver tissue where albumen and yolk precursors are made, and these are transferred to the egg. Yolk protein and lipid are laid down slowly at first, then more rapidly, and as the lipid needs halt when the ova are completed, the protein deposition in the form of albumen rises, then falls rapidly. In these birds, providing the amounts of nutrients of the two eggs appears not to be burdensome to the female, which begins its fast with large reserves.

The daily increments of food needed to provide egg nutrients vary widely among seabirds. For example, two species which lay eggs of the same yolk mass and total mass are the Tufted Puffin, an alcid, and the Grey-faced Petrel, a procellariiform (Figure 6). The nutrient needs of the petrel for yolk formation are minute during the first 10 days, and growth is also slow during the remainder of the 23-day period. In the puffin yolk growth is rapid throughout the 10-day period. These two seabirds represent extremes of yolk-growth rates. Unfortunately we do not know the lengths of their lag times, hence we cannot compare their total nutrient needs for egg production during each day of egg formation.

The nutrient needs of seabirds for egg formation appear generally to be less critical factors in reproduction than are The needs of supplying food for nestlings (Simons and Whittow, this volume). However, under conditions of marginal or deficient supplies or through interruption of normal nutrition by environmental factors such as disease or pollution, nutrient needs for reproduction may well become critical.

ACKNOWLEDGEMENTS

I thank Herman Rahn and G. Causey Whittow for their constructive suggestions. I am deeply indebted to Lee Astheimer for data, photographs, drawings, and valuable constructive criticisms. Beverly Wells assisted in preparation of the manuscript.

REFERENCES

Astheimer, L.B., and Grau, C.R., 1983, Timing of egg formation in the Adélie Penguin (Pygoscelis adeliae), Presented at Amer. Ornith. Union Meeting, New York, Sept. 26-Oct.1.

Astheimer, L., Roudybush, T.E., and Grau, C.R., 1980, Timing and energy requirements of egg synthesis in Cassin's Auklet, Pac. Seabird Group Bull., 6(2):29.

Bacon, W.L., and Cherms, F.L., 1968, Ovarian follicular growth and maturation in the domestic turkey, Poult. Sci., 47:1303.

Baker, C.M.A., and Manwell, C., 1975, Penguin proteins: biochemical contributions to classification and natural history, in:"The Biology of Penguins", B. Stonehouse, ed., Academic Press, London.

Bakken, G.S., Vanderbilt, V.C., Buttemer, W.A., and Dawson, W.K., 1978, Avian eggs: thermoregulatory value of very high near-infrared reflectance, Science, 200:321.

Bakst, M.R., and Howarth, B., Jr., 1977, Hydrolysis of the hen's perivitelline layer by cock sperm in vitro, Biol. Reprod., 17:370.

Bluhm, C.K., Philips, R.E., and Burke, W.H., 1983, Serum levels of luteinizing hormone, prolactin, estradiol and progesterone in laying and nonlaying mallards (Anas platyrhynchos), Biol. Reprod., 28:295.

Bradfield, J.R.G., 1951, Radiographic studies on the formation of the hen's egg shell, J. Exp. Biol., 28:125.

Carey, C., Rahn, H., and Parisi, P., 1980, Calories, water, lipid, and yolk in avian eggs, Condor, 82:335.

Collins, C.T., and LeCroy, M., 1972, Analysis of measurements, weights, and composition of common and roseate tern eggs, Wilson Bull., 84:187.

Conrad, R.M., and Warren, D.C., 1939, The alternate white and yellow layers of yolk in hen's ova, Poult. Sci., 18:220.

Day, S.L., and Nalbandov, A.V., 1977, Presence of prostaglandin f(PGF) in hen follicles and its physiological role in ovulation and oviposition, Biol. Reprod., 16:486.

Dixon, C., 1888, "Our Rarer Birds", Bentley and Sons, London.

Etches, R.J., and Cunningham, F.J., 1976, The interrelationship between progesterone and luteinizing hormone during the ovulation cycle of the hen (Gallus domesticus), J. Endocr., 71:51.

Evans, A.J., Perry, M.M., and Gilbert, A.B., 1979, The demonstration of very low density lipoprotein in the basal lamina of the granulosa layer in the hen's ovarian follicle, Biochim. Biophys. Acta, 573:184.

Everson, G.J., and Souders, H.J., 1957, Composition and nutritive importance of eggs, J. Am. Diet. Assn., 33:1244.

Feeney, R.E., and Allison, R.G., 1969, "Evolutionary biochemistry of Proteins", Interscience, New York.

Feeney, R.E., Osuga, D.T., Lind, S.B., and Miller, H.T., 1966, The egg-white proteins of the Adelie Penguin, <u>Comp. Biochem. Physiol</u>. 18:121

Fujii, S., Tamura, T., and Okamotoa, T., 1973, Studies on yolk formation in hen's eggs. 1. Light and scanning electron microscopy of the structure of yolk spheres, <u>J. Fac. Fish. Anim. Husb.</u>, Hiroshima Univ., 12:1.

Gaston, A.J., and Nettleship, D.N., 1981, "The Thick-billed Murres of Prince Leopold Island", Canadian Wildlife Serv. Monograph No. 6, Ottawa.

Gilbert, A.B., 1971, The egg: its physical and chemical aspects, in: "The Physiology and Biochemistry of the Domestic Fowl", Vol. 3, p. 1379., D.J. Bell and B.M. Freeman, eds., Academic Press, London.

Gilbert, A.B., 1972, The activity of the ovary in relation to egg production, p.3., in: "Egg Formation and Production", B.M. Freemann and P.E. Lake, eds., British Poultry Sciences, Ltd., Edinburgh.

Gilbert, A.B., 1979, Female genital organs, vol. 1, p. 237.60, in:"Form and Function in Birds", A.S. King and J. McLelland, eds., Academic Press, New York.

Grau, C.R., 1976, Ring structure of avian egg yolk, <u>Poult. Sci.</u>, 55:1418.

Grau, C.R., 1982, Egg formation in Fiordland Crested Penguins (<u>Eudyptes pachyrhynchus</u>), <u>Condor</u>, 84:172.

Grau, C.R., and Astheimer, L.B., 1982, Patterns of yolk growth in seabirds, <u>Pac. Seabird Group Bull.</u>, 9:65.

Harrison, C., 1978, "A Field Guide to the Nests, Eggs and Nestlings of North American Birds", Collins, New York,

Hatch, S.A., 1983, Mechanism and ecolgical significance of sperm storage in the Northern Fulmar with reference to its occurence in other birds, <u>Auk</u>, 100:183.

Hirsch, K.V., and Grau, C.R., 1981, Yolk formation ovipostiion in captive Emus, <u>Condor</u> 83:381.

Ho, C.Y-K., Prager, E.M., Wilson, A.C., Osuga, D.T., Feeney, R.E., 1976, Penguin evolution: protein comparisons demonstrate phylogenetic relationship to flying aquatic birds, J. Mol. Evol., 8:271.

Imber, M.J., 1976, Breeding biology of the grey-faced petrel Pterodroma macroptera gouldi, <u>Ibis</u>, 118:51.

Jones, P.J., 1978, Variability of egg size and composition in the Great White Pelican (<u>Pelecanus onocrotalus</u>), <u>Auk</u>, 96:407.

King, J.R., 1973, Energetics of reproduction in birds, in: "Breeding Biology of Birds", p. 78, D.S. Farner, ed., National Academy of Sciences, Washington, D.C.

Kinsky, F.C., 1971, The consistent presence of paired ovaries in the Kiwi (<u>Apteryx</u>) with some discussion of this condition in other birds, <u>J. Ornith.</u>, 112:334.

Kuroda, N., 1963, A comparative study on the chemical consitutions of some bird eggs and the adaptive significance, <u>Japanese Poult. Sci.</u>, 38:311.

Lack, D., 1968, The proportion of yolk in the eggs of water fowl, Wildfowl, 19:67.

Laishley, R., 1858, "British Birds' Eggs", Routledge, Warne, and Routledge, London.

Lake, P.E., 1975, Gamete production and the fertile period with particular reference to domesticated birds, Symp. Zoll. Soc. Lond., 35:225.

Lawrence, J.M., and Schreiber, R.W., 1974, Organic material and calories in the egg oif the Brown Pelican (Pelecanus occidentalis), Comp. Biochem. Physiol., 47A:435.

Leach, R.M., Jr., and Rucker, R.B., 1978, Studies on the major protein component of egg shell membranes, Poult. Sci., 57:1151.

Nalbandov, A.V., and James, M.F., 1949, The blood vascular system of the chicken ovary, Amer. J. Anat., 85:347.

Nelson, B., 1978, "The Gannet", Buteo Books, Vermillion, SD.

Packard, G.C., Tracy, C.R., and Roth, J.J., 1977, The physiological ecology of reptilian eggs ane embryos, and the evolution of viviparity within the Calss Reptilia, Biol. Rev., 52:71.

Paludan, K., 1951, Contributions to the breeding biology of Larus argentatus and Larus fuscus, Vidensk. Medd. fra Dansk. naturh. Foren, 114:1.

Perry, M.M., Gilbert, A.B., and Evans, A.J., 1978, Electron microscope observations on the ovarian follicle of the domestic fowl during the rapid growth phase, J. Anat., 125:481.

Perry, M.M., and Gilbert, A.B., 1979, Yolk transport in the ovarian follicle of the hen (Gallus domesticus): lipoprotein-like particles at the periphery of the oocyte in the rapid growth phase, J. Cell Sci., 39:257.

Perry, M.M., and Gilbert, A.B., 1981, Endocytosis in the avian oocyte, Trans. Biochem. Soc. 9(2):117P.

Reid, B., 1965, The Adelie Penguin (Pygoscelis adeliae) egg, N.Z. J. Sci., 8:503.

Ricklefs, R.E., 1974, The energetics of reproduction in birds, p. 152, in: "Avian Energetics", R.A. Paynter, Jr., ed., Publ. Nuttall Ornithol. Club, Cambridge, MA.

Ricklefs, R.E., 1977, Composition of eggs of several bird species, Auk, 94:350.

Ricklefs, R.E., and Montevecchi, W.A., 1979, Size, organic composition and energy content of North Atlantic Gannet Morus bassanus eggs, Comp. Biochem. Physiol. 64A:161.

Riddle, O., 1908, The rate of growth of the egg-yolk in the chick, and the significance of white and yellow yolk in the ova of vertebrates, Science, 27:945.

Riddle, O., 1910, Studies with Sudan III in metabolism and inheritance, J. Exp. Zool. 8:163.

Riddle, O., 1911, On the formation, significance and chemistry of the white and yellow yolk of ova, J. Morph. 22:455.

Romanoff, A.L., 1943, The Growth of the avian ovum, Anat. Rec., 85:261.

Romanoff, A.L., 1960, "The Avian Embryo", The Macmillan Company, New York.

Romanoff, A.L., and Romanoff, A.J., 1949, "The Avian Egg", J. Wiley & Sons, New York.

Rothchild, I., and Fraps, R.M., 1946, Induced ovipositions in relation to age of oviducal egg in the domestic hen, Proc. Soc. Exp. Biol. Med., 63:511.

Roudybush, T.E., Grau, C.R., Petesen, M.R., Ainley, D.G., Hirsch, K.V., Gilman, A.D., and Patten, S.M., 1979, Yolk formation in some charadriiform birds, Condor, 81:293.

Serventy, D.L., 1963, Egg-laying time table of the Slender-billed Shearwater, Proc. XIIIth Int. Ornith. Congr., 2:338.

Sharp, P.J., 1980, Female Reproduction, p. 435, in:"Avian Endocrinolgy", A. Epple and M.H. Stetson, ed., Academic Press, New York.

Shenstone, F.S., 1968, The gross composition, chemistry, and physio-chemical basis of organization of the yolk and white. Chapter 2, p. 26, in:"Egg Quality, A Study of the Hen's Egg", T.C. Carter, Ed., Oliver and Boyd, Edinburgh.

Sibley, C.G., 1960, The electrophoretic patterns of avian egg white proteins as taxonomic characters, Ibis, 102:215.

Siegfried, W.R., Williams, A.J., Burger, A.E, and Berruti, A., 1978, Mineral and energy contributions of eggs of selected species of seabirds to the Marion Island terrestrial ecosystem, S. Afr. J. Antarctic Res.8:75.

Simkiss, K., 1975, Calcium and avian reproduction. Symp. Zool. Soc., London, 35:305.

Simmons, G.S., Hetzel, D.J.S., 1983, Time relationships between ovipositions, ovulation and egg formation in Khaki Campbell ducks, British Poult. Sci., 24:21.

Smith, A.H., 1959, Follicular permeability and yolk formation, Poult. Sci., 38:1437.

Stieve, H., 1918, Über experimentell, durch veranderte aussere Bedingungen, hervorgerufene Ruckbildungsvorgange am Eierstock des haushuhnes (Gallus domesticus), Arch. EntwMech Org., 44:530.

Warham, J., 1974, The Fiordland Crested Penguin Eudyptes pachyrhynchus, Ibis, 116:1.

Warham, J., 1975, The crested penguins, p. 189, in: "The Biology of Penguins", B. Stonehouse, Ed., Macmillan, London.

Warham. J., 1983, The composition of petrel eggs, Condor, 85:194.

Warren. D.C., and Conrad, R.M., 1939, Growth of the hen's ovum. J. Agric. Res., 58:875.

Warren. D.C., and Scott, H.M., 1935, The time factor in egg formation, Poult. Sci., 14:195.

Williams, A.J., Siegfried, W.R., and Cooper, J., 1982, Egg composition and hatching precocity in seabirds, Ibis, 124: 456.

Woodard, A.E., and Mather, F.B., 1964, The timing of ovulation, movement of the ovum through the oviduct, pigmentation and shell deposition in Japanese Quail, Poult. Sci. 43:1427.

Wyburn, G.M., Johnston, H.S., Draper, M.H., and Davidson, M.F., 1973, The ultrastucture of the shell forming region of the oviduct and the development of the shell of Gallus domesticus, Quart. J. Exp. Physiol., 58:143.

Yoshimura, Y., and Koga, O., 1983, Ultrastructural changes in follicular stigma during the ovulation process in the hen, p. 107, in: "Avian Endocrinology: Environmental and Ecological Perspectives", S. Mikuma et al., eds., Japan Sci. Soc. Press, Tokyo.

ENERGY COST OF INCUBATION TO THE PARENT SEABIRD

Gilbert S. Grant

North Carolina State Museum of Natural History
Raleigh, NC 27611

INTRODUCTION

Our knowledge of incubation energetics has increased dramatically in the last decade since King (1973) and Kendeigh (1963, 1973) first presented their cost-of-incubation theories. Kendeigh's model assumes that the heat lost from the egg must be balanced by extra heat production by the parent bird while King argues that the heat produced as a by-product of metabolism could substitute at least part of the heat needed to maintain egg temperature. Much of the recent data on this subject stem from elaborate and extensive studies combining both field and laboratory methods conducted on seabirds nesting on remote oceanic islands. The tameness of the birds, the size of nesting colonies, and the ability to carry Scholander micro-gas analyzers, Haldane apparatus, sensitive balances, and radioactive water to nesting colonies have made such studies possible. Long fasting periods and long incubation periods of seabirds have facilitated these studies.

In this Chapter I will introduce and briefly discuss the methods (and results obtained) used to measure, calculate, or estimate incubating and resting metabolic rates of petrels, penguins, terns, and other birds. I adopt the usage of the term "petrel" to collectively apply to albatrosses, shearwaters, storm-petrels, and other members of the order Procellariiformes. Some of these methods have not been employed (and some cannot be) on seabirds. The primary concern is incubation metabolism within the thermal neutral zone (TNZ) as a point of comparison - it obviously goes up as ambient temperature drops below TNZ.

METABOLISM METHODOLOGY

Indirect

Indirect estimates of incubation costs are based on heat loss from the eggs, clutch mass method, time budget analysis, heat budget modeling, and weight loss of fasting birds.

Heat loss from eggs. - Incubation costs have been estimated indirectly by calculation of the rate of heat loss from the clutch of eggs (Kendeigh, 1963). Kendeigh's formula requires knowledge of the number of eggs in the clutch, mean weight of eggs, specific heat of eggs, their rate of cooling, egg and nest air temperature, proportion of time bird is on the nest and proportion of surface area of eggs covered by the incubating bird. The heat requirement for incubation of five species estimated by Kendeigh's formula ranged from 15 to 120% of BMR (Drent, 1972; Ricklefs, 1974). The incubation costs for the Herring Gull (Larus argentatus) was calculated to be only about 15% of the basal metabolic rate (BMR).

Clutch mass method. - This method was pioneered by West (1960) and involves the assumptions that the incubation patch permits heat transfer to the eggs as rapidly as heat transfer occurs within the body of the adult, and that the nest and integument of the incubator are equivalent in terms of insulative properties. The cost of incubation is taken to be the same as that of maintaining the temperature of the same weight of tissue. Incubation cost calculated by the clutch mass method for the same five species (estimated by Kendeigh's formula above) ranged from 1-153% of BMR (Ricklefs, 1974). The cost to the Herring Gull was 23% of BMR.

Time budget studies. - Quantitative incubation costs have been investigated by the partitioning of activities throughout the year. One simply times various activities at different air temperatures and applies appropriate laboratory data to estimate the energy cost of a particular activity. In the four species studied, the time spent in flight per day was reduced substantially during incubation (Black, 1975; Mugaas, 1976; Walsberg, 1977; Withers, 1977). On an annual basis, daily energy expenditure was lowest during incubation in the Black-billed Magpie, Pica pica (Mugaas, 1976; Mugaas and King, 1981) and the Phainopepla, Phainopepla nitens (Walsberg, 1977).

Weight loss of fasting birds. - Data on weight loss during prolonged fasts by incubating petrels and penguins can be used to calculate incubation energy costs. From the estimated composition of the material lost, the energy content can be calculated. I have followed Croxall (1982) in using the values obtained by Groscolas and Clement (1976) on fasting Emperor Penguins, Aptenodytes forsteri: 55.5% fat, 9.2% protein, and the remainder, water. The energy equivalents of fat and protein are taken as 39.7 kJ/g and 16.7 kJ/g,

respectively. Weight losses during incubation fasts have been
measured in a number of petrels and penguins. Calculated incubation
and predicted basal metabolic rates (most data are from the review by
Croxall, 1982) are summarized in Results.

Heat budget modeling. - Heat exchange modeling dependent upon
microclimatic measurements and thermal resistances of the nest, eggs
and bird's body have been employed by Mertens (1977) and Walsberg and
King (1978a,b) to estimate incubation costs in four passerines. The
resting energy expenditure of incubating White-crowned Sparrows
(Zonotrichia leucophrys), Red-winged Blackbirds (Agelaius phoenicus),
and Willow Flycatchers (Empidonax traillii) was reduced 15, 16, and
18%, respectively, compared to birds perched nearby (Walsberg and
King, 1978a,b). According to their calculations, incubating White-
crowned Sparrows spend 10 hours per day at the basal level of energy
expenditure. Energy savings are achieved by insulation characteristics of the nest and microclimate amelioration associated with the
location of the nest. Walsberg and King (1978a,b) found that the
effects of the nests' microclimate and insulation more than compensate for heat loss through the brood patch. It would be interesting
to compare calculated values for birds in nests with eggs to birds in
nests without eggs. Mertens (1977) found that an incubating Great
Tit (Parus major) incubating 10 eggs had to increase her heat production when the ambient temperature dropped below 27°C, and that
overall metabolic rate was 1.5 times the basal rate over the entire
incubation period. Such modeling applies only to incubation costs
while the bird is inactive on the nest. Further savings are achieved
through embryonic metabolism during the last portion of the incubation period. Unfortunately, heat budget modeling has not been undertaken on any seabird species.

Direct

Food consumption, heat production measurements, oxygen consumption and/or carbon dioxide production, and heavy water techniques
have been employed to measure incubation costs in birds.

Food consumption. - Incubation costs have been estimated by
monitoring food intake during the incubation period in Ringed Doves,
Streptopelia risoria, and Rock Doves, Columba livia (Riddle and
Braucher, 1934), Ringed Doves (Brisbin, 1969), and Zebra Finches,
Poephila guttata (El Wailly, 1966). This method, like that of the
heavy water technique, is not fine-tuned enough to separate "pure"
incubation costs from peripheral costs such as postural changes,
alertness, preening, etc. The increment in food intake over control
levels was taken as the incubation cost. Zebra Finches incurred an
incubation cost at all but the highest temperature regime (34.4°C, El
Wailly, 1966). There was no increase in food consumption by doves
and pigeons during the incubation period (Riddle and Braucher, 1934;
Brisbin, 1969).

Heat production measurements. - Mertens (1980) measured the
energy requirements of Great Tits, Parus major, while roosting and
while incubating eggs in a modified nestbox. By inserting 35 heat-
flux disks in the walls of the nestbox, he was able to quantify
incubation costs at various ambient temperatures. Heat loss at an
ambient temperature of $8°C$ was about three times greater than the
heat loss incurred at the resting metabolic level. Mertens (1980)
pointed out that the heat loss through the bottom of the nest/nestbox
increased five-fold after the fifth egg was laid. These values are
much higher than the predicted resting and incubating costs (Mertens,
1977). It is instructive that the incubating and resting metabolic
rate regressions approach each other at $24°C$ (ambient temperature).

O_2 and CO_2 measurements. - A variety of gas exchange methods
have been employed on several species of seabirds and other species
to obtain metabolic costs during incubation. Norton (1973), Mertens
(1977), Biebach (1979, 1981), Gessaman and Findell (1979), Hamilton
and Gessaman (1981), Vleck (1981), Grant and Whittow (1983), Brown
(in press), and Brown and Adams (in press) have measured oxygen
consumption and/or carbon dioxide production of incubating birds.
Some studies (Norton, 1973; Mertens, 1977) did not include measurements of non-incubating birds; thus a cost of incubation above
resting, non-incubating levels is not available for these species.
Birds nesting in nestboxes are relatively easy to make measurements
upon, while gas exchange of open nesting species may be technically
difficult. Seabirds are typically very tame and tolerate
experimental manipulation well. Recent studies have capitalized on
these conditions and produced several informative studies.
Incubating and resting birds can be easily obtained and induced to
incubate on the nest site in a metabolic nest chamber and/or with
face mask in place. Such studies have now been obtained on two
penguins (Brown, in press), two albatrosses (Grant and Whittow, 1983;
Brown and Adams, in press), and a petrel (Grant and Whittow, 1983).
Others will undoubtedly be studied in the near future. One caution
may be to use the same method on both resting (non-incubating) and
incubating birds - this eliminates method biases. Four of five
seabird species measured by this method had incubation rates lower
than resting rates. In the exception, the Wandering Albatross
(Diomedea exulans; Brown and Adams, in press), a chamber was used for
resting metabolism and a face mask was used for incubating
metabolism. As suggested by the authors, the stresses imposed on the
incubating bird by the face mask may have resulted in the elevated
levels of metabolism in this albatross (see Results).

Heavy water technique. - Doubly-labeled water has received wide
attention recently as an accurate means of measuring total CO_2
production and water flux in animals free to behave normally in their
natural environments (Lifson et al., 1955; Lifson and Lee, 1961;
LeFebvre, 1964; Nagy and Costa, 1980; Nagy, 1980; Flint and Nagy, in
press). Oxygen is lost from the body in both water and carbon

Table 1. Incubation costs in petrels as calculated by the weight-loss method. Data for the sexes and different sources were averaged and are from references in Croxall (1982) unless otherwise noted.

Species	Weight[a] (g)	Mean Weight Loss (g/g/day)	Calculated IMR[b] (kJ/day)	Calculated BMR[c] (kJ/day)	IMR/BMR
D. exulans	8955	0.0092	2015	1627	1.24
D. melanophrys	3808	0.0120	1072	864	1.24
D. chrysostoma	3688	0.0110	1039	843	1.23
D. immutabilis	2867	0.0115	776	700	1.11
P. macroptera	627	0.0129	192	227	0.85
P. phaeopygia	368	0.0343	297	153	1.94
P. inexpectata	325	0.0214	163	139	1.17[d]
P. hypoleuca	202	0.0322	151	98	1.54
P. pycrofti	197	0.0256	113	96	1.17
P. puffinus	401	0.0250	234	163	1.44
P. lherminieri	168	0.0359	142	85	1.66
F. tropica	52	0.0479	59	36	1.65
O. castro	37	0.0453	38	28	1.36
O. oceanicus	39	0.0663	67	29	2.31

[a] average of initial and final weights of an incubation fast.
[b] assuming 55.5% fat, 9.2% protein--values averaged from those in Croxall (1982) and converted to kJ/day. IMR = incubation metabolic rate.
[c] BMR equation from Lasiewski and Dawson (1967) where BMR = 76.7 $W^{0.741}$. BMR = basal metabolic rate.
[d] from Grant and Whittow (1983).

dioxide while hydrogen is lost only in water. The difference in turnover rates is proportional to CO_2 production. Birds are captured, weighed, and injected with doubly-labeled water, and held for a brief interval to allow thorough mixing of the isotopes with body fluids. A blood sample is withdrawn and the birds are released. A few days later the bird is recaptured and a second blood sample is withdrawn and the bird is weighed again. Final water content must be measured or estimated if water flux is to be determined. A number of assumptions must be made and methodological errors eliminated in order to minimize experimental error (Nagy, 1980).

The method has been used only once to estimate the cost of incubation: Flint and Nagy (in press) measured incubation and flight metabolism of Sooty Terns (Sterna fuscata) on Tern Island, French Frigate Shoals, Hawaii.

Table 2. Measured incubation costs and resting metabolic rates in birds. Measurements are those thought to be within their thermal neutral zone.

Species	Ref.	Incubation Metabolic Rate (IMR) (kJ/day)	Resting Metabolic Rate (RMR) (kJ/day)	IMR/RMR
Macaroni Penguin	1	1032	1161	0.89
Rockhopper Penguin	1	701	863	0.81
Wandering Albatross	2	2415	1755[a]	1.38
Laysan Albatross	3	618	645	0.96
Bonin Petrel	3	90	109	0.82
Sooty Tern	4	140[b]	87[b]	1.62
American Kestrel	5	–	–	ca. 1.0
Barn Owl	6	–	–	ca. 1.0
Great Tit	7	–	–	ca. 1.0
Starling	8,9	–	–	ca. 1.0
Zebra Finch	10	–	–	ca. 1.0

Ref. 1 Brown (in press)
2 Brown and Adams (in press)
3 Grant and Whittow (1983)
4 Flint and Nagy (in press)
5 Gessaman and Findell (1979)
6 Hamilton and Gessaman (1981)
7 Mertens (1977)
8,9 Biebach (1979, 1981)
10 Vleck (1981)

[a] different methods – mask for IMR, box for RMR.
[b] see text for discussion of this value. BMR from MacMillen et al. IMR by heavy water – integrates all activities over ca. 2 days at the nest.

RESULTS

Incubation costs in petrels. – Incubation costs calculated by the weight loss method for 14 species of Procellariiformes appear in Table 1. The ratio of incubation to basal metabolic rate (IMR/BMR) ranged from 0.85 in the Gray-faced Petrel (Pterodroma macroptera) to 2.31 in Wilson's Storm-Petrel (Oceanites oceanicus). Half of the species are within 25% of unity. The agreement is reasonable, given that we are comparing calculated IMR to predicted BMR, the wide latitudinal range (equator to Antarctica), and the consequent thermal variation encountered by nesting birds.

In Table 2 measured incubation and resting metabolic rates (both measured by the same investigator) of three species of petrels are presented. IMR/RMR ratios ranged from 0.82 to 1.38. All measurements were probably made within the respective thermal neutral zones. There is no incubation cost in the Laysan Albatross, Diomedea immutabilis, and Bonin Petrel, Pterodroma hypoleuca (Grant and Whittow, 1983) while the small incubation cost for the Wandering Albatross (Brown and Adams, in press) may be due to the different methods employed for incubating and resting birds. Incubation costs of Wandering Albatrosses were measured with a mask over the birds' bill, while resting or basal measurements were undertaken in a metabolic chamber. It seems to me that a mask over the face of an albatross delivering ambient air of very low water vapor pressure at a high velocity would greatly increase evaporative water loss and increase metabolism. The evaporative cooling in the nasal chamber and throat required to elevate incoming vapor pressure would, in turn, increase heat loss from the head area. To compensate, metabolism must increase and selective shunting of blood probably occurs. Alternately, increased evaporative cooling could lower body temperature and therefore heat transfer to the egg.

Incubation costs in penguins. - Incubation costs in six species of penguins have been calculated by the weight loss method (Table 3). IMR/BMR ranged from 1.08 in the Emperor Penguin to 1.69 in Gentoo Penguins (Pygoscelis papua). Incubation costs have been measured in only two species. IMR/RMR in the Macaroni Penguin (Eudyptes chrysolophus) was 0.89 while that of the Rockhopper Penguin (Eudyptes chrysocome) was 0.81 (Table 2). Calculated costs were, therefore, 36% greater than measured costs in the Macaroni Penguin, the only penguin for which both sets of data are available.

Incubation costs in other species. - Incubation costs have been measured in only one other seabird, the Sooty Tern (Flint and Nagy, in press; Table 2). Unfortunately, they did not measure resting or basal metabolic rate in Sooty Terns, but there are two other studies of BMR or SMR available. MacMillen et al. (1977) gave a standard metabolic rate of 87 kJ/day while Ricklefs et al. (1980) obtained a value of 123 kJ/day (my conversions to kJ/day assume that the average weight of Sooty Terns is 187g - Flint and Nagy, in press). Therefore, incubation costs can be great (140 IMR/87 SMR = 1.62) or minor (140 IMR/123 BMR = 1.14) depending on which baseline study we use (Table 2). It is instructive to examine Fig. 1 of Flint and Nagy (in press), where they regressed field metabolic rate against proportion of time flying. Three of nine values for incubating Sooty Terns cluster around the SMR of MacMillen et al. (1977). The regression of Flint and Nagy (in press) intercepts the y-axis (incubating birds) very near the BMR measured by Ricklefs et al. (1980). It should be noted that the birds used for SMR and BMR were smaller than those studied by Flint and Nagy (B. Flint, pers. comm.). At any rate, some Sooty Terns appeared to incubate their eggs without

Table 3. Incubation costs in penguins as calculated by the weight-loss method. Data for the sexes and different sources averaged from references in Croxall (1982).

Species	Weight[a] (g)	Mean Weight loss (g/g/day)	Calculated IMR[b] (kJ/day)	Calculated BMR[c] kJ/day	IMR/BMR
A. forsteri	23900	0.0065	3653	3369	1.08
A. patagonicus	13643	0.0110	3689	2223	1.66
P. papua	5848	0.0145	2002	1187	1.69
P. adeliae	3999	0.0136	1273	895	1.42
P. antarctica	3515	0.0160	1310	814	1.61
E. chrysolophus	4124	0.0146	1399	916	1.53

[a] average of initial and final weights of an incubation fast.
[b] assuming 55.5% fat and 9.2% protein, values averaged from those in Croxall (1982) and converted to kJ/day. IMR = metabolic rate during incubation.
[c] basal metabolic rate (BMR = $76.7 \ W^{0.741}$) equation from Lasiewski and Dawson (1967).

expending additional energy. As Flint and Nagy (in press) recorded weight loss of incubating birds, it is instructive to calculate incubation metabolism by this method. Assuming 55.5% fat and 9.2% protein metabolized during a fast (Groscolas and Clement, 1976), calculated incubation cost is 231 kJ/day or 65% greater than that measured by Flint and Nagy (in press) and 2.66 times greater than the standard metabolic rate measured by MacMillen et al. (1977).

Incubating and resting metabolic rates have been measured in five other species (Table 2). Even though these are not seabirds, they are instructive because of the larger clutch sizes and the size range of the birds. In these species incubation costs increased considerably as ambient temperature decreased. However, within the measured or presumed thermal neutral zone, regressions of IMR approached those of RMR. Specific IMR and RMR values for these species are not presented in Table 2 because of the need to extract specific numbers from the regression lines.

DISCUSSION OF METHODS AND RESULTS

Indirect or calculated values of incubation costs are, by definition, less precise. Inherent in the heat loss and clutch mass methods are that costs are additive to basal or resting levels. There is no way (when using the formulas) to obtain negative costs - or energy savings - while incubating an egg or a clutch of eggs.

Time budget analyses illustrate reduced flying time during the incubation period due to parent(s) spending large quantities of time on the nest. Obtaining an energetic cost value for incubation depends on what multiple of BMR is chosen.

Heat budget modeling could be relatively easily applied to seabirds nesting from the tropics to the poles. Measurements and/or estimates of short- and long-wave radiation, air temperature, wind velocity, and resistances of eggs, nests, brood patch and body could be obtained on a number of seabirds in relatively quick order. Comparison of results with measured (oxygen consumption or doubly-labeled water) values would be most instructive.

To evaluate the fasting weight loss method I compared values of species where both the calculated and measured ratios are available. Good agreement exists in the Wandering Albatross (1.24 vs 1.38) and the Laysan Albatross (1.11 vs 0.96), but the calculated ratio (1.54) is nearly twice that measured (0.82) in the Bonin Petrel. Fasting weight losses are ideally begun on seabirds a couple of days after they return from the sea and relieve their mate. We noted that Laysan Albatrosses essentially ceased defecating within a couple of days after the beginning of an incubation bout. A more steady-state weight loss occurs thereafter. Weight losses during the middle of Laysan Albatross incubation bouts averaged 24.3 g/day (Grant and Whittow, 1983), somewhat lower than the 33 g/day measured by Fisher (1967) and Rice and Kenyon (1962). This reduced weight loss value (without defecation losses) gave a calculated incubation cost of 573 kJ/day, or 0.82 of predicted basal metabolic rate.

Weight loss of postabsorptive birds occurs through three routes: gaseous exchange (RQ), fecal and urinary losses, and evaporative water loss (EWL). Fasting petrels have an RQ of about 0.71 (no weight gain or loss due to O_2 intake or CO_2 outflow) and typically do not defecate or urinate during the mid-portion of a fast. Therefore, weight loss should be equal to EWL. With an R.Q. of 0.71, oxygen consumption of 89.9 kJ/day, and weight loss of 6.4 g/day in the Bonin Petrel (Grant and Whittow, 1983), we can calculate that 35% of the weight lost was fat. To further evaluate metabolic measurements by the weight loss method (EWL), I have used measurements of simultaneous O_2 consumption and EWL in the Emu (Dromaius novaehollandiae) and Rhea, Rhea americana (Crawford and Lasiewski, 1968). If we assume (as we did in penguins and petrels) that 55.5% fat and 9.2% protein are used while fasting in the Emu and Rhea, weight loss (EWL) metabolism exceeds actual measured metabolism by 1.40 (Emu) and 1.64 (Rhea). These ratios are similar to many of those in Tables 1 and 3. As discussed previously with the Sooty Tern and Macaroni Penguin, weight loss measurements also gave elevated metabolic rate values. The conclusion here is that the estimates of incubation costs based on the weight loss method are only as good as the estimates of the substrates consumed. The composition of substrates consumed by petrels probably changes during a long incubation fast depending on the relative contribution of stomach oil vs stored fat. The composition of material lost during a fast has been given as 55.5% fat and 9.2% protein (fasting Emperor Penguins; Groscolas and Clement, 1976), 38% fat and 6% protein (fasting and molting Rockhopper and Macaroni Penguins; Williams et al., 1977), 23% fat and 10% protein (fasting and molting Jackass Penguins, Spheniscus demersus; Cooper, 1978), 60% fat (fasting House Sparrows, Passer domesticus, Dol'nik and Gavrilov, 1975), and 46% fat (molting and fasting House Sparrows, Dol'nik and Gavrilov, 1975). Composition data on fasting, non-molting petrels are desired. However, by measuring the R.Q. one does not have to make assumptions about the % of fat and protein.

Direct measures of energy metabolism during incubation are technically more difficult to obtain. Measurements of food intake have been made on captive birds (Zebra Finch, Ringed Dove, and Rock Dove). This technique integrates all activity over the time span measured and is probably not adaptable to seabirds (unless they can be induced to breed and take food in a captive, controlled setting). The heat production technique of Mertens can be applied to burrow-nesting seabirds if the sufficiently modified burrows are accepted by incubating petrels. Heavy water techniques, although expensive and technically difficult, are potentially the best way to obtain direct incubation costs on many species in a relatively brief time interval. The fact that most species can be recaptured easily, facilitates this method of measurement.

Oxygen consumption and/or carbon dioxide production can, with a little ingenuity, be measured on incubating petrels and penguins rather easily. Similar studies on pelagic terns will be more difficult because they seem to remain more concerned about human activities near their nests. Equipment to measure oxygen or carbon dioxide concentration can be transported to the field with relative ease.

SUMMARY

Incubation costs to the parent seabird have been estimated or measured using a variety of techniques. Calculated incubation cost values include a number of assumptions and vary from 0.85 to 2.31 times the predicted basal metabolic rate of penguins and petrels. Measured incubation costs in the Macaroni and Rockhopper Penguins, Laysan Albatross and Bonin Petrel were lower than resting levels of metabolism while that of the Wandering Albatross was greater than resting levels. Some Sooty Terns incubated at basal or standard metabolism levels while others showed significant incubation costs. In addition, measurements within the thermoneutral zone for the American Kestrel, Barn Owl, Great Tit, Starling and Zebra Finch showed no cost of incubation above resting levels. It seems that many species do not have an added metabolic cost during incubation bouts.

REFERENCES

Biebach, H., 1979, Energetik des Brutens beim Star (Sturnus vulgaris), J. Ornithol., 120:121.

Biebach, H., 1981, Energetic costs of incubation on different clutch sizes in Starlings (Sturnus vulgaris), Ardea, 69:141.

Black, C. P., 1975, The ecology and bioenergetics of the Northern Black-throated Blue Warbler (Dendroica caerulescens caerulescens), Unpubl. Ph.D. dissertation. Darmouth College, Hanover, N.H.

Brisbin, I. L. Jr., 1969, Bioenergetics of the breeding cycle of the Ring Dove, Auk, 86:54.

Brown, C. R., In press, Resting metabolic rate and energetic cost of incubation in Macaroni Penguins (Eudyptes chrysolophus) and Rockhopper Penguins (E. chrysocome), Comp. Biochem. Physiol.

Brown, C. R. and N. J. Adams, In press, Basal metabolic rate and energetic cost of incubation in the Wandering Albatross, (Diomedea exulans), Condor.

Cooper, J., 1978, Moult of the Black-footed Penguin Spheniscus demersus, International Zoo Yearbook, 18:22.

Crawford, E. C., Jr. and R. C. Lasiewski, 1968, Oxygen consumption and respiratory evaporation of the Emu and Rhea, Condor, 70:333.

Croxall, J. P., 1982, Energy costs of incubation and moult in petrels and penguins, J. Animal Ecol., 51:177.

Dol'nik, V. R. and V. M. Gavrilov, 1975, A comparison of the seasonal and daily variations of bioenergetics, locomotor activities and major body composition in the sedentary House Sparrow (Passer d. domesticus (L.) and the migratory 'Hindian' sparrow (P. d. bactrianus Zar et Kudasch), Ekologia Polska, 23:211.

Drent, R., 1972, Adaptive aspects of the physiology of incubation, Proc. XVth Intern. Ornithol. Congr., p. 255.

El-Wailly, A. J., 1966, Energy requirements for egg-laying and incubation in the Zebra Finch, Taeniopygia castanotis, Condor, 68:582.

Fisher, H. I., 1967, Body weights in Laysan Albatrosses, Diomedea immutabilis, Ibis, 109:373.

Flint, E. N. and K. A. Nagy, In press, Flight energetics of free-living Sooty Terns, Auk.

Gessaman, J. A. and P. R. Findell, 1979, Energy cost of incubation in the American Kestrel, Comp. Biochem. Physiol., 63A:57.

Grant, G. S. and G. C. Whittow, 1983, Metabolic cost of incubation in the Laysan Albatross and Bonin Petrel, Comp. Biochem. Physiol., 74A:77.

Groscolas, R. and C. Clement, 1976, Utilisation des reserves energetiques au cours de jeune de la reproduction chez le manchot empereur Aptenodytes fosteri, Comptes Rendus, Academic des Sciences, Paris. Serie D, 282:297.

Hamilton, K. L. and J. A. Gessaman, 1981, Energetic cost of incubation of the Barn Owl: a preliminary report, Am. Zool., 21:964.

Kendeigh, S. C., 1963, Thermodynamics of incubation in the House Wren, Troglodytes aedon, Proc. Intern. Ornithol. Congr., 13:884.

Kendeigh, S. C., 1973, Discussion, In Breeding Biology of Birds (D. S. Farner, ed.), pp. 311, Natl. Acad. Sci., Washington, D.C.

King, J. R., 1973, Energetics of reproduction in birds. In Breeding Biology of Birds (D. S. Farner, ed.), pp. 78, Natl. Acad. Sci., Washington, D.C.

Lasiewski, R. C. and W. R. Dawson, 1967, A re-examination of the relation between standard metabolic rate and body weight in birds, Condor, 69:12.

LeFebvre, E. A., 1964, The use of D_2O^{18} for measuring energy metabolism in Columba livia at rest and in flight, Auk, 81:403.

Lifson, N., G. B. Gordon, and R. McClintock, 1955, Measurement of total carbon dioxide production by means of D_2O^{18}, J. Appl. Physiol., 7:704.

Lifson, N. and J. S. Lee, 1961, Estimation of material balance of totally fasted rats by doubly labeled water, Am. J. Physiol., 200:85.

MacMillen, R. E., G. C. Whittow, E. A. Christopher, and R. J. Ebisu, 1977, Oxygen consumption, evaporative water loss, and body temperature in the Sooty Tern, Auk, 94:72.

Mertens, J. A. L., 1977, The energy requirements for incubation in Great Tits, Parus major L., Ardea, 65:184.

Mertens, J. A. L., 1980, The energy requirements for incubation in Great Tits and other bird species, Ardea, 68:185.

Mugaas, J. N., 1976, Thermal energy exchange, microclimate analysis, and behavioral energetics of Black-billed Magpies, Pica pica hudsonia, Unpubl. Ph.D. dissertation. Washington State Univ., Pullman.

Mugaas, J. N. and J. R. King, 1981, Annual variation of daily energy expenditure by the Black-billed Magpie: A study of thermal and behavior energetics, Studies in Avian Biology No. 5, Cooper Ornithol. Society.

Nagy, K. A., 1980, CO_2 production in animals: analysis of potential errors in the doubly labeled water method, Am. J. Physiol., 238:R466.

Nagy, K. A. and D. P. Costa, 1980, Water flux in animals: analysis of potential errors in the tritiated water method, Am. J. Physiol., 238:R454.

Norton, D. W., 1973, Ecological energetics of calidridine sandpipers breeding in northern Alaska, Unpubl. Ph.D. dissertation, Univ. of Alaska.

Rice, D. W. and K. W. Kenyon, 1962, Breeding cycles and behavior of Laysan and Black-footed Albatrosses, Auk, 79:517.

Ricklefs, R. E., 1974, Energetics of reproduction in birds, In Avian Energetics (R. A. Paynter, ed.) pp. 152, Nuttall Ornithol. Club, Cambridge, MA.

Ricklefs, R. E., S. C. White, and J. Cullen, 1980, Energetics of postnatal growth in Leach's Storm-Petrel, Auk, 97:566.

Riddle, O. and P. F. Braucher, 1934, Studies on the physiology of reproduction in birds. XXXIII. Body size changes in doves and pigeons incident to stages of the reproductive cycle, Am. J. Physiol., 107:343.

Vleck, C. M., 1981, Energetic cost of incubation in the Zebra Finch, Condor, 83:229.

Walsberg, G. E., 1977, Ecology and energetics of contrasting social systems in Phainopepla nitens (Aves: Ptilogonatidae), Univ. Calif. Publ. Zool., No. 108.

Walsberg, G. E. and J. R. King, 1978a, The heat budget of incubating mountain white-crowned sparrows (Zonotrichia leucophrys oriantha) in Oregon, Physiol. Zool., 51:92.

Walsberg, G. E. and J. R. King, 1978b, The energetic consequences of incubation for two passerine species. Auk, 95:644.

West, G. C., 1960, Seasonal variation in the energy balance of the Tree Sparrow in relation to migration, Auk, 77:306.

Withers, P. C., 1977, Energetic aspects of reproduction by the Cliff Swallow, Auk, 94:718.

Williams, A. J., W. R. Siegfried, A. E. Burger, and A. Berruti, 1977, Body composition and energy metabolism of moulting eudyptid penguins, Comp. Biochem. Physiol., 56A:27.

PARENT-EGG INTERACTIONS: EGG TEMPERATURE AND WATER LOSS

Ralph A. Ackerman and Richard C. Seagrave

Departments of Zoology and Chemical Engineering
Iowa State University
Ames, Iowa

INTRODUCTION

The avian egg is encased in a hard, calcareous shell and
deposited outside the body of the parent. The embryo develops and
grows, separated by the shell from the ambient atmosphere surrounding
the egg. The chief role of the parent appears to be to attend the
egg, moderating and modifying the thermal environment around the egg
and controlling the energy exchanged between the egg and
surroundings. Control may be exercised in several ways. Eggs are
typically deposited during the most appropriate season. The
micro-climate of the egg may be influenced by the selection of an
appropriate site for egg deposition or by the construction of a nest
which acts to separate the egg or some part of it physically from the
surroundings. Finally, most birds intervene directly in the process
of egg energy exchange by applying a specialized area of skin, the
brood patch, to the surface of the egg. Since adult birds maintain
body temperatures constant at 38-41 °C and brood patch temperatures
are slightly cooler than this (Drent, 1970), energy may be
transferred to the egg when the egg is losing heat to the
surroundings and absorbed from the egg when the egg is gaining heat
from the surroundings.

The interaction between parent and egg acts to produce stability
in the energy content of the egg with the result that egg temperature
is relatively constant throughout incubation (Drent, 1970; White and
Kinney, 1974). It is clear, however, that significant excursions of
egg temperature may occur during successful incubation for some
species (Boersma and Wheelright, 1979; Vleck, 1981). The stability
of egg temperature during incubation means that the vapor pressure
of water inside the egg is also stable since the saturation vapor

pressure of water is directly related to temperature and since the vapor pressure inside the egg is very close to saturation (Rahn and Ar, 1974; Rahn et al., 1976; Rahn, Ackerman and Paganelli, 1977). Thus, if egg temperature is constant so too is egg water vapor pressure. This is an important relationship because bird eggs lose about 15% of their initial mass in water during incubation (Ar and Rahn, 1980). The rate of water loss by the egg is proportional to the water vapor partial pressure difference between the inside and outside of the egg across the eggshell. Thus the water lost by the egg will be sensitive to the temperature of the egg.

Egg temperature is a variable of substantial biological and physiological interest. This interest is reflected in the numerous and often detailed accounts of egg temperatures during natural incubation (Drent, 1970, 1975; Kinney and White, 1974; Howell, 1979; Vleck, 1981; Grant, 1982). However, relatively few of these reports are for living eggs and those that do report the temperatures of living eggs demonstrate that while egg temperature is relatively stable there are interesting and systematic changes in temperature of eggs during incubation. The main difficulty in analyzing egg temperature data lies in the absence of any conceptual framework defining the relationships between simultaneous heat and water exchange or the relationships between the egg, the parent and the nest. For this reason, it is often difficult to perceive and quantify the adaptation of incubation physiology and biology to environmental circumstance (Walsberg, 1980). We present here an analysis of heat and energy exchange by avian eggs. Our intention is to develop a quantitative model of the exchange between the avian egg and its surroundings. In doing so we make simplifying assumptions and approximations, leaving out some biological variables that are of real significance. We stress that our analysis does not represent the last word on the exchange of heat and water by avian eggs but does serve, we hope, to identify and quantify those physical factors that are likely to be of importance in the process.

HEAT AND WATER EXCHANGE BETWEEN EGGS AND SURROUNDINGS

We proceed by dividing an egg into three compartments. One compartment represents the volume of egg defined by the egg surface contacting the brood patch of the incubating adult and a slice through the egg subtending that surface area. A second compartment represents the volume of the egg defined by the surface area of the shell contacting a substrate (which may be nest or soil) and a slice subtending that surface. The third compartment is the remainder of the egg surface and volume and is presumed to be exposed only to the gas phase inside the nest. We also assume, for computational simplicity, that the egg is a sphere and we then proceed by performing energy and mass balances on each compartment. Exchange between compartments is considered.

74

Energy Balance

An energy balance requires that equations be written describing the heat exchanged between the compartment and the external surroundings, the heat exchanged by the compartment as water changes state, the heat transferred between compartments and heat produced in the compartment by embryonic energy conversion. The energy balance for the compartment contacting the brood patch may be written as:

$$C_P M G_V dT_m/dt = H_M G_A A(T_B - T_M) + H_V K_M G_A A(P_B - P_M)$$
$$+ h_E A_1/R(T_U - T_M) + M_m G_V \qquad \qquad 1.)$$

where G_A and G_V represent the shell area and egg volume fractions of the compartment. The energy balance for the compartment contacting the substrate may be written as:

$$C_P M F_V dT_L/dt = H_S F_A A(T_S - T_1 L) + H_V K_S F_A A(P_S - P_L)$$
$$+ h_E A_1/R(T_M - T_U) + h_E A_1/R(T_U - T_L) + M_m F_V \qquad \qquad 2.)$$

where F_A and F_V represent the shell area and egg volume fractions of the compartment. The energy balance for the remaining compartment may be written as:

$$(1 - G_V - F_V) C_P M dT_U/dt = H_a (1 - G_A - F_A) A(T_A - T_U)$$
$$+ H_V K_A (1 - G_A - F_A) A(P_A - P_U) + h_E A_1/R(T_M - T_U)$$
$$+ h_E A_1/R(T_L - T_U) + M_m (1 - G_V - F_V) \qquad \qquad 3.)$$

Symbols are summarized in Table 1. Three heat transfer coefficients are required, one for the convective-radiative exchange between the egg and the surrounding atmosphere (H_a) and one each for the exchange between the egg and brood patch (H_M) and between the egg and substrate (H_S). The magnitudes of H_m and H_s are approximated (Seagrave, 1971) by the relationship

$$H_m = 2.0 \ (h_M/D) \qquad \qquad 4.)$$

$$H_s = 2.0 \ (h_S/D) \qquad \qquad 5.)$$

where h_M is the thermal conductivity of tissue, h_s is the thermal conductivity of the substrate and D is the characteristic linear dimension. Conduction of heat between compartments is approximated using the thermal conductivity of tissue (h_E) and a geometric term (A_1/R) where A_1 is the surface area dividing the compartments and R is the radius of the compartment. The influence of embryonic blood circulation on transport of heat within the egg is simulated by allowing h_E to increase by some factor. Tazawa (1980) reported that

Table 1. Heat and mass variables uses in equations 1-6.

M = egg mass (g)
A = shell surface area (cm^2)
A_1 = internal surface area between buried and unburied fractions of the egg (cm^2)
R = egg radius (cm)
D = egg diameter (cm)
G_A = fraction of egg surface contacting brood patch
G_V = fraction of egg volume contacting brood patch
F_A = fraction of egg surface contacting substrate
F_V = fraction of egg volume contacting substrate
P_A = partial pressure of water vapor in atmosphere (torr)
P_B = partial pressure of water vapor in the brood patch (torr)
P_M = partial pressure of water vapor in egg contacting brood patch (torr)
P_S = partial pressure of water vapor in substrate gas phase (torr)
P_U = partial pressure of water vapor in fraction of the egg contacting neither brood patch nor substrate (torr)
P_L = partial pressure of water vapor in fraction of the egg contacting substrate (torr)
K_M = mass transfer coefficient for water vapor exchange between shell and brood patch (mg $H_2O \cdot cm^{-2} \cdot h^{-1} \cdot torr^{-1}$)
K_A = mass transfer coefficient for water vapor exchange between shell and atmosphere (mg $H_2O \cdot cm^{-2} \cdot h^{-1} \cdot torr^{-1}$)
K_S = mass transfer coefficient for water vapor exchange between shell and substrate (mg $H_2O \cdot cm^{-2} \cdot h^{-1} \cdot torr^{-1}$)
T_B = brood patch temperature (^oC)
T_A = atmospheric temperature (^oC)
T_S = substrate temperature (^oC)
T_M = internal temperature of the fraction of the egg contacting brood patch (^oC)
T_L = internal temperature of the fraction of the egg contacting substrate (^oC)
T_U = internal temperature of the fraction of the egg contacting neither brood patch nor substrate (^oC)
H_M = heat transfer coefficient between shell and brood patch (kcal $\cdot m^{-2} \cdot h^{-1} \cdot {}^oC^{-1}$)
H_a = heat transfer coefficient from shell to atmosphere (kcal $\cdot m^{-2} \cdot h^{-1} \cdot {}^oC^{-1}$)
H_s = heat transfer coefficient from shell to substrate (kcal $\cdot m^{-2} \cdot h^{-1} \cdot {}^oC^{-1}$)
H_V = heat of vaporization of water (kcal $\cdot g^{-1}$)
C_P = heat capacity of tissue (kcal $\cdot g^{-1} \cdot {}^oC^{-1}$)
H_m = metabolic heat generation by the embryo (kcal $\cdot h^{-1}$)
h_M = thermal conductivity of the brood patch (kcal $\cdot cm^{-1} \cdot h^{-1} \cdot {}^oC^{-1}$)
h_E = thermal conductivity of egg contents (kcal $\cdot cm^{-1} \cdot h^{-1} \cdot {}^oC^{-1}$)
h_S = thermal conductivity of substrate (kcal $\cdot cm^{-1} \cdot h^{-1} \cdot {}^oC^{-1}$)
M_M = metabolic heat generation by the embryo (Kcal $\cdot h^{-1}$)
t = time (h)

the chorio-allantoic circulation increased by a factor of 5 during incubation and since the increase looked exponential we used a five-fold exponential increase in h_E with incubation time. Heat loss from the egg due to embryonic energy conversion is estimated (empirically) by fitting a series of curves to data for the heat loss of eggs or to data for the O_2 consumption (after conversion to heat) of eggs. Values for the O_2 consumption of eggs are available for about 35 species but all are reported at incubator temperature. The eggs are 1-3 $^\circ$C warmer than the incubator by the end of incubation (Drent, 1970; Romijn and Lokhorst, 1956; Kashkin, 1961). Since avian egg metabolism is influenced by changes in temperature (Drent, 1970; Barott, 1932) some correction (Q_{10}) may be required. The heat transfer coefficient for a chicken egg (H_A) has recently been estimated to be 8.60 (kcal·m^{-2}·h^{-1}·$^\circ$C^{-1}) by Briedis and Seagrave (1984) and we use this value. The temperatures of the brood patch (T_B), substrate (T_C) and nest air (T_A) are used as input variables and may be assigned values as desired. The selection of a nest air temperature, in particular, represents the simplification of a complex relationship. Little information is available for the temperature of substrate under eggs although there are data for the temperature at the eggshell-substrate interface and for substrate away from nests (Drent, 1970; Grant, 1982). We fix brood patch temperature at 39 $^\circ$C throughout incubation. As upper and lower values for the thermal conductivity of substrate we use the thermal conductivity of sand (Hillel, 1978) and the thermal conductivities of air. The thermal conductivity of sand will vary as a function of its water content as well as other factors. We used a value for a 20% volumetric water content. Walsberg (1980) reports that the thermal conductivity of some avian nests are about 3 times that of a similar thickness of still air.

Mass Balance

A mass balance is constructed for each of the three compartments. The mass balance may be written formally as:

$$dM/dt = G_A K_M A (P_B - P_M) + F_A K_S A (P_S - P_L)$$

$$+ (1 - F_A - G_A) K_A A (P_A - P_U) \tag{6.}$$

The symbols are in table 1. The mass transfer coefficients, K_m, K_s and K_A are calculated from values for shell water vapor conductance available in the literature (G_{H2O}) after taking into account shell area. The partial pressure of water vapor inside the egg, P_m, P_u, P_l are calculated as a function of temperature after correcting for the osmotic effects. The external water vapor pressures, P_A, P_S are set as desired. The vapor pressure of water in brood patch is calculated for tissue water concentration at brood patch temperature. The conductance of the brood patch is assumed to much greater than that of the egg shell thus all the resistance to vapor movement resides in

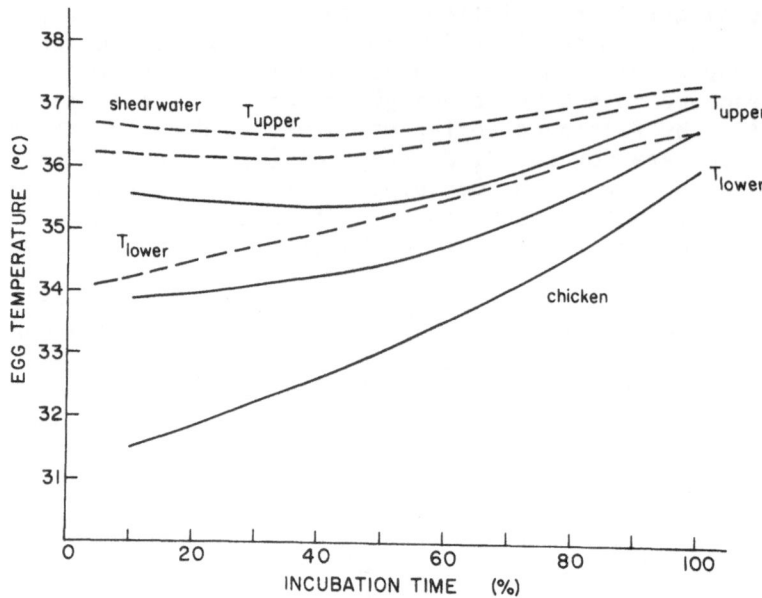

Fig. 1. Egg temperatures of chicken and shearwater eggs during
incubation. Temperatures are shown for the compartment
(10%) contacting the brood patch (T_{upper}), the
compartment (20%) contacting the substrate (T_{lower}) and
the average egg temperature (unlabeled). The substrate
temperature is 25 °C for the chicken egg and 28 °C for the
shearwater egg. The substrate thermal conductivity is
0.0002 Kcal·cm^{-1}·h^{-1}·°C^{-1}. The brood patch temperature is
39 °C while the nest temperature is set at 36 °C.
Initial mass is set at 60 g.

the egg shell. We also assume that the conductance of substrate
contacting the egg shell is much greater than that of the shell and
thus too can be neglected. These assumptions are reasonable in some
cases (Walsberg, 1980) but in other cases there may be appreciable
resistance to water vapor movement. Sand or soil, for example, is
likely to present a significant resistance to water vapor movement.

The equations relating the heat and mass transfer variables were
solved simultaneously and integrated in time using digital
computation and numerical integration techniques. Our principal
objective is to examine the general behavior of the model so rather
than make a species by species analysis we limit our analysis to
chicken sized eggs and use thermal and mass data for chicken eggs.
We have in addition solved the model using variables taken from
studies on Wedge-tailed Shearwater eggs (Puffinus pacificus; Ackerman
et al., 1981; Whittow et al., 1982).

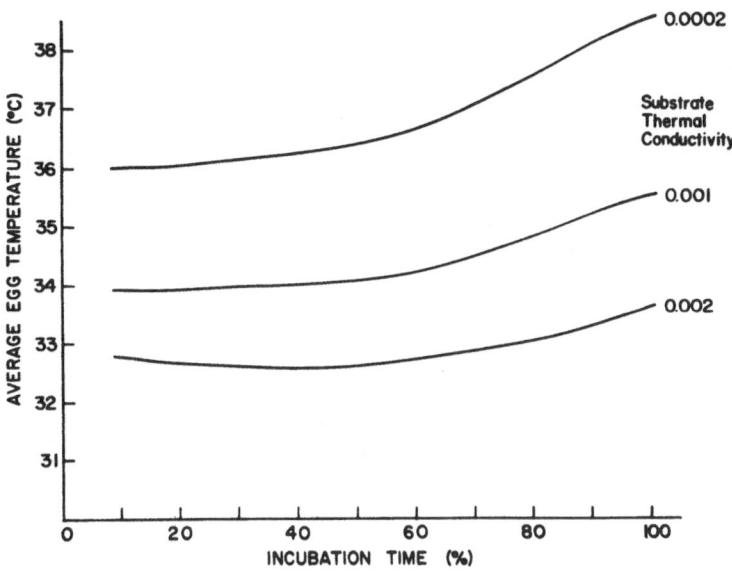

Fig. 2. The effect of thermal conductivity on the average egg temperature of eggs during incubation. Other variables are as described in Figure. 1. The average egg temperature is calculated as the weighted mean of the three compartments. The increase in average egg temperature is due to embryonic heat production.

RESULTS AND DISCUSSION

Egg Temperature

The simulated changes in the egg temperatures of two 60g eggs (chicken and Wedge-tailed Shearwater) during incubation are shown in Figure 1. Illustrated are the temperatures in the egg compartments contacting the brood patch, the substrate and the average egg temperature computed as the weighted mean of the three compartmental temperatures. Temperatures in all compartments increase during incubation. This increase is due to the heat produced by metabolic energy conversion inside the egg. The chicken egg experiences an increase in average temperature of about 2.5 $^\circ$C while the shearwater egg experiences a smaller increase. The difference between these two species is related to the much lower rates of oxygen uptake reported for the shearwater egg (Ackerman et al., 1981). The shearwater egg temperatures are greater than those of the chicken because the substrate temperatures we used (Whittow et al., 1982) were 3 $^\circ$C higher than the substrate temperature used for the chicken egg. The changes in egg temperature simulated in Figure 1 do not take into

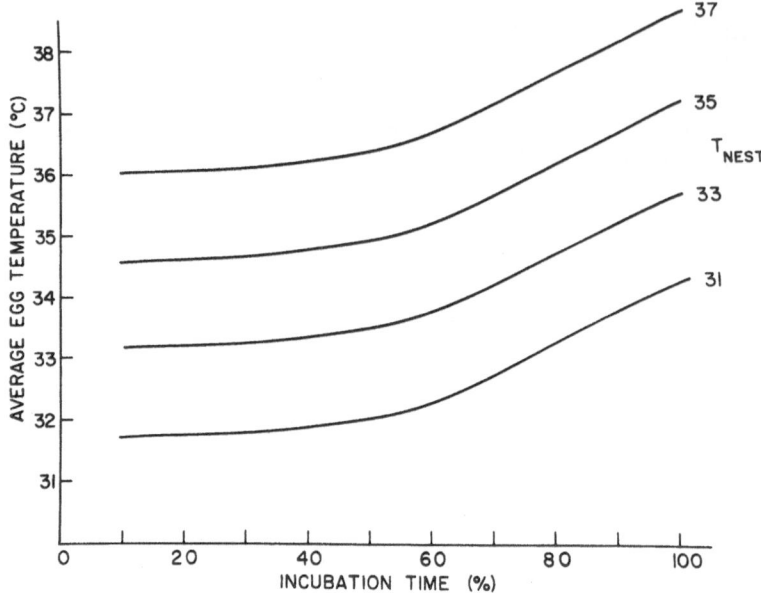

Fig. 3. The effect of nest temperature on the average temperature of eggs during incubation. Substrate thermal conductivity is set at 0.0002 Kcal·cm^{-1}·h^{-1}·$^{\circ}$C^{-1} and other variables are as described in Figure 1.

account changes in incubation behavior by the parent. Parental behavior will affect egg temperature (Drent, 1970; White and Kinney, 1974; Vleck, 1982) and can be quantitatively simulated when sufficient data become available. The narrowing of the temperature difference between the top and bottom of the egg during incubation is due principally to an increase in embryonic heat production. The temperature at the bottom of the egg increases because the temperature at the top of the egg is stabilized by the constant, high brood patch temperature. In addition, the increase in circulation has a smaller but noticable role in this process. A decrease or increase in the circulation variable by a factor of two does produce small changes in the pattern of temperature change. Data on egg temperature are often reported for the interface between the egg and brood patch and the interface between the egg and substrate. These temperatures are determined by the temperatures on either side of the interface and are not strictly comparable to our simulated compartmental temperatures. However, the patterns of change in temperature appear to be quite similar and represent temperature changes experienced by fertile eggs. Temperatures measured inside dead or infertile eggs will be comparable but only for that period of incubation when the heat production and circulation of the embryo are negligible. Data collected from non-viable or artificial eggs whose thermal properties are known will yield useful information if the other variables in the exchange process have been measured but will be much less useful in describing temperature changes actually experienced by fertile eggs.

A number of factors influence egg temperature. The effect of two factors, substrate thermal conductivity and nest air temperature, which are likely to be of biological interest are illustrated in Figures 2 and 3. The change in average egg temperature is simulated at 10% brood patch and 20% substrate coverage of the egg and at a constant brood patch (39 $^{\circ}$C) and substrate (25 $^{\circ}$C) temperature. The values for the substrate thermal conductivity encompass the range of values likely to be encountered by avian eggs. The eggs of ground nesting species, as many marine species are, are likely to encounter substrates of much greater thermal conductivity than the eggs of species nesting above the ground. Eggs contacting sand at 25 $^{\circ}$C or any substrate with a similarly high conductivity will have an average temperature of around 33 $^{\circ}$C. The average egg temperature of the egg increases as the thermal conductivity of the substrate is decreased. If it is important, as it appears to be, to have a high egg temperature (35-38 $^{\circ}$C; White and Kinney, 1974) then the problem of contact with the substrate may be solved in one of two ways. Eggs may be laid only when the substrate is hot and preferably near desired egg temperature. However, if the substrate temperature exceeds egg temperature heat will flow into the egg raising egg temperature. In this circumstance, the incubating adult may absorb heat from the egg via the brood patch but this is not a quantitatively effective means of controlling egg temperature. Egg temperature is altered only by about 0.5 $^{\circ}$C when brood patch coverage varies between 1% and 30%. The adult must, somehow, keep the surroundings of the rest of the egg cool. Water may be added to the region of the egg and substrate and evaporative cooling utilized to limit the temperature increase. This appears to be the case for a number of species which deposit their eggs directly on the ground surface (Grant, 1982; Howell, 1979). If the substrate is cold, as is the case in northern nesting areas, then the solution is to interpose a layer of air between the egg and the substrate; that is, to build a nest. The nest need not necessarily be very substantial, perhaps only enough to break the physical contact between the egg and the ground surface. On the other hand when nesting on very cold substrates, it may be necessary to interpose a thick layer of a good insulator (such as down) between the egg and the substrate as is the case for the eider duck (Rahn et al., 1984). The nest air temperature also has a significant influence on the average egg temperature (Figure 3). Nest temperature as a variable represents a complex relationship between the adult, the nest and the substrate. The effect illustrated is due to the 70% of the egg which is exposed to the nest atmosphere. Even though the insulative value of the air is very high (low thermal conductivity), the large volume and area of the egg exposed to the nest atmosphere insures that it will have an

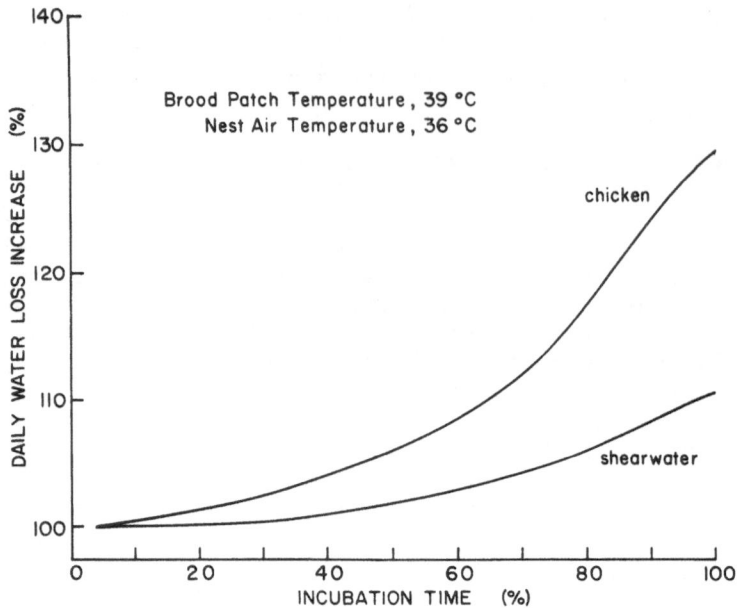

Fig. 4. The simulated increase in daily water loss of chicken and
Wedge-tailed Shearwater eggs during incubation. Results
are shown as the % of the initial daily water loss.
Variables are as described in Figure 1.

important effect on egg temperature. Thus parent birds must not only
control the exposure of the egg to the substrate, they must also
limit heat loss from the nest interior. This implies that nest
structure will have important thermal function.

Egg Water Exchange

The temperature of the egg will influence the water lost from
the egg to the drier environment. Even in an environment close to
water vapor saturation, such as the nest of mound-builders or the wet
nest of grebes is likely to be, eggs will lose water as internal heat
generation increases egg temperature and consequently egg water vapor
pressure above ambient. The effect of the increase in embryonic heat
generation on daily water loss of the chicken and the shearwater egg
is shown in Figure 4. The difference in incubation period of these
two species with similar egg mass is a factor of almost three or so
we express time as the % of incubation. The daily water loss of the
chicken egg increases during incubation by about 30%, while that of
the shearwater increases only by about 10%. The difference is due
to the smaller temperature increase experienced by the shearwater egg
during incubation (Figure 1). Since the daily rate of water loss
production is related to embryonic heat production, we may expect to

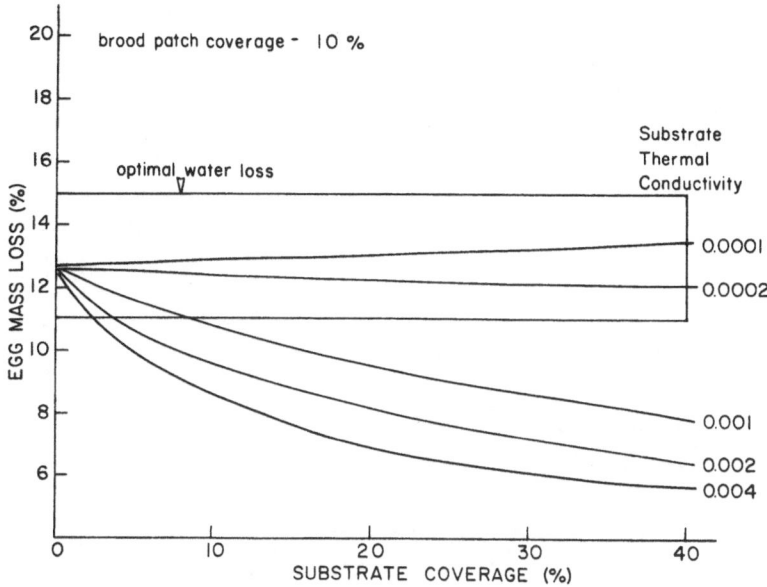

Fig. 5. The fractional water loss (expressed as %) of eggs as a
function of the fraction (%) of the egg contacting the
nesting substrate and the thermal conductivity of the
nesting substrate. The optimal water loss is defined as
the water loss between 11% and 15% of initial egg mass.
Other variables are as described in Figure 1.

find that the chicken egg represents the typical avian egg while the
shearwater egg represents those eggs with unusually long incubation
periods and, consequently, low rates of embryonic heat generation.
The small petrel egg should demonstrate an even smaller increase in
daily water loss during incubation than the shearwater egg while the
albatross egg with an incubation period closer to normal should
experience a greater increase in daily water loss as incubation
progresses. Several recent reports indicate that this is the case
(Grant et al., 1982). Circulation inside the egg has been suggested
as a mechanism for increasing the daily rate of water loss (Grant et
al., 1982). Our simulation indicates that the effect, while present,
is small. A four-fold change in the rate of internal circulation
changes the daily water loss by only a few percent. It is clear from
our earlier discussion that a number of factors influence egg
temperature and must therefore influence egg water exchange. One of
these factors is likely to be the fraction of the egg contacting the
substrate. We simulate the effect of varying the surface area of the
egg contacting the substrate on the fractional water loss of the egg
in Figure 5. Recall, that we assume that water vapor exchange is not
limited by the substrate so that only the shell effects exchange.

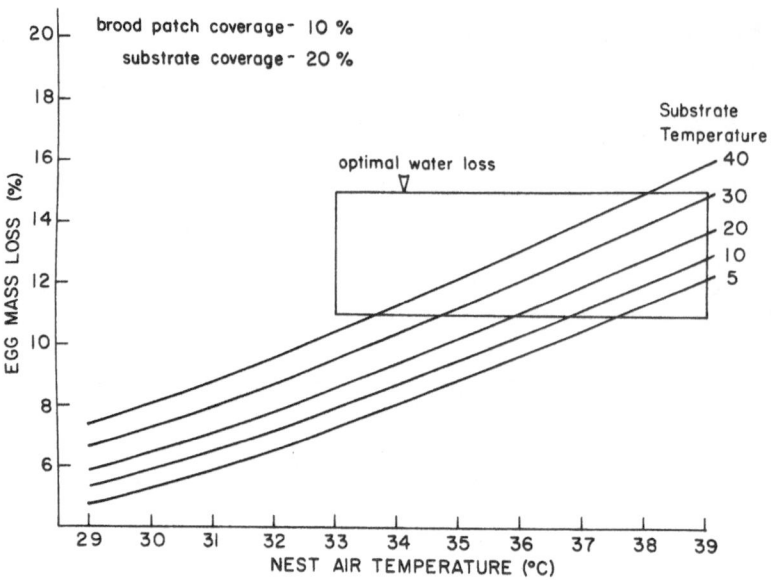

Fig. 6. The fractional water loss (expressed as %) of eggs as a function of nest air temperature and substrate temperature. Variables are as described in Figure 1 and the thermal conductivity of the substrate is set at 0.0002 $Kcal \cdot cm^{-1} \cdot h^{-1} \cdot {}^{o}C^{-1}$.

The simulation is carried out at several substrate thermal conductivities. The fractional water loss from 11 to 15% is enclosed in a box to represent the water loss typical of most bird eggs (Ar and Rahn, 1980). We can state at the outset that average egg temperatures must exceed $35^{o}C$ if egg water loss is to exceed 11% of initial mass. Whenever, the average egg temperature of the simulated egg fell below $35^{o}C$, water loss fell below 11% of initial mass. Egg temperature is influenced by a number of variables and since vapor pressure inside the egg is established by temperature, all thermal effects will also influence egg water loss. It is particularly interesting that if the thermal conductivity of the substrate is low, the degree of contact between the egg and substrate is unimportant. That is, if the parent constructs the nest of materials containing mostly air then contact is irrelevant. However, if the nest material conducts heat well then the degree of contact becomes significant and if a fractional mass loss of the appropriate magnitude is to be achieved the degree of contact must be limited. Another factor in this relationship is the substrate temperature. A hot, dry substrate will counteract the effect of elevated thermal conductivity. In cold climates and on solid substrates the contact between the egg and the substrate must be limited but in warm climates other means of

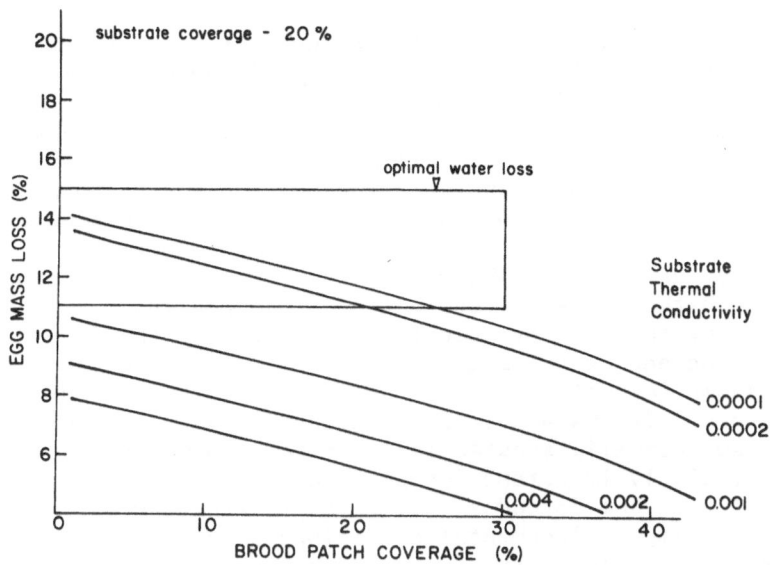

Fig. 7. The fractional water loss (expressed as %) of eggs as a
function of the fraction (%) of the egg contacting the
brood patch and at different substrate thermal
conductivities. Other variables are as described in
Figure 1. The thermal conductivity ranges from that for
air (0.0001) to that for sand (0.004).

controlling egg water exchange are likely to be effective. We can
also simulate the effect of nest air temperature on the fractional
water loss of eggs (Figure 6) and this can be done at a variety of
substrate temperatures. Once again the expected or typical
fractional water loss is shown. As nest temperature decreases
fractional water loss decreases by about 0.7% per $^\circ$C. Thus, for
every 1°C drop in nest temperature, egg water loss during incubation
falls by nearly 1% and as nest temperature decreases so too does egg
temperature and this occurs despite a constant brood patch
temperature and on a substrate with low thermal conductivity. As the
substrate temperature increases, the fractional water loss
experienced by the egg at a given nest air temperature also
increases. The limiting nest air temperature is about 38 $^\circ$C on a 5
$^\circ$C substrate and 33.5 $^\circ$C on a 40 $^\circ$C substrate if a water loss of at
least 11% is to be achieved. Thus, the temperature maintained by the
parent around the part of the egg not contacting the brood patch is
very important both for thermal and water exchange. The parent bird
can perhaps most directly influence the thermal exchange of the egg
by varying the degree of contact between the brood patch and the
surface of the egg. We have not examined this relationship
explicitly but it should be clear that the more surface covered by

the warm brood patch the warmer the egg will be. However, the effect on the egg water loss will be just the opposite of that expected for egg heat exchange. This is shown in figure 7 where we simulate the fractional water loss of eggs covered to varying degree by the brood patch. The effect of varying substrate thermal conductivity is also shown on this process. The overall effect is due to the way we have structured the relationship between the egg and the brood patch. The vapor pressure in the skin of the brood patch is calculated at the osmotic pressure of blood and brood patch temperature and this means that it always very high. Therefore, only very little exchange occurs between the egg and brood patch (and depending on temperature, water uptake by the egg may occur) and the greater the degree of coverage of the egg by the brood patch the smaller the area of egg shell available for water loss to the environment. It would appear that under most circumstances brood patch coverages in excess of 20% of the egg surface will substantially impede egg water loss. This may be particularly important for particular groups of birds such as penguins which may be forced to cover large fractions of their eggs because of extreme environmental conditions. Rahn and Hammel (1982) reported relatively low water loss for the eggs of Adelie penguins in the Antarctic. This reduced water loss is likely to be due to the thermal requirement that the Adelie cover its egg rather thoroughly. We might predict on the same grounds that the egg water loss of the Emperor Penguin egg will be even lower since this species incubates during the middle of the Antarctic winter and covers its egg completely.

SUMMARY AND RECAPITULATION

Egg temperature and water loss are interrelated. If water losses of about 15% of initial mass are to be achieved then the temperature of the egg must be maintained in excess of 35 $^{\circ}$C. This can only be done by limiting the heat lost by the egg to the surroundings. Nest sites, nest structures and incubation behavior by the adult will reflect these constraints. Our analysis has been concerned with the steady-state heat exchange of avian eggs. The transfer of heat directly between the egg and the brood patch is quantitatively only a small fraction of the heat exchange experienced by the egg and has a minimal effect on average egg temperature. It is not likely that sufficient quantities of heat can be transferred directly between the brood patch and egg so as to effectively keep eggs cool in a hot environment or hot in a cool environment. This conclusion should apply to non-steady state heat transfer as well. Further, coverage of the egg by the brood patch will impede egg water loss. The principal thermal function of the brood patch may be to control the temperature of the air surrounding the section of the egg not contacting the brood patch at a constant, elevated level. The role of the nest would then be to minimize heat transfer between nest air and the environment and between the egg (where contact occurs) and the environment.

ACKNOWLEDGEMENTS

Our research has been supported by NSF grant PCM 79-16256 and by grants from Iowa State University.

REFERENCES

Ackerman, R.A., Whittow, G.C., Paganelli, C.V. and Pettit, T.N., 1980, Oxygen consumption, gas exchange and growth of embryonic Wedge-tailed Shearwaters (Puffinus pacificus chlororhynchus), Physiol. Zool., 53:210.

Ar, A. and Rahn, H., 1980, Water in the avian egg: overall budget of incubation, Amer. Zool., 20:373.

Barott, H.G., 1937, Effects of temperature, humidity, and other factors on hatch of hen's eggs and on energy metabolism of chick embryos, USDA Tech. Bull., #553:1.

Briedis, D. and Seagrave, R.C., 1984, Energy transformation and entropy production in living systems I. Applications to embryonic growth, Journ. Theoret. Biol., in press.

Boersma, P.D. and Wheelright, N.T., 1979, Egg neglect in the Procellariiformes: reproductive adaptations in the Fork-tailed Storm Petrel, Condor, 81:157-165.

Drent, R., 1970, Heat exchange during incubation, Behaviour Supplement, 17:58.

Drent, R., 1975, Incubation, in: "Avian Biology," D.S. Farner and King, J.R., ed., Academic Press, New York.

Grant, G.S. 1982, Avian incubation: egg temperature, nest humidity, and behavioral thermoregulation in a hot environment, Ornithol. Monog., #30:2-75.

Grant, G.S., Pettit, T.N., Rahn, H., Whittow, G.C. and Paganelli, C.V., 1982, Water loss from Laysan and Black-footed Albatross eggs, Physiol. Zool., 55:405.

Grant, G.S., Pettit, T.N., Rahn, H., Whittow G.C. and Paganelli, C.V., 1982, Regulation of water loss from Bonin Petrel (Pterodroma hypoleuca) eggs, Auk, 99:236.

Hillel, D., 1978, "Soil and Water", Academic Press, New York.

Howell, T.R., 1979, Breeding biology of the Egyptian Plover, Pluvianus aegyptius (Aves: Glareolaridae), Univ. Calif. Publ. Zool., 113:1.

Khaskin, V.V., 1961, Heat exchange in bird's eggs on incubation, Biophysics, 6:97.

Rahn, H., Paganelli, C.V., Nisbet, I.C.T. and Whittow, G.C., 1976, Regulation of incubation water loss in eggs of seven species of terns, Physiol. Zool., 49:245.

Rahn, H. and Ar, A., 1974, The avian egg: incubation time and water loss, Condor, 76:147.

Rahn, H., Ackerman, R.A., and Paganelli, C.V., 1977, Humidity in the avian nest and the egg water loss during incubation, Physiol. Zool., 50:269.

Rahn, H. and Hammel, H.T., 1982, Incubation water loss, shell conductance, and pore dimensions in Adelie Penguin eggs, Polar Biol., 1:91.

Rahn, H., Krog, J. and Mehlum, F., 1984, Microclimate of the nest and egg water loss of the Eider and other water fowl in Spitsbergen, Polar Research, in press.

Romijn, C. and Lokhorst, W., 1956, The caloric equilibrium of the chicken embryo, Poultry Science, 35:829.

Seagrave, R.C., 1971, "Biomedical applications of heat and mass transfer", Iowa State Univ. Press, Ames, IA.

Tazawa, H., 1980, Oxygen and CO_2 exchange and acid-base regulation in the avian embryo, Amer. Zool., 20:395.

Vleck, C.M., 1981, Hummingbird incubation: Female attentiveness and egg temperature, Oecologia, 51:199.

Walsberg, G.E., 1980, The gaseous microclimate of the avian nest during incubation, Amer. Zool., 20:363.

White, F.N. and Kinney, L.L., 1974, Avian Incubation, Science, 186:107.

Whittow, G.C., Ackerman, R.A., Paganelli, C.V. and Pettit, T.N., 1982, Pre-pipping water loss from the eggs of the Wedge-tailed Shearwater, Comp. Biochem. Physiol., 72A:29.

EGGS, YOLK, AND EMBRYONIC GROWTH RATE

H. Rahn[1], R. A. Ackerman[2], and C. V. Paganelli[1]

[1]Department of Physiology
State University of New York at Buffalo
Buffalo, New York   14214
[2]Department of Zoology
Iowa State University
Ames, Iowa   50011

INTRODUCTION

In this chapter we describe various strategies that seabirds have adopted in their reproductive process such as the relative size of their eggs, the relative amount of yolk they contain, the caloric content of their eggs, the cost of egg formation, and the embryonic growth rate. In addition we document energetic strategies of some species whose eggs have larger than normal yolk content.

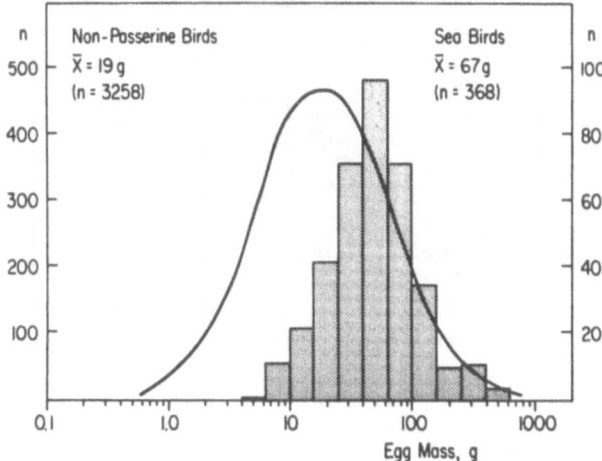

Fig. 1. Log-normal egg mass distribution curve of non-passerine birds (left ordinate). Bar graph distribution of seabird eggs (right ordinate).

How big are seabird eggs and how do they compare in size with those of other birds? Fig. 1 shows the log-normal distribution of egg mass for 3258 species and subspecies of non-passerine birds derived from the tables of Schonwetter (1960). The total range extends from the 0.3 g eggs of Trochilids to the 1500 egg of the African Ostrich. The mean egg mass is 19 g. The right ordinate indicates the number of seabird species and subspecies (total = 368) plotted as a bar graph against their egg mass. The range extends from 5.5 g for the smallest Storm Petrel egg to about 450 g for the largest eggs, those of the Wandering Albatross and the Emperor Penguin. Their mean egg mass is 67 g and 90% of their eggs are larger than 19 g, the mean value for birds in general.

Relationship of Hatchling Mass to Adult Mass

In figure 2 we have plotted hatchling mass against adult body mass (Rahn, 1982). To obtain hatchling mass for figure 2 we multiplied egg mass by 0.67, a factor established from values of egg and hatchling mass in 63 species reported by Heinroth (1922). It is of particular interest that the slope of the regression line, 0.70, is not significantly different from the slope of basal metabolic rate, 0.72, when regressed against adult body mass (Lasiewski and Dawson, 1967). This suggests a general rule: the size of bird eggs is proportional to the adult metabolic rate. As a consequence, egg mass as a fraction of adult body mass (henceforth called relative egg mass) is proportional to the mass-specific metabolic rate of the adult. This is shown in figure 3 in which we have plotted relative

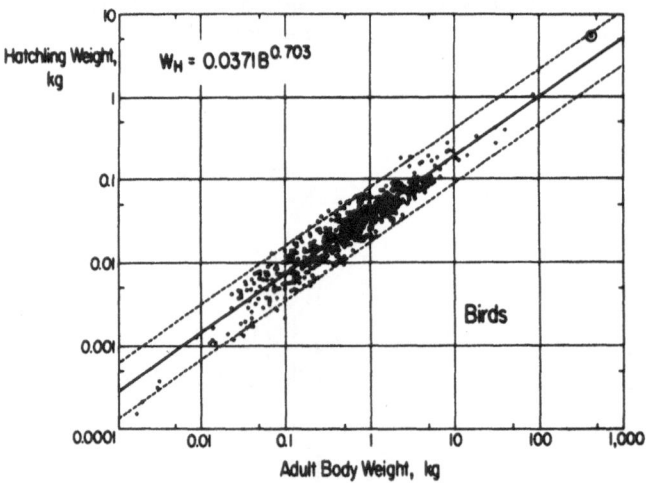

Fig. 2. Hatchling mass (= 0.67 egg mass) plotted against adult body mass of non-passerine birds (n = 514 species). Slope 0.70 (Rahn, 1982).

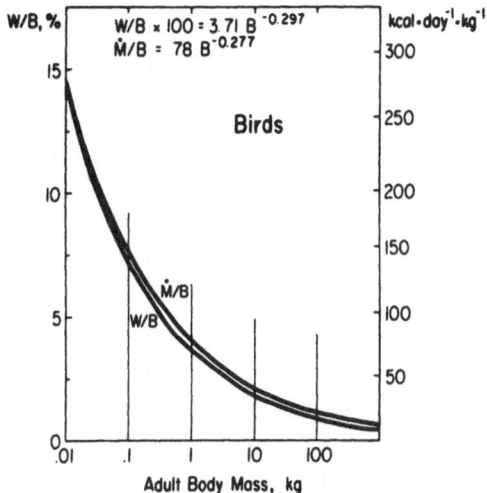

Fig. 3. Relative hatchling mass (% of adult body mass, left ordinate) on a linear scale plotted against adult body mass (kg) on a log scale. Mass specific metabolic rate (kcal·d^{-1}·kg^{-1}) of adult birds on right ordinate (from Lasiewski and Dawson, 1967).

hatchling mass and mass-specific metabolic rate of the adult on semi-log coordinates against adult body mass. Figure 3 shows clearly that relative egg or hatchling mass declines as body mass increases, a general phenomenon which is also seen for the relative litter mass of mammals when plotted against adult body mass (Rahn, 1982).

Relative Egg Mass of Seabirds

Fig. 4 shows the regression of egg mass against body mass for 36 species of Procellariiformes. The allometric equation is $g = 0.67 A^{0.74}$, where g is egg mass and A = adult body mass, g. Dividing both sides of the equation by A, we obtain an equation relating relative egg mass as % of adult mass (Y) to adult mass: $Y = 67A^{-0.16}$. Similar regressions for five other seabird taxa are shown in Table 1 and are graphed on a semi-log plot in Fig. 4.

To compare relative egg mass in various taxa one must choose a common body mass, since the slopes of the regression lines for the various groups (Table 1) are not identical. All groups in Fig. 5 have representatives at 1000 g body mass or very near this value, and thus serve as a convenient point for such comparisons, as shown in Table 2. For seabirds the smallest eggs, 3% of adult mass, are found among the Phalacrocoracidae, while the largest eggs are found among the Alcidae and Procellariiformes. For comparison, the right side of the table gives relative egg mass for seven orders of land-birds, calculated from the tables of Rahn et al. (1975).

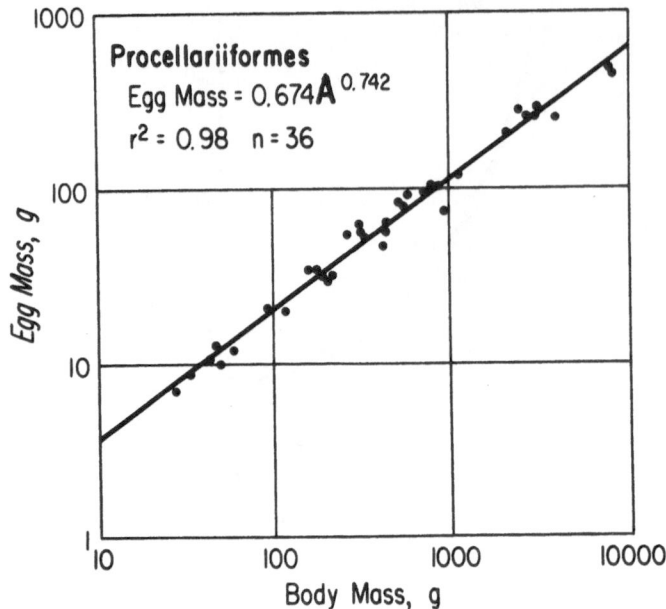

Fig. 4. Egg mass (g) as a function of adult body mass (g) for 36 species of Procellariiformes. Slope = 0.74. r^2 = coefficient of determination.

Table 1. Relationship of relative egg mass (% of adult mass) to adult body mass, where relative egg mass, $Y = a A^b$ and A = body mass, g; r^2 = coefficient of determination, n = number of species.

	a	b	r^2	n	Range, g
Procellariiformes	67	− 0.26	0.98	36	28 − 8,040
Alcidae	51	− 0.22	0.98	17	92 − 970
Laridae	80	− 0.32	0.98	17	40 − 1,800
Sulidae	82	− 0.40	0.80	8	950 − 3,100
Spheniscidae	50	− 0.34	0.96	16	1,100 − 30,000
Phalacrocoracidae	63	− 0.44	0.67	8	600 − 3,000

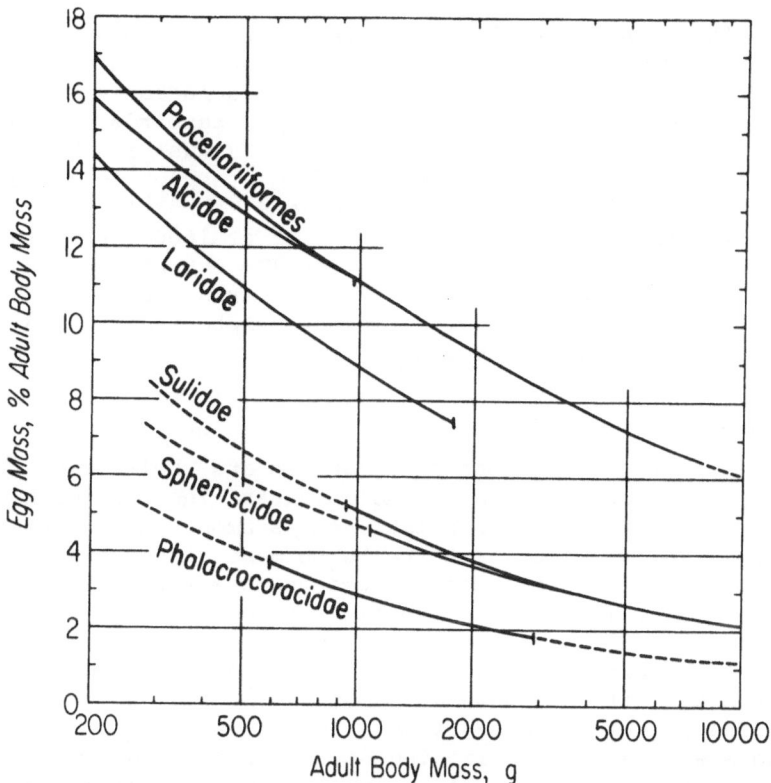

Fig. 5. Egg mass (% of adult mass) on a linear scale plotted against
adult body mass (g) on a log scale for 6 seabird families.
The heavy lines indicate range of body mass for each group.

Table 2. Comparison of relative egg mass of seabirds and landbirds
expressed as percent of 1000 g body mass. Values in parentheses indicate number of species used in the regression of
egg mass as a function of body mass. For landbird values
see Rahn, Paganelli and Ar (1975).

Seabirds			Landbirds		
Phalacrocoracidae	3.0	(8)	Psittaciformes	3.5	(23)
Spheniscidae	4.8	(16)	Galliformes	4.0	(52)
Sulidae	5.2	(8)	Strigiformes	4.6	(15)
Laridae	8.8	(17)	Gruiformes	5.6	(18)
Alcidae	11.2	(17)	Falconiformes	5.9	(54)
Procellariiformes	11.2	(36)	Anseriformes	6.7	(149)
			Charadriiformes[*]	9.2	(60)

[*]Includes 12 Larid and 7 Alcid species.

It is important to realize that the regression equations shown in Table 1 and Fig. 5 represent approximations at best and to note that certain individual exceptions were not included in the regressions. For example, the only truly precocial seabirds are found among a few of the Alcidae (Sealy, 1975b). Xantus' and Ancient Murrelets (Appendix) weighing 167 and 206 g, respectively, have a relative egg mass of 22%. Among the Sulidae, Abbott's Booby has an exceptionally large relative egg mass of 7% at 1600 g body mass. The relative egg masses for three Tropicbirds (Phaethontidae) and two Frigatebirds (Fregatidae) fall between the Laridae and Sulidae, and for two Pelicans (Pelecanidae), between the Spheniscidae and Phalacrocoracidae. (See Appendix.)

Shell Characteristics

Characteristic dimensions of eggshells are their mass, thickness, and density. Because shell mass and thickness increase with egg mass, we must first establish relationships among these variables for each taxon if valid comparisons are to be made at comparable egg masses. For example, shell mass increases more rapidly than egg mass (Paganelli et al., 1974), while shell thickness increases approximately as the square root of egg mass (Ar et al., 1979). To compare shell characteristics among seabird taxa, allometric equations for shell mass and shell thickness as functions of egg mass are shown in Table 3. The equations werer calculated from the tables of Schönwetter (1960) and are based on his listing of the species and subspecies for the various orders. The values of the 175 species and subspecies of the Charadriiformes represent only the family Laridae and the family Alcidae. The allometric equations allow one to predict meal values of shell mass and thickness for any egg mass.

Table 3. Egg mass range and shell characteristics of seabird eggs calculated from the data of Schönwetter. n = number of species and subspecies, r^2 = coefficient of determination

| | | range of egg mass | $Y = aW^b$, W = egg mass, g | | | | | |
| | | | shell mass, g | | | shell thickn., mm | | |
	n	g	a	b	r^2	a	b	r^2
Spheniscif.	19	52 - 425	0.052	1.16	0.98	0.042	0.53	0.92
Procellariif.	88	6 - 455	0.047	1.09	0.99	0.055	0.40	0.95
Pelecanif.	87	19 - 185	0.073	1.07	0.93	0.080	0.39	0.62
Charadriif.	175	8 - 115	0.038	1.15	0.99	0.045	0.45	0.88

Table 4. Shell dimensions of 100 gram eggs

	Shell mass	thickness	density
	g	mm	$g \cdot cm^{-3}$
Sphenisciformes	10.9	0.49	2.20
Procellariiformes	7.1	0.34	2.04
Pelecaniformes	10.1	0.48	2.02
Charadriiformes	7.6	0.36	1.95

In Table 4 we compare predicted values for shell mass and shell thickness for 100 g eggs of each taxon. Shell density was obtained by dividing shell mass by shell volume, the latter calculated from the product of shell thickness and egg surface area (Paganelli et al., 1974). This calculation was carried out for each species and the values in Table 4 represent averages with coefficients of variations between 3.4 and 4.2%. Note in particular that Sphenisciformes and Pelecaniformes have appreciably heavier and thicker eggshells than the Procellariiformes and Charadriiformes. The shell density was significantly greater in the Sphenisciformes than in the other taxa.

Relative Yolk Mass and Caloric Egg Density

A recent review of the relative yolk content (% of egg content) of 60 seabird eggs has recently become available (Williams et al., 1982) which also includes an analysis of the lipid, protein, and water contents of yolk and albumen of 25 species of seabirds. In Table 5 we present their average values for six groups of seabirds, which varied from 19% in the Sulidae to 38% in the Alcidae.

Table 5. Relative amount of yolk in seabird eggs, expressed as percent of egg content. (From tables of Williams et al., 1982.)

	No. Species	% Yolk	S.D.	Range
Sphenisciformes[*]	9	28.2	2.2	25 – 30
Procellariiformes	19	37.6	4.6	26 – 44
Phalacrocoracidae	7	20.9	4.0	15 – 28
Sulidae	3	19.0	1.9	18 – 21
Laridae	18	30.7	3.5	27 – 38
Alcidae	4	38.2	3.0	35 – 41

[*]Values of the first and second eggs were averaged in 3 species of Eudyptes.

Members of the Procellariiformes lay a single egg. However, their relative yolk content varies considerably, and Warham (1983) showed a surprising inverse correlation between relative yolk content and egg mass in 23 species ranging in egg mass from 11 to 400 g. Additional values for 10 more species have recently become available (Williams et al., 1982; Pettit et al. (this volume); Boersma, 1982; Montevecchi et al., 1983; and Rahn and Huntington, unpubl.). These have been added to Warham's data in Fig. 6 where the yolk value, % of yolk content (not egg mass), has been plotted against egg mass.

Does the range of yolk values from ca 30% in the largest eggs to ca 40% in the smallest eggs reflect differences in hatchling maturity from semiprecocial to precocial as noted in other birds (Nice, 1962; Ar and Yom-Tov, 1978; Carey et al., 1980), or must one look for other explanations as discussed by Williams et al. (1982), Warham (1983), Boersma (1982), and Montevecchi et al., (1983). According to Carey et al. (1980), 71 species of precocial birds (Nice's classification) have a mean yolk content of 40% $\pm$ 0.6 (S.E.). The average value for 11 petrel eggs weighing less than 43 g is nearly the same (40.7% $\pm$ 0.6 (S.E.)), yet their hatchling maturity is hardly comparable to that of precocial species.

Below we also discuss yolk values found in 1-clutch nests of five species of terns which are appreciably larger than those found among the 3-clutch tern eggs. Thus, the relative yolk content within certain taxonomic groups can vary greatly. This is true even within a given genus. Williams et al. (1982) have shown that among the Spheniscidae three species of Eudyptes lay two eggs, the first one of which is appreciably smaller (65, 71, and 80% of the second egg) but has a relatively greater yolk content (7.3, 4.8, and 2.2%, respectively). The relatively greater yolk content among the terns and Procellariiformes appears to be linked to prolonged incubation periods and relatively greater cost of hatchling production. This will be discussed under the heading In-shore, Off-shore, and Pelagic Feeders.

COST OF EGG PRODUCTION

Caloric Egg Density. In Fig. 7 caloric density (kcal·g^{-1} of wet mass egg content) is plotted against relative yolk content (% of egg content) for 21 species of seabirds taken from data presented by Carey et al. (1980) and Pettit et al. (this volume). The linear regression, kcal·g^{-1} = .0332 (% yolk) + .602, provides a reasonable prediction of caloric density if only the relative yolk content is known.

Net Efficiency of Egg Production. Brody (1945, p. 882) established the net efficiency of egg production as 77% for the chicken. Thus (1/.77) or 1.3 kcal is required by the hen to produce 1.0 kcal

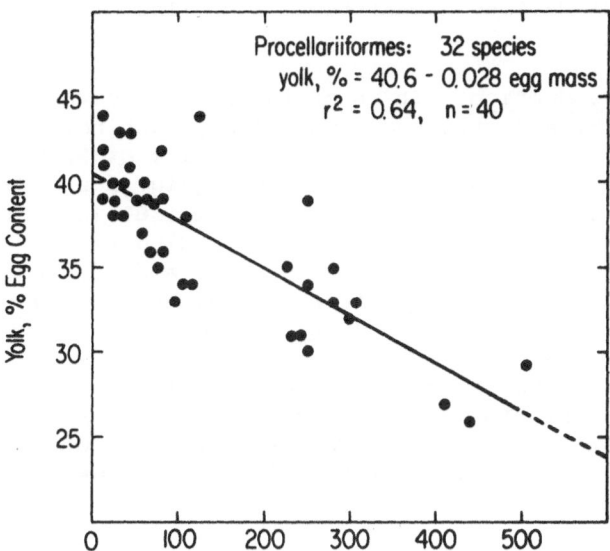

Fig. 6. Yolk (% of egg content) plotted against egg mass (g) for
32 species of Procellariiformes, extending the data of Warham (1983).

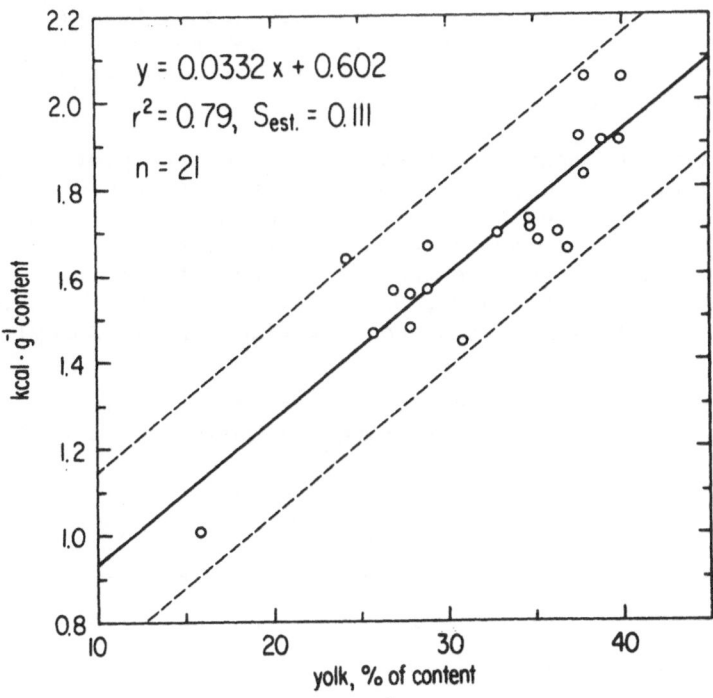

Fig. 7. Caloric density (kcal·g^{-1} of egg content) as a function of
yolk (% of egg content) in 21 seabird species. From Carey
et al. (1980) and Pettit et al. (this volume).

of egg content. El-Wailley (1966) found a similar efficiency (75%) for the Zebra Finch.

Cost of Egg Production. It is of interest to compare the cost of egg production of a petrel and a cormorant, each having a body mass of 1000 g and a predicted resting metabolism of 92 $kcal \cdot day^{-1}$. Their respective egg masses are 111 and 30 g (Fig. 5 and Table 1). The relative yolk content of the petrel eggs is 38% (see Fig. 6) and that of the cormorant is 21% (Table 5). According to Fig. 7 these yolk contents predict caloric contents of 1.85 and 1.30 $kcal \cdot g^{-1}$, respecitvely, or 205 and 39 kcal per egg (Table 6). When these values are multiplied by 1.3, the cost of producing these eggs is 267 and 51 kcal, respectively. Thus, it costs five times more to produce a petrel egg than a cormorant egg. On the other hand, the clutch size for cormorants is typically four and for the petrel one. Therefore, the overall energy requirements for the clutch in this case are similar, being equivalent to 2.9 and 2.2 times the basal metabolic rate.

IN-SHORE AND OFF-SHORE FEEDERS

Lack (1968) pointed out that for terns and gulls the proximity of feeding grounds and quantity of available food may affect clutch size, incubation time, fledging period, incubation spells, and feeding visits and that longer fledging periods were associated with longer incubation periods. Lack has designated these groups as in-shore and off-shore feeders, fully realizing that similar adaptations might also arise from lack of available food.

To expand the examples which Lack cited we have plotted incuba-tion period against egg mass for members of the Sterninae in Fig. 8. The solid dots are species which lay a clutch of 2 or 3 eggs (see Appendix). The open circles represent terns which lay a single egg; in order of increasing egg mass they are *Gygis alba* (White Tern), *Anous tenuirostris* (Black Noddy), *Sterna lunata* (Grey-backed Tern), *Sterna fuscata* (Sooty Tern), and *Anous stolidus* (Brown Noddy).

Table 6. Cost of egg production of a petrel and a cormorant, each weighing 1000 g and with a resting metabolism of 92 $kcal \cdot day^{-1}$.

	Egg Mass	Yolk	Energy Content		Effic. Factor	Total Cost
	g	%	$kcal \cdot g^{-1}$	$kcal \cdot egg^{-1}$	--	$kcal \cdot egg^{-1}$
Petrel	111	38	1.85	205	1.3	267
Cormorant	30	21	1.30	39	1.3	51

98

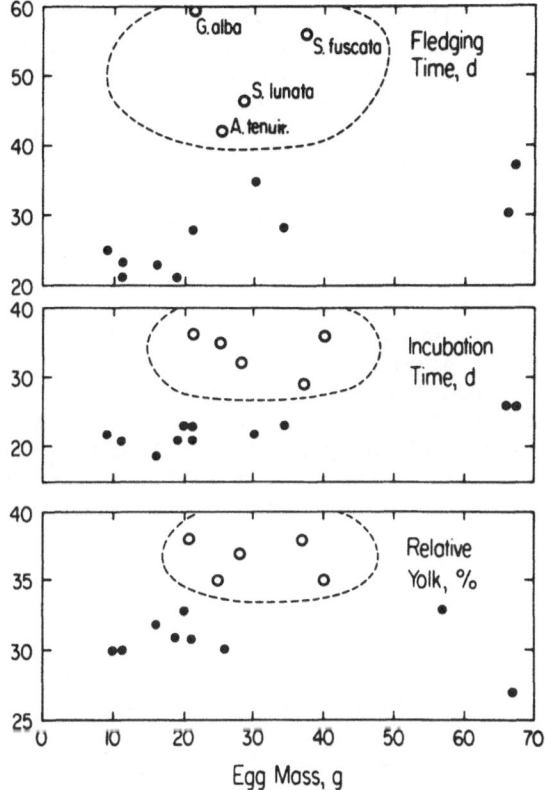

Fig. 8. Fledging period (days), incubation period (days), and yolk
(% of egg content) plotted against egg mass (g) of subfamily
Sterninae, Terns. Solid points represent terns with a
clutch of 2 or 3 eggs. Circles represent 5 species which
lay a single egg.

Their incubation periods are considerably longer than eggs of similar
mass that are laid in clutches of 2 or 3. As Lack (1968) pointed
out, increased incubation time is linked with increased fledging
time, as shown in the upper part of Fig. 8. (Fledging time for the
Brown Noddy is not available.)

An additional correlate of one-clutch eggs is a larger
relative yolk content (percent of egg content) as shown in the lower
part of Fig. 8. The average value for the 5 species is 36.6%, which
can be compared with 35.3% for the average value of the domestic
chicken based upon 7 independent reports (Carey et al., 1980).
Furthermore, as Pettit et al. (this volume) have pointed out, the
caloric egg density of these five tern species is similar to that of

other eggs with equal yolk content. The larger yolk content of these five tern species is possibly related to the greater maintenance cost of the embryo as reflected in the energetic cost of producing a hatchling which averages 3.6 kJ·g^{-1} for the Brown Noddy, Black Noddy, and White Tern (Pettit and Whittow, 1983; Pettit et al., 1984) and can be compared with ca. 3.0 kJ·g^{-1} hatchling, which is typical for for other birds (Hoyt and Rahn, 1980). It may also provide a larger yolk reserve for the hatchling. It can be viewed as an additional adaptation to off-shore feeding or lack of available food supply.

Similar examples can be found among the gulls. As Lack (1968) pointed out, *Creagrus furcatus*, the Swallow-tailed Gull of the Galapagos, has a 32-day incubation and a 60-day fledging period. It lays a single egg. *Rissa brevirostris*, the Red-legged Kittiwake of the Pribilof Islands in the Bering Sea, also lays a single egg with average incubation and fledging periods of 31 and 37 days (Hunt, G.L., Jr., pers. comm.). Whether these species will eventually show a relatively larger yolk than other gull eggs of similar size remains to be seen.

PELAGIC FEEDERS

As noted in Fig. 6 there is an inverse relationship between relative yolk content and egg mass among the Procellariiformes. Is there also an inverse relationship between incubation time and egg mass? Whittow (1980) was the first to point out that, indeed, such a relationship exists when one compares their incubation time with that of "other birds." This is shown in Fig. 9 from the study of Grant et al. (1982) where the regression of incubation time of Procellarii- formes (shaded area) approaches the regression of "other birds" as egg mass increases. The ratio of the observed incubation to that predicted for "other birds" of similar egg mass is a measure of rela- tive prolongation of incubation time among Procellariiformes and this decreases as egg mass increases. For example, relative incubation period for a 7 g petrel egg is 2.2 times greater than that predicted, while for a 500 g egg the ratio is 1.5 times the predicted value for other birds. This would suggest a relationship similar to that ob- served among the one-clutch terns (see above), i.e., relatively longer incubation periods are associated with relatively greater yolk content

When one compares the total energy cost per gram of hatchling mass during development in eggs ranging from 10 to 300 g as shown in Table 7, it becomes apparent that relatively longer incubation periods require greater maintenance energy. As egg mass diminishes there is an approximate parallelism between the increase in relative incubation time, energy cost per hatchling mass, and relative yolk content. These observations suggest that smaller petrel eggs require rela- tively longer incubation periods, which require greater maintenance

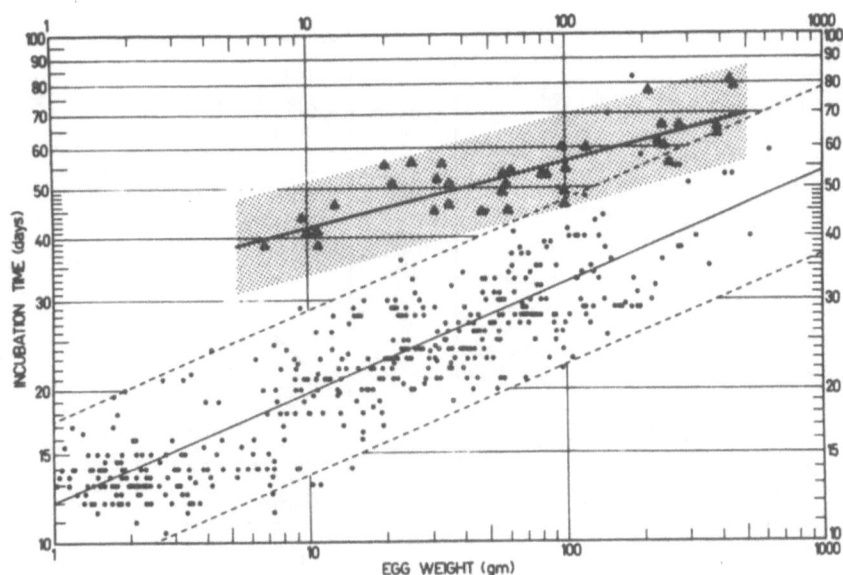

Fig. 9. Incubation time (days) as a function of egg mass (g) for 38 species of Procellariiformes (upper regression) and 460 species of "other birds" (lower regression), indicating limits of ± 2 SEE. From Grant et al. (1982).

Table 7. Egg mass, incubation time, energy cost per g hatchling mass, and relative yolk content of Procellariiformes eggs ranging from 10 to 300 g. For details see text. Ref.: (1) Pettit et al., 1982a; (2) Ackerman et al., 1980; (3) Pettit et al., 1982b; (4) Rahn and Huntington (unpubl.)

| | Egg Mass | Incub | Incub Ob/Pred | Hatchl. Cost | Yolk | Ref. |
	g	d	–	kJ·g^{-1}	%	
Diomedea nigripes Blk-ftd. Albatross	305	66	1.6	2.9	33	1
Diomedea immutabilis Laysan Albatross	285	75	1.6	3.1	35	1
Puffinus pacificus Wedge-t. Shearwater	58	52	1.8	3.9	40	2
Pterodroma hypoleuca Bonin Petrel	39	49	1.8	4.2	40	3
Oceanodroma leucorhoa Leach's Storm-Petrel	10.5	42	2.1	5.4	39	4

101

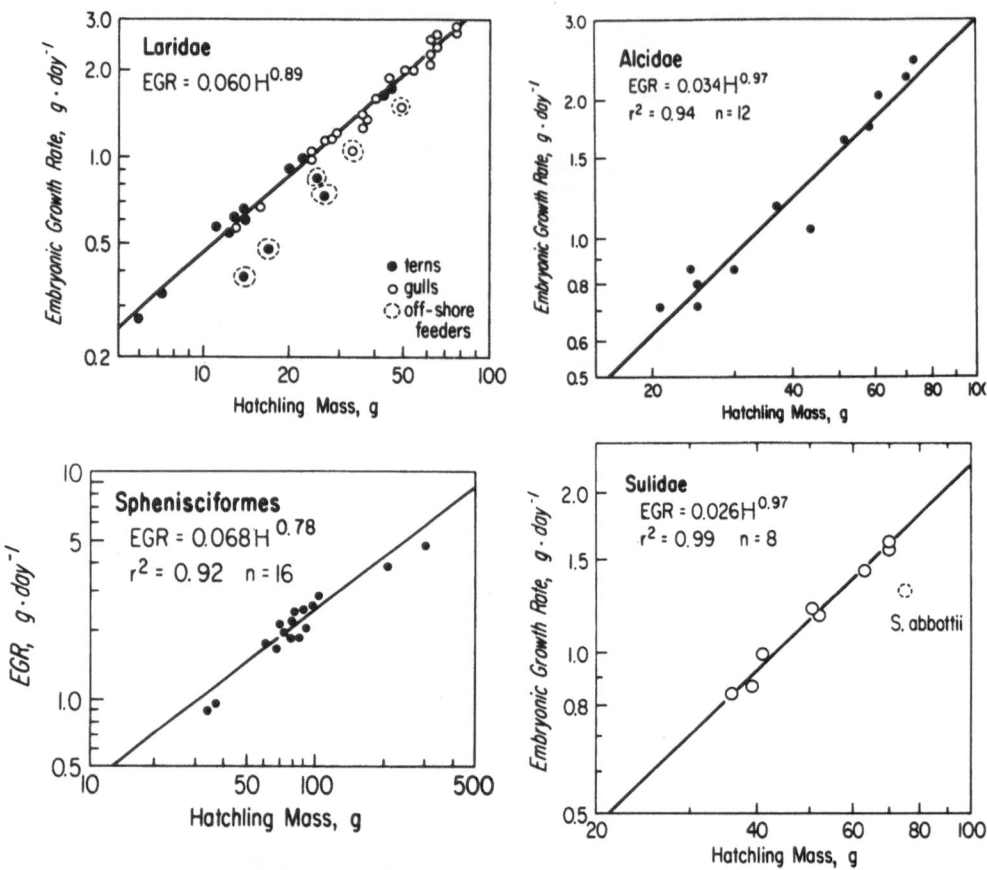

Fig. 10 - 13. Embryonic growth rates, EGR (g·day^{-1}), as a function of hatchling mass, H, (g) for Laridae, Alcidae, Sphenisciformes, and Sulidae. r^2 = coefficient of determination; n = number of species.

energy, and that this, in turn, is provided by relatively larger yolk contents. To what extent smaller hatchlings require a relatively larger yolk reserve upon hatching still needs to be explored.

EMBRYONIC GROWTH RATE

At what rate does the avian embryo develop and how does this rate vary among different groups of seabirds? We define absolute mean embryonic growth rate, EGR, as hatchling mass divided by incubation time and express this ratio in units of g·day^{-1}. Hatchling mass is defined as 67% of initial egg mass, as noted previously, from the observations of Heinroth (1922). Recent measurements in

30 species (A. Ar, pers. comm.) also gave a mean value of 67%. The mass also includes part of the remaining yolk sac which is taken up by the embryo during the hatching act. This yolk mass averages 7% of hatchling mass in altricial hatchlings (Schmekel, 1960) and about twice this value in precocial species (A. Ar, pers. comm.). No allowance was made for yolk-sac mass in our calculations. In Figs. 10 - 13 we have plotted EGR against hatchling mass taken from the values listed in the Appendix. Each group has its own growth rate characteristics, and the slopes of the regression lines vary from 0.78 to 0.97. Among the Laridae the circled points were not included in the regression. These have a lower growth rate and represent off-shore feeders as discussed above.

The EGR's of the Procellariiformes are shown in solid circles in Fig. 14. The open circles represent the juvenile growth rates for the same species. Juvenile growth rate was defined as the difference between adult and hatchling mass divided by the fledging period. Here the assumption was made that the fledgling mass was equal to the adult mass. Even though nestling mass in these birds tends to exceed adult mass during later stages of development, it decreases again just prior to fledging and appears to be close to that of the adult. With the exception of the two heaviest albatrosses, which were not included in the regression, it appears that juvenile and embryonic growth rates are well described by a single regression. One may also note the overlap between the growth rates of the larger hatchlings and the smaller fledglings. One can now compare EGR for different seabirds (Fig. 15) by choosing a common hatchling mass, such as 50 g, where each family has a representative. Embryos of the Laridae develop nearly twice as fast as those of the Procellariiformes. The Alcidae, Sulidae, and Spheniscidae (not shown) fall in between. The top regression is an extrapolation beyond 22 g hatchling mass for the order

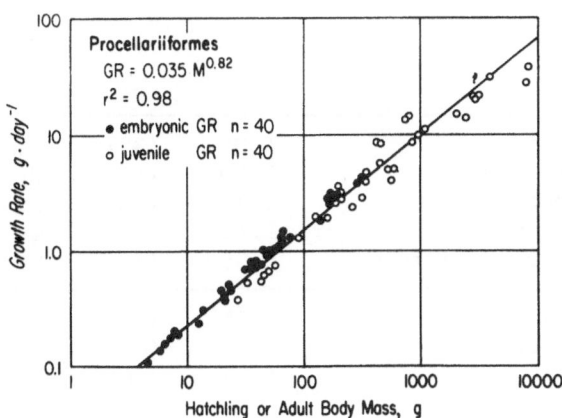

Fig. 14. Embryonic and juvenile growth rates $(g \cdot day^{-1})$ plotted against hatchling and adult body mass, M, (g). See text.

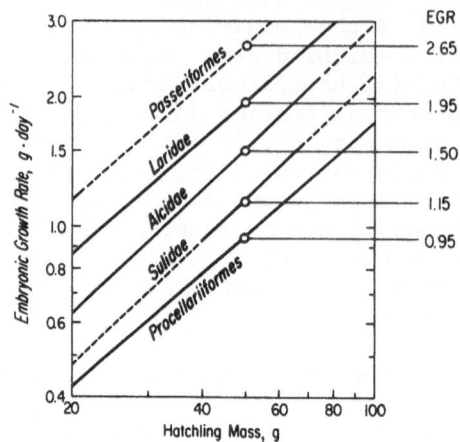

Fig. 15. Comparison of EGR (g·day^{-1}) for various taxa plotted against
hatchling mass (g). Heavy line indicates range of hatchling
mass for each group. Dotted lines denote extrapolations of
regressions.

Table 8. Comparison of embryonic growth rate, EGR, (g·d^{-1}), incubation time, I, (days), and adult body mass, A, (g) among seabirds and landbirds with a hatchling mass of 50 g, equivalent
to an egg mass of 75 g. Data for landbirds unpubl. Values in
parentheses are number of species used in regression.
Incubation time calculated from relationship I = 50 g/EGR.
Adult body mass derived from equations of Table 1 for seabirds
and for landbirds from Rahn et al. (1975).

		EGR	n	I	A
		g·d^{-1}	-	days	g
Seabirds	Laridae	1.95	(32)	26	800
	Alcidae	1.50	(12)	33	640
	Spheniscidae	1.44	(16)	35	2,000
	Sulidae	1.15	(8)	44	1,800
	Procellariiformes	0.95	(40)	53	460
Landbirds	Passeriformes	2.65[*]	(92	-	-
	Anseriformes	1.79	(82)	28	1,200
	Falconiformes	1.32	(40)	38	1,500

[*]Extrapolated to a hatchling mass of 50 g.

Passeriformes (n = 92, unpublished). Because the slopes are not identical, slightly different ratios between taxa will be found when similar comparisons are made at other hatchling masses.

In Table 8 are listed EGR values for five seabird families and three landbird orders for a hatchling mass of 50 g. It also predicts incubation period = 50 g/EGR as well as the adult body mass. It can be seen that each group has developed its own strategy for producing a 75 g egg or a 50 g hatchling. From the few data of the Phalacrocoracidae (Appendix) one can estimate the EGR from the four largest eggs which have an average hatchling mass of 33 g and an EGR of 1.19 $g \cdot d^{-1}$. This value can be compared with the predicted EGR of a 33 g Larid hatchling, which is 1.35 $g \cdot day^{-1}$. If we omit the Passeriformes, Larids have the fastest and Procellariiformes the lowest growth rate, while all other groups so far investigated fall between these two extremes. These observations provide a first approximation of embryonic growth rates and suggest that each group is genetically endowed with a given rate of development which distinguishes it from other groups and that in the Procellariiformes the juvenile growth rate regression is similar to that for embryonic growth rate (Fig. 14).

While EGR as defined here (0.67 W/I) may appear as an oversimplified expression of <u>absolute</u> mean embryonic growth rate, it should be noted that <u>relative</u> mean growth rate (W/I) is highly correlated (r^2 values 0.93 to 0.98) with various eggshell structures and functions (Rahn and Ar, 1980). For example, the number of pores per egg, eggshell conductance, daily rate of water loss, and oxygen consumption at the pre-internal pipping stage when regressed against W/I have slopes between 0.90 and 1.04. As a first approximation all these values can be considered as being directly proportional to the absolute or relative mean growth rate.

REFERENCES

Ackerman, R. A., G. C. Whittow, C. V. Paganelli and T.N. Pettit, 1980, Oxygen consumption, gas exchange, and growth of embryonic Wedge-tailed Shearwaters (*Puffinus pacificus chlororhynchus*), *Physiol. Zool.* 53: 210.

Ar, A., H. Rahn and C. V. Paganelli, 1979, The avian egg: mass and strength, *Condor* 81: 331.

Ar, A., and Y. Yom-Tov, 1978, The evolution of parental care in birds, *Evolution* 32: 655.

Blomquist, S., and M. Elander, 1981, Sabine's Gull (*Xema sabini*), Ross's Gull (*Rhodostethia rosea*) and Ivory Gull (*Pagophila eburnea*) -- Gulls in the Arctic: A review, *Arctic* 34: 122.

Boersma, P. D., 1982, Why some birds take so long to hatch, *Amer. Natural.* 120: 733.

Brody, S., 1945, "Bioenergetics and Growth," Reinhold Publ. Co., New York. (Reprinted 1964, Hafner Publ. Co., New York).

Burger, J., 1980, The transition to independence and postfledging parental care in seabirds. In: "Behavior of Marine Animals," Vol. 4: Marine Birds, J. Burger, B. L. Olla and H. E. Winn, ed., Plenum Press, New York.

Carey, C., H. Rahn and P. Parisi, 1980, Calories, water, lipid and yolk in avian eggs. *Condor* 82: 335.

Cramp, S., 1977, "Handbook of Birds of Europe, the Middle East and North Africa: The birds of the Western Palearctic," Vol. I; Vol. II (1980), Oxford University Press, Oxford.

Croxall, J. P., and P. A. Prince, 1981, A preliminary assessment of the impact of seabirds on marine resources in South Georgia, Colloque sur les Ecosystems Subantarctiques, Paimpont, C.N.F.R.A., n° 51, 501.

El-Wailley, A. J., 1966, Energy requirements for egg-laying and incubation in the Zebra Finch, *Taeniopygia castanotis, Condor* 68: 582.

Gaston, A. J., and D. N. Nettleship, 1981, The Thick-billed Murres of Prince Leopold Island, *Canad. Wildlife Service Monograph S.* No. 1

Grant, G. S., T. N. Pettit, H. Rahn, G. C. Whittow, and C. Paganelli, 1982, Water loss from Laysan and Black-footed Albatross eggs, *Physiol. Zool.* 55: 405

Harrison, C. S., M. B. Naughton and S. I. Fefer, The status and conservation of seabirds in the Hawaiian Archipelago and Johnston Atoll. Internal Report. U. S. Fish and Wildlife Service, P.O. Box 50167, Honolulu, Hawaii.

Heinroth, O., 1922. Die Beziehungen zwischen Vogelgewicht, Eigewicht Gelegegewicht und Brutdauer, *J. Ornithol.* 70: 172.

Hoyt, D. F., and H. Rahn, 1980, Respiration of avian embryos: a comparative analysis, *Respir. Physiol.* 39: 255.

Hunt, G. L., Jr., B. Burgeson and G. A. Sanger, 1981, Feeding ecology of seabirds of the Eastern Bering Sea, In: The Eastern Bering Sea Shelf: Its Oceanography and Resources, D. W. Hood and J. Calder, ed., N.O.A.A.

Lack, D., 1968, Ecological adaptations for breeding in birds, Methuen, London.

Lasiewski, R. C., and W. R. Dawson, 1967, A re-examination of the relation between standard metabolic rate and body weights in birds, *Condor* 69: 13.

Makatsch, W., 1974, Die Eier der Vögel Europas, Vol. I, Verlag J. Neumann-Neudamm, Melsungen.

Manuwal, D. A., 1974, The natural history of Cassin's Auklet (*Ptychoramphus aleuticus*), *Condor* 76: 421.

Montevecchi, W.A., I. R. Kirkham, D. D. Roby, and K. L. Brink, 1983, Size, organic composition, and energy content of Leach's Storm Petrel (*Oceanodroma leucorhoa*) eggs with reference to position in the precocial-altricial spectrum and breeding ecology, *Canad. J. Zool.* 61: 1456.

Murray, K. G., K. Winnett-Murray, Z. A. Eppley, G. L. Hunt, Jr., and
    D. B. Schwartz, 1983, Breeding biology of the Xantus'
    Murrelet, *Condor* 85: 12.
Nelson, J. B., 1978, The Sulidae, Oxford Univ. Press, Oxford.
Nice, M. M., 1962, Development of behavior in precocial birds.
    Trans. Linn. Soc. N. Y. 8: 1.
Paganelli, C. V., A. Olszowka and A. Ar, 1974, The avian egg:
    surface area, volume and density, *Condor* 76: 319.
Palmer, R. S., 1962, Handbook of North American Birds, Vol. I, Yale
    Univ. Press, New Haven, Conn.
Pettit, T. N., G. S. Grant, G. C. Whittow, H. Rahn and C.V. Paganelli,
    1982a, Embryonic oxygen consumption and growth of Laysan and
    Black-footed Albatross, *Am. J. Physiol.* 242: R121.
------, 1982b, Respiratory gas exchange and growth of Bonin Petrel
    embryos, *Physiol. Zool.* 55: 162.
Pettit, T. N., and G. C. Whittow, 1983, Embryonic respiration and
    growth in two species of Noddy Terns, *Physiol. Zool.* 56: 455.
Pettit, T. N., G. S. Grant, and G. C. Whittow, 1984, Nestling meta-
    bolism and growth in the Black Noddy and White Tern, *Condor*
    86: 83.
Rahn, H., 1982, Comparison of embryonic development in birds and
    mammals: birth weight, time and cost. In: "A Companion to
    Animal Physiology," C. R. Taylor, K. Johansen, and L. Bolis,
    ed., Cambridge Univ. Press, Cambridge.
Rahn, H., and A. Ar, 1980, Gas exchange of avian eggs: Time,
    structure and function, *Amer. Zool.* 20: 477.
Rahn, H., C. V. Paganelli and A. Ar, 1975, Relation of avian egg
    weight to body weight, *Auk* 92: 750.
Schmekel, L., 1960, Daten über das Gewicht des Vogeldottersackes vom
    Schlupftag biszum Schwinden., *Rev. Suisse Zool.* 68: 103.
Schönwetter, M., 1960, "Handbuch der Oologie," W. Meise, ed.,
    Akademie Verlag, Berlin.
Sealy, S. G., 1973a, Breeding biology of the Horned Puffin on St.
    Lawrence Island, Bering Sea, with Zoogeographical Notes on the
    North Pacific Puffins, *Pacific Science* 27: 99.
------, 1973b. Adaptive significance of post-hatching developmental
    patterns and growth rates in the Alcidae, *Ornis Scand.* 4: 113.
------, 1975a, Aspects of the breeding biology of the Marbled Murrelet
    in British Columbia, *Bird-Banding* 46: 141.
------, 1975b, Egg size of Murrelets, *Condor* 77: 500.
------, 1976, Biology of nesting Ancient Murrelets, *Condor* 78: 294.
Sealy, S. G., and J. Bédard, 1973, Breeding biology of the Parakeet
    Auklet (*Cyclorrhyncus psittacula*) on St. Lawrence Island,
    Alaska, *Astarte* 6: 59.   ·
Stempniewicz, L., 1981, Breeding biology of the Little Auk, *Plautus
    alle*, in the Hornsund region, Spitsbergen, *Acta Ornithol.*
    18: 142.
Warham, J., 1977, Wing loadings, wing shapes and flight capabilities
    of Procellariiformes, *New Zealand J. Zool.* 4:73.

-----, 1983, The composition of petrel eggs, *Condor* 85: 194.

Watson, G. E., 1975, Birds of the Antarctic and Sub-Antarctic, Amer. Geophysic. Union, Washington, D.C.

Whittow, G. C., 1980, Physiological and ecological correlates of prolonged incubation in sea birds, *Amer. Zool.* 20: 427.

Williams, A. J., W. R. Seigfried and J. Cooper, 1982, Egg composition and hatchling precocity in seabirds. *Ibis* 124: 456.

APPENDIX

The following tables give estimates of the adult body mass, egg mass, hatchling mass = 0.67 W, incubation time, and fledging time for various seabird families. For the family Laridae relative yolk mass and clutch size are also given. These values were obtained from many sources. The major sources are listed below as well as the particular values that may be found with each reference, where W = egg mass, I = incubation period, F = fledging period, A = adult mass, C = clutch size, H = hatchling mass, and Y = relative yolk content. **All values in the Appendix should be regarded as tentative.**

	W	I	F	A	C	Y	
Cramp (1977)	+	+	+	+	+	−	Vol. 1-3
Palmer (1962)	+	+	+	+	+	−	Procell., Pelec.
Makatsch (1974)	−	+	+	−	+	−	All birds
Schönwetter (1960)	+	−	−	+	−	−	All birds
Burger (1980)	−	+	+	−	+	−	Seabirds
Lack (1968)	+	+	+	+	+	−	Seabirds
Watson (1975)	+	+	+	−	+	−	Antarctic Birds
Warham (1977)	+	−	−	+	−	−	Procellar.
Nelson (1978)	+	+	+	+	+	−	Sulidae
Croxall & Prince (1981)	−	+	−	+	−	−	Sphenis., Procel.
Hunt et al. (1981)	−	−	−	+	−	−	Alaska Seabirds
Harrison et al.	−	−	−	+	−	−	Hawaiian Seabirds
Williams et al. (1982)	+	−	−	+	−	+	Seabirds

Alcidae -- additional references: Sealy (1973, a, b), Sealy and Bédard (1973), Sealy (1975 a, b, 1976), Manuwal (1974), Gaston and Nettleship (1981), Stempniewicz (1981), and Murray et al (1983).

Laridae -- additional references: Pettit and Whittow (1983), Pettit et al. (1984), G. Hunt (pers. communication), Blomquist and Elander (1981), G. Grant (pers. communication).

Procellariiformes -- additional references: Ackerman et al. (1980), Pettit et al. (1982 a, b), Grant et al. (1982), Williams et al. (1982), Warham (1983), Boersma (1982), Montevecchi et al. (1983).

Order, Family, Species		A	W	H	F	I
Procellariiformes		g	g	g	days	days
Diomedeidae						
Diomedea exulans	Wandering Albatross	8,040	488	323	280	78
" epomophora	Royal Albatross	8,290	445	298	236	79
" nigripes	Black-footed Albatross	3,090	291	195	140	65
" immutabilis	Laysan Albatross	2,450	279	187	165	64
" melanophris	Black-browed Albatross	2,911	258	173	135	56
" chlororhynchos	Yellow-nosed "	2,068	206	138	130	78
Phoebetria palpebrata	Light-mantled "	3,000	257	172	140	65
Procellariidae						
Macronectes giganteus	Giant Fulmar	4,000	250	168	119	60
Daption capense	Cape Petrel	450	67	45	49	44
Fulmarus glacialis	Northern Fulmar	720	98	66	49	56
" glacialoides	Southern Fulmar	800	103	69	51	46
Pagadroma nivea	Snow Petrel	425	47	31	46	45
Pachyptila desolata	Antarctic Prion	200	30	20	50	45
" vittata	Broad-billed Prion	196	32	21	55	56
Puffinus assimilis	Little Shearwater	222	32	21	72	52
" tenuirostris	Short-tailed Shearwater	530	82	55	94	54
" gravis	Greater "	870	102	68	92	53
" lherminieri	Audubon's "	163	35	23	75	49
" puffinus	Manx "	450	58	39	72	53
" navitatis	Black "	324	63	42	70	54
Calonectris diomedea	Cory's "	950	74	50	90	54
Bulweria bulwerii	Bulwer's Petrel	92	21	14	62	45
Procellaria aequinoct.	White-chinned Petrel	1,130	120	80	95	60
Pterodroma macroptera	Gray-faced "	570	80	54	130	53
" alba	Phoenix "	270	56	38	96	53
" lessoni	White-headed "	590	96	64	105	58
" brevirostris	Kerguelen "	330	56	38	60	49
" inexpectata	Scaled "	320	58	39	98	51
Halobaena caerulea	Blue "	180	35	23	58	46
Hydrobatidae						
Oceanites oceanicus	Wilson's Storm-Petrel	34	9	6.0	52	43
Pelagodroma marina	White-faced " "	47	13	8.7	60	46
Hydrobates pelagicus	Storm Petrel	28	7	4.7	63	41
Oceanodroma castro	Harcourt's Storm-Petrel	44	11	7.4	67	42
" leucorhoa	Leach's " "	50	10	6.7	66	42
Fregetta grallaria	White beld. " "	58	12	8.0	68	41
Pelecanoididae						
Pelecanoides urinatrix	Kerguelen Diving Petrel	124	20	13	56	54
Sphenisciformes						
Spheniscidae						
Aptenodytes forsteri	Emperor Penguin	30,000	450	302	165	62
" patagonicus	King "	15,000	305	204	360	53
Pygoscelis adeliae	Adelie "	5,000	124	83	52	35
" papua	Gentoo "	6,200	136	91	75	37
" antarctica	Chinstrap "	4,500	110	74	51	37
Eudyptes chrysolophus	Macaroni "	4,200	142	95	65	35
" schlegeli	Royal "	4,500	150	101	62	35

		A	W	H	F	I
		g	g	g	days	days
Spheniscidae (cont.)						
Eudyptes crestatus	Rockhopper Penguin	2,500	92	62	70	34
" sclateri	Erect-crested Penguin	3,600	120	80	63	35
" pachyrhynchus	Crested "	3,000	108	72	65	34
Eudyptula albosignata	White-flippered "	1,500	55	37	56	38
" minor	Little blue "	1,100	52	35	55	39
Megadyptes antipodes	Yellow-eyed "	5,200	138	92	106	43
Spheniscus demersus	Jackass "	2,900	103	69	90	40
" humboldti	Humboldt "	4,200	119	80	77	42
" magellanicus	Magellanic "	4,900	120	80	77	43
Pelecaniformes						
Sulidae						
Sula b. bassana	North Atlantic Gannet	3,100	105	70	91	44
" " capensis	Cape Gannet	2,640	105	70	97	44
" " serrator	Australasian Gannet	2,350	94	63	102	44
" dactylatra	Masked Booby	1,890	78	52	120	44
" nebouxii	Blue-footed Booby	1,800	61	41	102	41
" abbotti	Abbott's Booby	1,600	112	75	168	57
" variegata	Peruvian "	1,520	76	51	78	42
" leucogaster	Brown Booby	1,260	54	36	95	43
" sula	Red-footed Booby	950	58	39	120	45
Phaethontidae						
Phaethon aethereus	Red-billed Tropicbird	750	54	36	80	43
" lepturus	White-tailed "	450	45	30	75	41
" rubricauda	Red-tailed "	665	71	48	80	44
Pelecanidae						
Pelecanus erythrorhyn.	White Pelican	8,000	150	101	60	29
" rufescens	Pink-backed Pelican		116	78	72	34
" occidentalis	Brown Pelican	3,500	112	75	63	30
Phalacrocoracidae						
Phalacrocorax carbo	Great Cormorant	2,500	58	39	50	30
" urile	Red-faced Cormorant	2,500	48	32		
" penicillatus	Brandt's "	2,400	51	34		
" auritus	Double-crested "	2,300	48	32	40	25
" pelagicus	Pelagic Cormorant	2,000	40	27	–	26
" olivaceous	Neotropic "	1,800	37	25		
" aristotelis	Shag	1,800	51	34	53	30
" pygmaeus	Little Cormorant	600	23	15	–	29
Charadriiformes						
Alcidae						
Uria aalge	Common Murre	970	109	73	18	30
" lomvia	Thick-billed Murre	900	106	71	20	32
Lunda cirrhata	Tufted Puffin	797	91	61	50	30
Alca torda	Razorbill	770	88	59	13	34
Fratercula corniculata	Horned Puffin	594	80	54	38	41
Cerorhinca monocerata	Rhinoceros Auklet	518	77	52	38	32
Fratercula arctica	Puffin	476	65	44	14	42

		A	W	H	F	I
		g	g	g	days	days
Alcidae (cont.)						
Cepphus columba	Pigeon Guillemot	440	55	37	–	31
" grylle	Black "	410	51	34	35	24
Cyclorrhyn. psittacula	Parakeet Auklet	288	38	25	35	35
Aethia cristatella	Crested "	286	41	27	–	–
Brachyramphus marmor.	Marbled Murrelet	224	36	24	30	28
Aethia pygmaea	Whiskered Auklet	224	34	23	–	–
Brachyramphus breviros.	Kittlitz's Murrelet	222	34	23	–	–
Synthliboramphus antiq.	Ancient "	206	45	30	–	35
" hypoleuca	Xantus's "	167	37	25	–	31
Ptychoramphus aleut.	Cassin's Auklet	167	28	19	45	37
Plautus alle	Dovekie	163	31	21	27	29
Aethia pusilla	Least Auklet	92	17	11	–	–

Laridae   (See next page.)

		A	W	F	I	C	Y
		g	g	days	days	n	%

Charadriiformes

Laridae

		A	W	F	I	C	Y
Larus hyperboreus	Glaucous Gull	1,400	118	40	28	3	–
" marinus	Gr. Black-bkd. G.	1,500	117	56	27	3	28
" glaucescens	Glaucous-wng. G.	1,800	98	42	27	3	–
" occidentalis	Western Gull	900	97	50	24	3	–
" dominicanus	So. Black-bkd. G.	--	92	49	29	3	31
" . argentatus	Herring Gull	1,000	92	45	27	3	26
" fuscus	Les. Blk-bkd. G.	840	81	32	27	3	27
" californicus	California G.	--	75	–	25	3	–
Creagrus furcatus	Swallow-tail. G.	--	73	60	32	1	–
Larus crassirostris	Black-tailed G.	--	66	–	–	–	29
" audouinii	Audouin's Gull	--	65	–	23	3	–
Pagophila eburnea	Ivory Gull	--	60	–	25	2	–
Larus delawarensis	Ring-billed Gull	455	57	–	22	3	–
" canus	Mew Gull	420	54	30	25	3	28
Rissa tridactyla	Black-lgd. Kittiw.	455	54	43	28	1.4	–
' brevirostris	Red-legged Kittiw.	380	49	37	31	1	–
Larus atricilla	Laughing Gull	325	43	32	24	3	32
" genei	Slender-billed G.	--	42	–	24	2.5	–
" melanocephalus	Mediteranean Gull	--	42	–	24	3	–
" novae hollandiae	Silver Gull	--	39	42	24	3	31
" ridibundus	Black-headed Gull	250	36	42	23	3	29
" pipixcan	Franklin's Gull	--	36	39	24	3	–
Xema sabini	Sabine's Gull	--	24	–	24	2	–
Larus minutus	Little Gull	125	19	24	23	3	–
Sterna caspia	Caspian Tern	767	67	37	26	2.5	27
" maxima	Royal Tern	491	66	26	30	1.5	30
" bergii	Great crested Tern	--	57	–	–	–	33
Anous stolidus	Brown Noddy	210	40	–	36	1	35
Sterna fuscata	Sooty Tern	175	37	56	29	1	38
" sandvicensis	Sandwich Tern	250	34	28	25	2	33
Gelochelidon nilotica	Gull-billed Tern	--	30	35	22	3	–
Sterna lunata	Grey-backed Tern	150	28	46	32	1	37
" striata	White-fronted Tern	--	26	–	–	–	30
Anous tenuirostris	Black Noddy	120	25	42	35	1	35
Gygis alba	White Tern	138	21	60	36	1	38
Sterna hirundo	Common Tern	135	21	28	21	3	31
" forsteri	Forster's Tern	153	21	–	23	3	–
" dougallii	Roseate Tern	95	20	–	23	2	33
" paradisaea	Arctic Tern	110	19	21	21	3	31
Chlidonias hybrida	Whiskered Tern	--	16	23	19	3	32
" leucoptera	White Wng. Tern	--	11	23	–	3	30
" niger	Black Tern	46	11	21	21	3	30
Sterna albifrons	Least Tern	40	9	25	22	3	–

CALORIC CONTENT AND ENERGETIC BUDGET OF TROPICAL SEABIRD EGGS

T.N. Pettit[1], G.C. Whittow[1] and G.S. Grant[2]

[1]Department of Physiology
John A. Burns School of Medicine and
P.B.R.C. Kewalo Marine Laboratory
University of Hawaii
Honolulu, Hawaii
[2]North Carolina State Museum of Natural History
Raleigh, North Carolina

INTRODUCTION

Although avian egg composition and caloric density have been related to the degree of developmental maturity shown by the hatchling (Romanoff and Romanoff, 1949; Nice, 1962; Ricklefs, 1977; Carey et al., 1980), recent evidence suggests that there is no simple relationship between egg composition and hatchling precocity (Williams et al., 1982). Among tropical seabirds, incubation periods are characteristically prolonged; the incubation period is much greater than predicted on the basis of egg mass (Whittow, 1980). In addition, the energetic cost of prolonged embryonic development is high (Whittow, 1980; Ackerman et al., 1980; Pettit and Whittow, 1983). Accordingly, the energetic content of the freshly-laid egg would be expected to be high since embryonic development can utilize only those nutrients which are incorporated in the egg at the start of incubation, and the oxygen which diffuses through the pores of the eggshell, to meet metabolic requirements.

The purpose of this chapter is (a) to present information on the energy content of the eggs of a number of tropical species of seabirds, (b) to relate the energy content of the egg to the precocity of development and duration of incubation, (c) to attempt to allocate the energy resources of the freshly-laid egg to embryonic tissue synthesis and maintenance, and storage in embryonic tissue and yolk sacs, and (d) to estimate the efficiency of embryonic development in tropical seabirds.

METHODS

Fresh eggs were collected within 24 hours of laying on Manana Island, Oahu (21°30'N, 157°40'W), Tern Island, French Frigate Shoals (23°52'N, 165°18'W), Laysan Island (25°46'N, 171°40'N), and Midway Islands (28°13'N, 177°23'W) in the Hawaiian Islands. The fresh eggs were wrapped in plastic film and refrigerated prior to processing. The initial egg mass was determined by injecting distilled water into the air cell (if any) with a syringe to replace water lost by evaporation (Grant et al., 1982) and weighing the egg to the nearest 0.001 g with a Mettler balance. The shell was then gently cracked open, and the contents drained into a previously weighed glass beaker. The egg contents were homogenized by rapid stirring with a metal rod. The eggshell and shell membranes were rinsed with distilled water and dried for 48 hr over silica gel. The difference between the initial egg mass and the dry shell mass is the mass of the egg contents. Except for albatross eggs, the entire egg contents were dried to constant mass in an oven at 45°C. An aliquot of approximately 40 g from the albatross egg homogenate was dried to constant mass. The water content of the egg was determined by subtracting the dry mass fraction from the initial egg content. The caloric content of approximately 10 mg samples was analyzed in triplicate using a Phillipson Microbomb Calorimeter, and benzoic acid as a standard. The percentage ash of the egg contents was determined by burning replicate dry samples of each egg in a muffle furnace at 600°C for 6 hr.

Fresh eggs of each species were collected for determination of the relative yolk content; the eggs were boiled and the yolk:albumen ratio determined from the weights of the solidified yolk and albumen. Eggs of known age from seven species were collected during the internally-pipped phase (penetration of the air cell by the bill), wrapped in plastic film and frozen. Chicks were collected within 24 hr of hatching. The hatchling yolk sac was dissected free of the chick and the chick and yolk sac were weighed separately on a torsion balance (± 0.01g), prior to freezing. After thawing at room temperature, embryos, hatchlings and yolk reserves were dried to a constant mass i an oven at 45°C, homogenized with a mechanical tissue homogenizer, an ground by hand with mortar and pestle. Water, ash and caloric conten were determined as described above. Water, ash and caloric content i a series of noddy embryos of known age were measured by the same tech niques. This assessment allows an evaluation of the conversion of eg matter into embryos over the entire incubation period.

RESULTS

Table 1 presents the mean values (±SD) for initial egg mass; egg contents; relative yolk, water and ash content; and the caloric density (kcal.g^{-1}) of wet and dry egg content. The relative yolk

Table 1. Calories, water, ash, and yolk in Hawaiian seabird eggs.
Values are presented as means ± S.D.; sample size in
parentheses.

Order/Species	Fresh Egg Mass (g)	Egg Contents (g)	Yolk Content (% of egg content)	Yolk Albumen Ratio	Water Content (% wet)	Ash Content (% dry)	Caloric Density of Egg Contents (kcal · g^{-1})	
							dry	wet
PROCELLARIIFORMES								
Black-footed Albatross (Diomedea nigripes)	306.88 ± 25.12 (6)	283.81 ± 24.37 (6)	33.0 [1] ± 1.4 (6)	0.49 [1] (6)	74.92 ± 1.12 (6)	3.77 0.54 (6)	6.770 +0.112 (6)	1.697
Laysan Albatross (D. immutabilis)	281.72 ± 18.75 (6)	261.25 ± 17.00 (6)	34.8 [1] ± 1.6 (6)	0.50 [1] (6)	74.16 ± 1.25 (6)	3.69 +0.27 (6)	6.695 +0.131 (6)	1.730
Wedge-tailed Shearwater (Puffinus pacificus)	57.76 ± 3.14 (6)	53.52 ± 3.14 (6)	40.0 [2]	0.69 [2] +0.12 (15)	73.17 ± 0.37 (6)	3.63 +0.11 (6)	7.111 +0.182 (6)	1.908
Bonin Petrel (Pterodroma hypoleuca)	37.92 ± 1.12 (6)	35.99 ± 1.07 (6)	40.0 [3] ± 1.2 (4)	0.67 [3] +0.02 (4)	70.55 ± 0.96 (6)	4.01 +0.46 (6)	6.991 +0.094	2.059
Bulwer's Petrel (Bulweria bulwerii)	21.75 ± 1.55 (10)	20.58 ± 1.11 (10)	38.0 (1)	0.65 (1)	72.15 (1)	---	6.559 +0.162 (1)	1.827
PELECANIFORMES								
Great Frigatebird (Fregata minor)	88.82 ± 4.09 (12)	82.21 ± 3.77 (4)	25.70 ± 1.26 (6)	0.365 +0.032 (3)	77.54 ± 1.61 (4)	4.23 +1.01 (4)	6.579 +0.144 (2)	1.478
Red-tailed Tropicbird (Phaethon rubricauda)	70.05 ± 5.03 (2)	63.83 ± 4.67 (2)	24.40 ± 1.76 (3)	0.375 +0.026 (3)	74.78 ± 0.90 (2)	4.80 +0.36 (2)	6.482 +0.067 (2)	1.635
Red-footed Booby (Sula sula)	57.93 ± 6.72 (12)	52.65 ± 6.11 (12)	15.77 ± 1.26 (6)	0.195 +0.019 (6)	83.61 ± 0.48 (6)	4.01 +0.25 (4)	6.156 +0.200 (6)	1.009
CHARADRIIFORMES								
Brown Noddy (Anous stolidus)	40.13 ± 1.61 (11)	37.36 ± 1.54 (11)	35.29 ± 2.76 (6)	0.564 +0.069 (6)	74.73 ± 0.43 (5)	4.25 +0.38 (1)	6.655 +0.099 (4)	1.682
Sooty Tern (Sterna fuscata)	36.73 ± 2.66 (6)	34.51 ± 2.51 (6)	37.71 ± 1.05 (8)	0.599 +0.021 (8)	71.73 ± 1.89 (6)	4.25 +0.18 (4)	6.778 +0.283 (6)	1.916
Grey-backed Tern (S. lunata)	28.08 ± 1.47 (9)	26.80 ± 0.59 (4)	36.46 ± 2.33 (4)	0.569 +0.077 (4)	74.55 ± 0.60 (5)	4.38 +1.02 (3)	6.669 +0.089 (3)	1.697
Black Noddy (A. minutus)	24.91 ± 1.63 (7)	23.45 ± 1.54 (7)	34.92 ± 1.12 (4)	0.560 +0.025 (4)	74.94 ± 0.57 (4)	5.09 +0.62 (3)	6.803 +0.089 (3)	1.705
White Tern (Gygis alba)	22.25 ± 1.41 (6)	20.98 ± 1.34 (6)	38.0 [4] ± 2.0 (5)	0.610 [4] (6)	71.11 ± 1.01 (6)	4.07 +0.50 (6)	7.115 +0.133- (6)	2.056

[1] Pettit et al. 1982a

[2] Ackerman et al. 1980

[3] Pettit et al. 1982b

[4] Pettit et al. 1981

content is of interest because the yolk provides most of the energy consumed for embryonic development. High relative yolk contents (33-40%) and high yolk:albumen ratios (0.49-0.69) were observed in small procellariiform eggs (petrels and shearwater) and charadriiform eggs. Lower yolk contents (16-26%), along with lower yolk: albumen ratios (0.20-0.37), were found in altricial pelecaniform eggs.

Table 2 presents the mean values ($\pm$SD) for the fresh, yolk-free mass and the relative water, ash and caloric content of internally-pipped embryos and hatchlings. Tern embryos were generally higher in water content and lower in caloric density ($kcal.g^{-1}$ wet) than were procellariiform embryos. At hatching, however, chicks of both groups were similar in body water content but the higher caloric density was maintained by procellariiform hatchlings (mean = 1.277 $kcal.g^{-1}$) over tern hatchlings (mean = 1.232 $kcal.g^{-1}$), although this differences was not statistically significant (p<.20).

Table 3 presents the fresh mass of the hatchling yolk reserve and the water, ash and caloric content of the yolk reserves from seven species of Hawaiian seabirds. The yolk reserve was a greater fraction of the hatchling mass in procellariiform (12%-17%) than in charadriiform chicks (6.5%). The caloric densities of the yolk sacs were similar in all species, and thus the caloric content of the yolk reserve remained a higher fraction of total hatchling caloric content among Procellariiformes. The caloric content of the yolk reserve was used to compute the time required for depletion of the energy available in the yolk sac, based upon the resting metabolic rate of the hatchling (see Discussion).

The caloric density of Brown Noddy (Anous stolidus) embryos increased from approximately 450 $cal.g^{-1}$ (wet) at age 4-8 days to 950 $cal.g^{-1}$ (wet) during the internal-pipping phase of incubation (Fig. 1). Caloric density continued to increase to a mean value of 1.152 $kcal.g^{-1}$ (wet) in the hatchling. The increase in caloric content was accompanied by a decrease in water content from approximately 90% at age 10 days to a mean of 78.3% in the hatchling chick. The caloric content of dry matter increased from approximately 4.9 $kcal.g^{-1}$ (dry) in the 11-17 day old embryo to 5.5 $kcal.g^{-1}$ (dry) in the internally-pipped embryo. During the hatching process, the caloric density of the dry embryo decreased from 5.547 $kcal.g^{-1}$ to 5.297 $kcal.g^{-1}$, a decrease of 5%. Energy-expensive hatching efforts and the work of pulmonary ventilation during this period may deplete energy stores within the embryonic tissue and contribute to the decrease in dry caloric density. The ash content of the embryo decreases during incubation from approximately 19% of the dry mass at age 11-14 days to 10% of the dry mass at hatching. The initially high proportion of ash content may be related to the deposition of minerals for skeletal and muscle tissue in the young embryo.

116

Table 2. Calories, water, and ash in embryos and hatchlings of Hawaiian seabirds. Values are presented as means ± S.D.; sample size in parentheses.

Species	Age (days)	INTERNALLY-PIPPED EMBRYOS					HATCHLINGS				
		Fresh Mass (yolk-free;g)	% Water (wet)	% Ash (dry)	kcal · g^{-1} dry	kcal · g^{-1} wet	Fresh Mass (yolk-free;g)	% Water (wet)	% Ash (dry)	kcal · g^{-1} dry	kcal · g^{-1} wet
Black-footed Albatross	61	202.47 +21.36 (6)	75.47 +1.22 (6)	7.50 +1.06 (3)	5.605 +.265 (4)	1.375	195.78 +4.55 (4)	79.23 +0.38 (4)	6.85 +0.57 (3)	6.022 +0.018 (3)	1.251
Laysan Albatross	61	163.22 +10.91 (4)	76.13 +2.51 (4)	7.81 +1.23 (4)	5.773 +.410 (4)	1.368	189.13 +6.70 (2)	78.89 +2.11 (2)	6.54 +1.19 (2)	5.770 +0.193 (2)	1.218
Wedge-tailed Shearwater	50	26.08 +2.51 (3)	78.74 +3.15 (3)	6.92 +1.73 (3)	5.646 +0.606 (3)	1.199	28.90 +2.21 (4)	75.34 +1.97 (4)	7.02 +1.46 (4)	5.251 +0.344 (4)	1.295
Bonin Petrel	45	21.91 +1.78 (6)	78.43 +1.31 (6)	6.54 +1.94 (4)	5.377 +0.185 (6)	1.160	25.42 +1.73 (6)	75.78 +1.20 (6)	7.22 +0.92 (6)	5.540 +0.334 (6)	1.342
Brown Noddy	33	22.73 +1.93 (5)	83.04 +1.76 (5)	11.37 +1.33 (5)	5.547 +0.042 (4)	0.941	25.65 +1.47 (5)	78.25 +0.69 (5)	10.06 +0.29 (4)	5.297 +0.101 (4)	1.152
Black Noddy	33	13.22 +1.76 (4)	82.99 +1.30 (4)	11.46 +1.51 (4)	5.549 +0.097 (4)	0.943	17.64 +1.67 (4)	78.35 +1.06 (4)	10.11 +0.53 (4)	5.482 +0.074 (4)	1.192
White Tern	33	12.16 +0.70 (5)	79.41 +0.60 (5)	9.99 +1.26 (5)	5.509 +0.120 (5)	1.134	14.60 +2.03 (3)	75.37 +1.92 (3)	10.03 +0.68 (3)	5.362 +0.137 (3)	1.321

Table 3. Calories, water and ash in hatchling yolk reserves.

Species	Fresh Mass (g)	% Total Hatchling Mass	% Water (wet)	% Ash (dry)	Caloric Density (kcal·g⁻¹) (dry)	Caloric Density (kcal·g⁻¹) (wet)	% Total Hatchling Caloric Content (wet)	Hatchling Metabolic Rate (kcal·day⁻¹)	Time Required For Depletion of Yolk Energy Reserve (days)
Black-footed Albatross	26.5	12.7	49.0	---	7.498	3.82	33.1	16.26[1]	6.2
Laysan Albatross	22.8	12.0	51.9	---	7.343	3.53	28.9	18.41[1]	4.4
Wedge-tailed Shearwater	7.0	17.4	54.3	8.48	7.246	3.31	37.8	4.05[2]	5.7
Bonin Petrel	3.4	11.7	55.0	9.55	7.434	3.35	25.6	3.65[3]	3.1
Brown Noddy	1.8	6.6	48.9	7.93	7.834	4.00	19.2	3.33[4]	2.2
Black Noddy	1.1	6.5	56.7	6.28	7.952	3.44	15.8	2.53[4]	1.51
White Term	1.0	6.5	52.0	7.04	8.084	3.88	17.0	2.27[5]	1.7

[1]Pettit et al., 1982a
[2]Ackerman et al., 1980
[3]Pettit et al., 1982b
[4]Pettit and Whittow, 1983
[5]Pettit et al., 1981

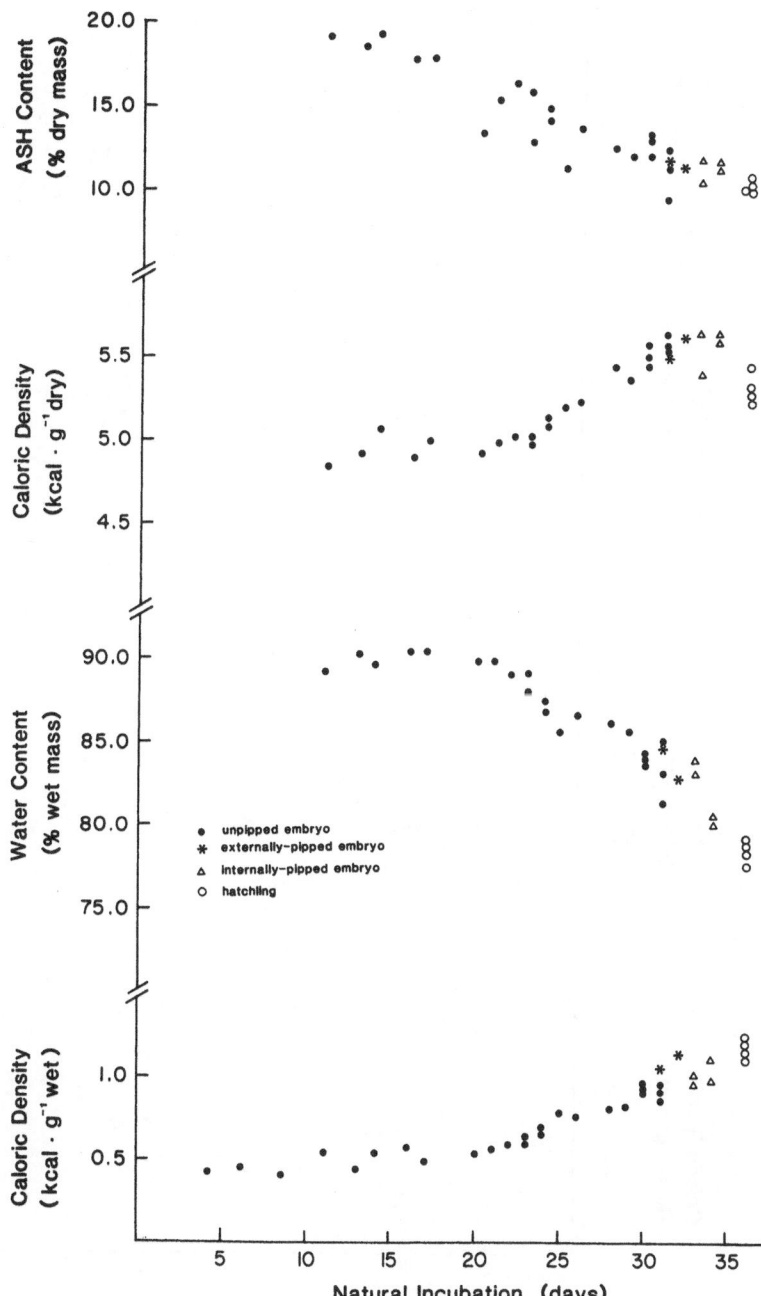

Fig. 1. Ash content, caloric density and water content of the
embryos and hatchlings of the Brown Noddy during incubation.

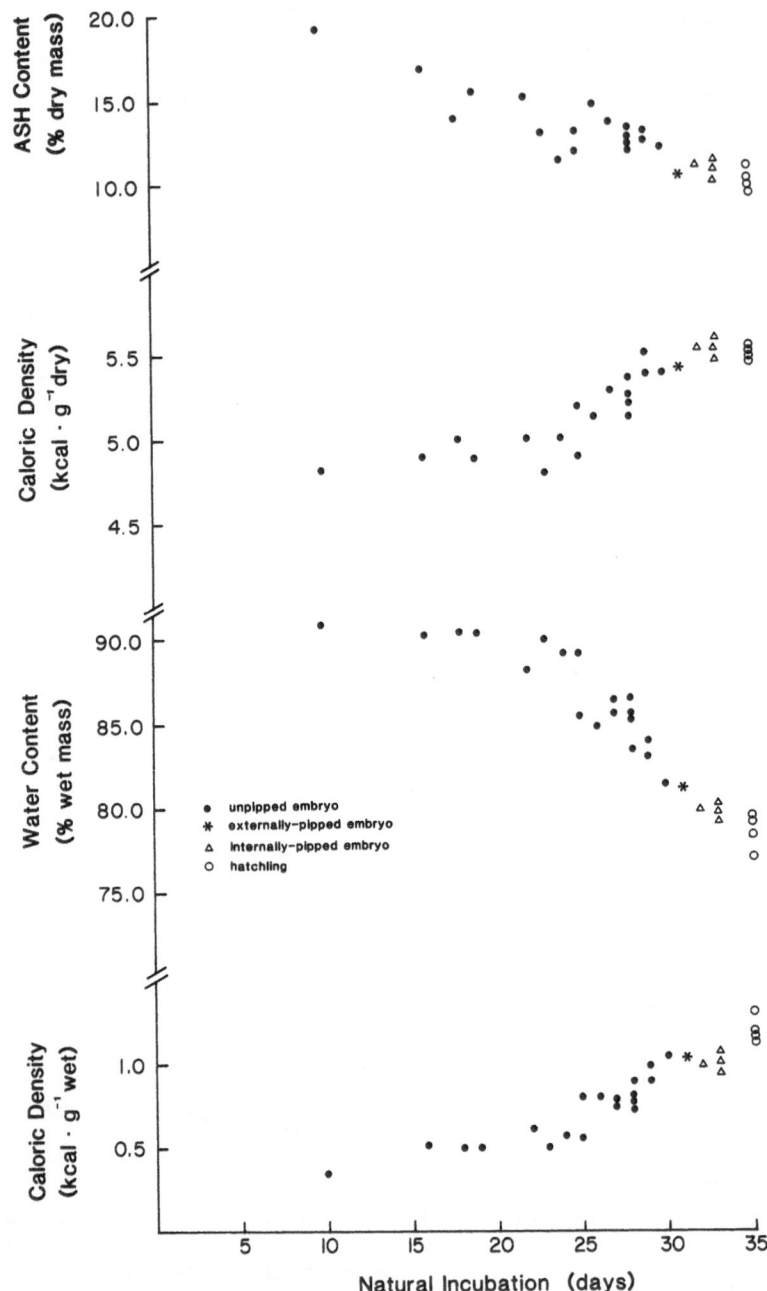

Fig. 2. Ash content, caloric density and water content of the embryos and hatchlings of the Black Noddy during incubation.

The Black Noddy (<u>Anous</u> <u>minutus</u>) displays a similar trend of decreasing ash and water content of the embryo during incubation (Fig. 2). The caloric density of dry and wet embryonic tissue increases in a sigmoidal fashion throughout development. A slight decrease in the caloric content of dry matter was observed during hatching. Caloric content decreased from 5.549 kcal.g^{-1} (dry) in the internally-pipped embryo to 5.482 kcal.g^{-1} (dry) in the hatching.

DISCUSSION

Yolk, Water and Caloric Content of Fresh Eggs

Procellariiform seabirds have been categorized as semi-altricial (Evans, 1980), either semi-altricial or semi-precocial (Nice, 1962), or as semi-precocial (Carey et al., 1980; Ricklefs, et al., 1980). The mean yolk content of 23 freshly-laid procellariiform eggs was recently reported to be 32.5% $\pm$ 4.1 (S.D.) by Warham (1983), close to the mean value of 33% for the eggs of 18 semi-precocial species (Carey et al., 1980). Procellariiform eggs, therefore, might be considered to have "semi-precocial" energy status. The relative yolk content of eggs from the five tropical tube-nosed species studied in the present investigation averaged 37.2% $\pm$ 3.2 (SD), a value which lies between the means for the semi-precocial (33%) and precocial 2,3,4 category (40%) reported by Carey et al. (1980). Similarly, the mean caloric density of tropical procellariiform eggs is 1.844 kcal.g^{-1} (wet) $\pm$ 0.146 (SD), a value which is also intermediate between semi-precocial and precocial eggs. The mean water content of procellariiform eggs in this study (72.99%) is also lower than that of either semi-precocial or precocial eggs (Carey et al., 1980) thus contributing to the high mean caloric density of fresh eggs from this group. The percentage water content of hatchlings is slightly greater than that of the fresh egg.

The yolk mass of procellariiform eggs is highly correlated with incubation period, having a correlation coefficient of 0.942 (P< 0.001; Warham, 1983). Similarly, the relative proportion of yolk in the egg is highly correlated with egg mass. Small petrel eggs have higher yolk contents than large petrel eggs. Based upon the egg composition and maturity of function in procellariiform chicks, Warham (1983) suggests that most tube-nosed chicks should have semi-precocial status. This suggestion is supported by the caloric density and water content of hatchlings (see below).

Tropical procellariiform eggs appear to be high in yolk content and calories to meet the energetic requirements of embryonic development among members of this Order (Pettit et al., 1982b). As egg mass decreases among Procellariiformes, the incubation period becomes relatively longer than that predicted on the basis of fresh-egg mass (Whittow, 1980; Grant et al., 1982) and the energy budget for embryonic development increases correspondingly. Thus smaller members of the Order (shearwaters and petrels) have eggs which are endowed with additional nutrients and calories to provide energy for prolonged embryogenesis. The egg of Leach's storm-petrel (Oceanodroma furcata) also has a high yolk content (41%) and high yolk:albumen ratio (0.71) apparently to meet the maintenance requirements of a semi-precocial embryo during a long incubation period (Montevecchi et al., 1983).

Among the three species of Hawaiian Pelecaniformes which also have prolonged incubation periods (Whittow, 1980), the relative yolk content of the egg is related to both the developmental status of the hatchling and the relative length of the incubation period. The altricial Red-footed Booby egg (Sula sula) has a low yolk content (15.8%) and a correspondingly low caloric density (1.009 $kcal.g^{-1}$, wet) due to the high water content (83.6%). Even lower values for the yolk content (15.8%) and energy content (0.897 $kcal.g^{-1}$ wet) of eggs from another altricial pelecaniform, the North Atlantic Gannet (Sula bassanus), have been reported by Ricklefs and Montevecchi (1979). The altricial Great Frigatebird egg has more yolk (25.7%), less water (77.5%) and a greater energy content (1.478 $kcal.g^{-1}$, wet) than either the booby or the gannet egg and its incubation period is relatively longer. The Red-tailed Tropicbird is considered to be semi-altricial (Nice, 1962) but the energetic profile of its egg in terms of water, yolk, and caloric content is closer to that for semi-precocial eggs. It also has a relatively shorter incubation period than do the Red-footed Booby and Great Frigatebird.

Among charadriiform eggs, the mean values (+SD) for yolk content (36.5% $\pm$ 1.4) and caloric density (1.811 $kcal.g^{-1}$ wet $\pm$ 0.167) are slightly greater, and the water content (73.4% $\pm$ 1.8) slightly less, than the mean values for semi-precocial eggs cited by Carey et al. (1980). Relatively high energy content in eggs of the White Tern (Gygis alba) and Black Noddy may be associated with the energetic demands of prolonged incubation (Pettit et al., 1981; Pettit and Whittow, 1983).

Caloric Content of Embryos and Hatchlings

The relatively high caloric density of hatchlings in this study supports the semi-precocial status of procellariiform and tern chicks. Romanoff (1967) reported a value of 1.2 $kcal.g^{-1}$ wet mass for the hatchling chicken, close to the values for albatross and noddy chicks (Table 2). The caloric density of shearwater, petrel and White Tern hatchlings exceeded 1.2 $kcal.g^{-1}$ and ranged from 1.295 $kcal.g^{-1}$ to 1.342 $kcal.g^{-1}$. The water content and caloric density of hatchlings may be an index of the precociousness and maturity of function in chicks. The caloric content of the hatchling among these species appears to be related also to the energy status of the freshly-laid egg (Table 1). A relatively long incubation period, slow rate of embryonic growth and greater maturity at hatching increase the energy requirements for development. Precocial development in other species requires a longer incubation time and an egg rich in energy stores (Vleck et al., 1980).

The caloric density of albatross hatchlings appears to be less than that of internally-pipped embryos. In the other species studied, the relationship is reversed such that the hatchlings have a greater caloric density. The reasons for this are unclear but may be related to the process of hatching. Albatross embryos follow the usual pattern of internal pipping (penetration of the air space by the beak) followed by external pipping (star-fracture of the shell; Pettit et al., 1982a). The shearwater, petrel and tern embryos display a different pattern of hatching. In these species, with relatively longer incubation periods, the first phase of hatching is the externally-pipped phase (star-fracture of the shell). This event occurs several days prior to internal pipping. After a few days of internal pipping and re-breathing of air cell gases, the embryo makes a pip-hole prior to complete emergence of the chick (Pettit et al., 1981, 1982b; Pettit and Whittow, 1983). Perhaps this lengthy pipping process (4-6 days) allows the embryo to break free of the shell without expending excessive energy from its tissue stores. The continued increase in caloric density of embryonic tissue may be an adaptive consequence of this particular pipping sequence. Alternatively, the pattern of pipping in terns and petrels may permit the increased oxygen consumption necessary for enhanced growth.

Hatchling Yolk Reserves

The relationship between the yolk reserve at hatching (% total body mass) and the independence of the newborn chick is such that the average precocial chick contains almost twice as much yolk (15.5%) as altricial hatchlings (8.0%; Ar and Yom-Tov, 1978). In the present study, however, the yolk reserve was greater in procellariiform chicks (mean = 13.5%) than in semi-precocial tern chicks (mean = 6.5%). This discrepancy, however, may be due to the larger size of procellariiform eggs in this study. As Romanoff (1944) pointed out, the weight of the yolk sac at hatching is closely related to the mass of the fresh egg. Thus, large eggs produce chicks with large yolk reserves.

During incubation, the mass of yolk in the egg slowly decreases as the embryo utilizes this lipid-rich source of nutrition. The yolk lipids are absorbed by the endodermal cells of the yolk sac, and subsequently transferred to the embryonic blood (Romanoff,

1967). However, a significant increase in yolk mass is observed in several species just prior to the start of yolk retraction (Romanoff, 1967; Vanheel et al., 1981). This increment in yolk mass may be caused by the entrance of proteins and/or water from the albumen. In the Laysan and Black-footed Albatross, the yolk mass increases from about 90 g to 100 g and coincides with the ingestion of albumen by the embryo at ages 45-50 days. An accelerated rate of yolk uptake by the embryo occurs in the week preceding hatching at 65 days. Prior to yolk retraction, lipid components - triglycerides, phospholipids and sterol esters - are resorbed preferentially (Vanheel et al., 1981); during the first few days of incubation mainly carbohydrates and proteins are assimilated (Hazelwood, 1965; Romanoff, 1967). It may be assumed that during the retraction process, as the yolk is rapidly assimilated, large amounts of yolk pass through the yolk sac into the intestine of the embryo. Mechanical pressure exerted on the yolk sac during retraction may enhance this process. Gradual absorption of yolk through the yolk sac epithelial cells may continue, but becomes quantitatively less important (Vanheel et al., 1981).

Comparison of the energetic content of the hatchling yolk reserve with the maintenance energetic requirements of hatchlings (Table 3) suggests that the yolk reserve can provide enough energy for about two days in newborn tern chicks, and 3-6 days in tube-nosed chicks, if growth, thermoregulatory and locomotion costs are minimal. The time required for depletion of yolk energy reserves (Table 3) is broadly correlated with the average incubation span near hatching and the first substantial meal. For example, the depletion time of yolk reserves in albatross chicks is 4.4 to 6.2 days (Table 3). Similarly, the incubation span near hatching averages 3-5 days and the feeding interval ranges from 2-5 days post-hatching. In the White Tern, the depletion time of 1.7 days for the chick's yolk reserves is matched by an incubation span of 2 days near hatching and daily feeding intervals after hatching. Thus, embryonic development and the provision of yolk energy reserves at hatching appears to match the timing of incubation bouts and feeding in these tropical seabirds. Procellariiformes are pelagic, deep-water feeders with long incubation spans; their newly-hatched chicks are endowed with yolk reserves to ensure 3-6 days without feeding. The terns feed closer to their colonies, have shorter incubation spans and feeding intervals, and produce chicks with yolk reserves to sustain them for 1-2 days.

Energy Budget for Embryonic Development

Budgeting of energy requirements is essential to successful embryonic development. The nutrients and energetic substrates for producing a chick must be incorporated into the egg from the beginning since no other materials (except oxygen) are available to the embryos. Little is known, however, about the bioenergetic

124

efficiency of chick production in avian eggs. Brody (1945) estimated the efficiency of biosynthesis ("gross production efficiency")
at 62% for the egg of the domestic fowl (<u>Gallus gallus</u>). He estimated the efficiency of embryonic growth as the energetic content of
the hatchling tissue divided by the initial energy of the egg contents. Thus, Brody assumed that the energetic content of unused
materials did not participate in the growth estimate. This method
of calculation, however, is likely to be a slight underestimate of
production efficiency since some unused materials such as extra-
embryonic membranes (e.g., chorioallantois) support the growth
process by providing respiratory exchange of gases. The caloric
content of extra-embryonic membranes, however, are ignored in the
estimate. Extra-embryonic membranes constitute about 5% of the
total neonate mass (Romanoff, 1967). No attempt was made in the
present study to determine the energetic content or estimate the
contribution of extra-embryonic tissue to the "work of embryonic
development". Measurements of consumed energy (see below) include
the work of producing and maintaining respiratory and excretory
extra-embryonic tissues and thus slightly overestimate the metabolic
requirements for producing embryonic tissue.

Analysis of energetic efficiency may be made from the caloric
values for the egg contents, internally-pipped embryos, hatchlings,
and yolk reserves. In addition, the consumed energy is estimated
from the total amount of oxygen consumed; one liter of oxygen being
equivalent to 4.8 kcal (Brody, 1945). Comparison of consumed energy
during the pre-internally pipped phase of incubation and over the
entire incubation period, allows one to compare metabolic needs
which are met by diffusive respiration with total energy consumption.
The consumed energy during the internally-pipped-to-hatching interval
includes the work of hatching efforts, the cost of pulmonary ventilation, and other costs associated with the transition from diffusive
respiration to pulmonary respiration, as well as the energy cost of
any growth. Some of the mean values for the mass of egg contents,
embryos, hatchlings, and yolk reserves (Table 4) are taken from other
studies which provide a larger sample size.

The basic relationships needed to explore the energetic efficiency of embryonic development (Table 5) are defined as follows:

Production Efficiency = [(yolk-free hatchling caloric content
    + yolk reserve caloric content) ÷ caloric content of fresh
    egg] X 100

Efficiency of Biosynthesis = (yolk-free tissue caloric content
    ÷ caloric content of fresh egg) X 100

Energetic Storage Efficiency = [yolk-free tissue caloric content
    ÷ (tissue energy + consumed energy)] X 100

Table 4. Energy Budget of Hawaiian Seabird Eggs.[1]

Species	EGG CONTENTS Mass (g)	EGG CONTENTS Calories kcal·g⁻¹ (wet)	EGG CONTENTS Calories Total (kcal)	EMBRYO Mass (yolk-free)	EMBRYO Calories kcal·g⁻¹ (wet)	EMBRYO Calories Total (kcal)	HATCHLING Mass (yolk-free)	HATCHLING Calories kcal·g⁻¹ (wet)	HATCHLING Calories Total (kcal)	YOLK RESERVE Mass (g)	YOLK RESERVE Calories kcal·g⁻¹ (wet)	YOLK RESERVE Calories Total (kcal)	CONSUMED ENERGY Pre-internal Pipping (kcal)	CONSUMED ENERGY Total (kcal)
Black-footed Albatross	281.6	1.697	477.9	156.0	1.368	213.4	182.5	1.218	222.3	26.5	3.82	101.2	84.0[1]	142.8[2]
Laysan Albatross	263.8	1.730	456.4	150.0	1.375	206.3	167.2	1.251	209.2	22.8	3.53	80.5	84.0[1]	142.8[2]
Wedge-tailed Shearwater	56.3	1.908	107.4	28.0	1.194	33.4	32.6	1.295	42.2	7.0	3.31	23.2	24.0[2]	38.4[3]
Bonin Petrel	37.5	2.059	77.2	21.4	1.160	24.8	25.7	1.342	34.5	3.4	3.35	11.4	16.4[3]	26.6[4]
Brown Noddy	37.7	1.682	63.4	22.7	0.941	21.4	25.6	1.152	29.5	1.8	4.00	7.2	13.5[4]	22.9[5]
Black Noddy	23.4	1.705	39.9	13.2	0.998	13.2	15.9	1.192	19.0	1.1	3.44	3.8	9.3[4]	15.2[5]
White Tern		2.056	45.4	12.2	1.134	13.8	14.1	1.321	18.6	1.0	3.88	3.9	9.2[5]	12.3[6]

[1] Some of the mean values for the mass of egg contents, embryos, hatchlings, and yolk reserves are taken from the studies footnoted which provide a larger sample size.

[2] Pettit et al., 1982a

[3] Ackerman et al., 1980

[4] Pettit et al., 1982b

[5] Pettit and Whittow, 1983

[6] Pettit et al., 1981

Table 5. Incubation Period and Energetic Efficiency of Embryonic Development in Hawaiian Seabirds.

Species	Incubation Period Days	Incubation Period % of Predicted	Production Efficiency (%)	Efficiency of Biosynthesis (%) Embryo	Efficiency of Biosynthesis (%) Hatchling	Total Consumed Energy (% of Fresh Egg Energy)	Unused Energy (% of Fresh Egg Energy)	Storage Efficiency (%) Embryo	Storage Efficiency (%) Hatchling
Black-footed Albatross	65	156	67.7	44.7	46.5	29.9	6.1	71.8	60.9
Laysan Albatross	65	158	63.5	45.2	45.8	31.3	7.6	71.1	59.4
Wedge-tailed Shearwater	52	178	60.9	31.1	39.3	35.8	3.4	58.2	52.5
Bonin Petrel	49	183	49.4	32.1	44.7	34.5	8.0	60.2	56.5
Brown Noddy	36	133	57.9	33.7	46.5	36.1	4.8	61.3	56.3
Black Noddy	35	145	56.1	33.0	47.6	38.1	4.7	58.7	55.6
White Tern	36	149	49.6	30.4	41.0	37.1	10.7	60.0	60.2

Consumed Energy = estimated from total amount of oxygen consumed during incubation; one liter of oxygen being equivalent to 4.8 kcal.

Unused energy = total fresh-egg energy - (hatchling tissue energy + yolk reserve energy + consumed energy)

Production Efficiency

Values for the production efficiency of procellariiform hatchlings ranged from 49.4% (Bonin Petrel) to 67.7% (Black-footed Albatross). Among this group the petrel egg has the longest relative incubation period (183% of the predicted value) and the associated high maintenance energy requirements may effectively reduce the overall energetic efficiency of production. The energetic efficiencies of tern eggs are less than those of procellariiform eggs, possibly due to the higher energetic requirements of producing a more mature chick over a prolonged incubation period.

Efficiency of Biosynthesis

The efficiency of tissue biosynthesis also suggests a reduction in efficiency due to high maintenance requirements of the embryo (Table 5). Thus, the shearwater, petrel and tern have lower efficiencies than other members of their respective Orders. Slow embryonic growth in eggs with prolonged incubation may be the result of semi-precocious development and early functional maturation in the incubation period. The allocation of a high proportion of tissue to mature function may not only slow the growth rate of the embryo, but the energetic efficiency in conversion of matter to embryonic tissue may also be reduced as maintenance costs increase.

Comparison of the efficiency of tissue biosynthesis of internally-pipped embryos with that of hatchlings reveals an energetic relationship to the dynamics of growth just prior to hatching. The two species of albatross display little change in efficiency during the last few days of incubation (45-46%) due to limited growth during this period. Furthermore, the caloric density of hatchling tissue is less than that of embryos (Table 2). Thus, the efficiency of biosynthesis for hatchling yolk-free tissue increases only slightly over the value for internally-pipped embryos. On the other hand, species with relatively longer incubation periods display a relatively longer pip-to-hatch interval during which growth continues to increase rapidly, and the water content of the embryo decreases substantially. At hatching, the caloric density of the chick is greater than the caloric density of the internally-pipped embryo (Table 5).

Ricklefs (1974) reviewed the efficiency of biosynthetic processes in various animals and concluded that the cost of biosynthesis is approximately one-third of the chemical potential energy in newly synthesized tissue. Thus, one-third of the stored energy or caloric density of the hatchling represents the cost of biosynthesis. Using this rough index, the calculated values range from 384 cal.g^{-1} in the Brown Noddy to 447 cal.g^{-1} in the Bonin Petrel. These values are considerably greater than the estimate of 292 cal.g^{-1} generated for 11 precocial and altricial species (Vleck et al., 1980). The higher cost of biosynthesis in tropical seabirds may be indicative of their relatively mature status at hatching.

Consumed Energy

Consumed energy, derived from total egg O_2 uptake over the entire incubation period, ranged from approximately 30% of initial energy content in the two species of albatross to 38% in the Black Noddy egg. The energy consumed or "lost" during development was calculated to be 27% of the initial energy stores in the domestic fowl's egg (Tangl, 1903, in Brody, 1945). Tangl considered this component of the energy budget to represent the "Entwicklungsarbeit" or the "work of development". However, some maintenance energy requirement is also involved and undoubtedly contributes to the high proportion of consumed energy observed in shearwater, petrel and tern eggs. Consumed energy in the small egg of Leach's Storm-Petrel (Oceanodroma leucorhoa) represents approximately 59% of the initial egg energy (calculated from Vleck and Kenagy, 1980, and Carey et al., 1980).

Unused Energy

The unused energy, representing waste materials such as uric acid, and extra-embryonic membranes, averaged about 6.5% of the initial egg energy content. This calculated value contains some margin of error since it was derived by subtracting hatchling tissue energy and yolk reserve energy and consumed energy from the mean fresh egg energetic content. The energy budget for embryonic development in the Bonin Petrel egg (Fig. 3) accounts for the utilization of 77.2 kcal contained in the fresh egg. At the internally-pipped phase of incubation, 24.8 kcal are contained in the embryonic tissues while 20.5 kcal are present in the yolk sac and 15.5 kcal are present in extra-embryonic membranes. Up to this phase of incubation, the consumed energy represents 16.4 kcal or 21.2% of the initial fresh egg energy. At hatching, 44.7% (34.5 kcal) of the energy is in the chic and 14.7% (11.4 kcal) in the yolk reserve. The unused energy is approximately 4.7 kcal and the total consumed energy is approximately 26.6 kcal or 34.5% of the fresh egg energy.

Energetic Storage Efficiency

The energy storage efficiency for embryonic tissue ranged from

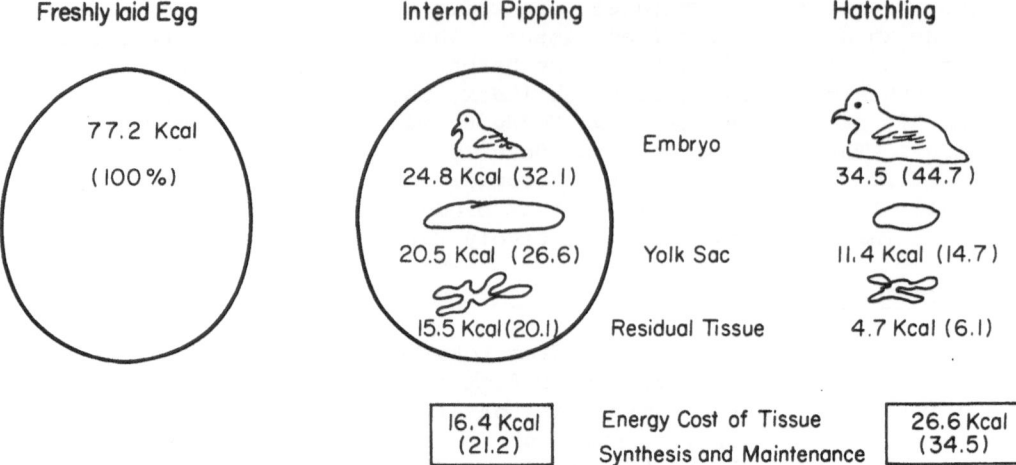

Freshly laid Egg	Internal Pipping		Hatchling
77.2 Kcal (100%)	24.8 Kcal (32.1)	Embryo	34.5 (44.7)
	20.5 Kcal (26.6)	Yolk Sac	11.4 Kcal (14.7)
	15.5 Kcal (20.1)	Residual Tissue	4.7 Kcal (6.1)

16.4 Kcal (21.2) Energy Cost of Tissue Synthesis and Maintenance 26.6 Kcal (34.5)

Fig. 3. Allocation of the energy contained in the freshly-laid egg of the Bonin Petrel to tissue synthesis, maintenance and storage. The numbers in parentheses are the percentage allocations (energy content of freshly-laid egg = 100%).

58.2% in the shearwater to 71.8% in the Black-footed albatross. It was lower in hatchling tissue, ranging from 46.8% in the shearwater to about 60% in the albatross. Storage efficiency in hatchlings includes the energetic costs during the pip-to-hatch interval. Nevertheless, the values are remarkably similar (53-60%) among all species regardless of the length of the incubation period and the pattern of pipping. It appears that prolonged embryonic development occurs in eggs which are high in initial energetic content, and the efficiency of biosynthesis is similar to the bioenergetic efficiency of the incubated egg of the domestic fowl. An enriched fresh egg initially high in calories compensates for the augmented cost of prolonged incubation for semi-precocial seabird embryos.

Needham (1931) reported that changes in the efficiency of tissue production during incubation are observed in the fowl's embryo, where the caloric density per gram of dry embryonic tissue increases in a sigmoidal fashion from 5.1 $kcal.g^{-1}$ at day 5 to 6.2 $kcal.g^{-1}$ at hatching (21 days). A similar increase in production efficiency was observed in embryos of both the Black and Brown Noddy. Noddy embryos also display a sigmoidal increase in caloric density from about 4.9 $kcal.g^{-1}$ (dry mass at day 10 to 5.6 $kcal.g^{-1}$ just prior to hatching (Figs. 1 & 2).

The efficiency of embryonic energy storage during the incubation period is represented in Figure 4 for the Brown Noddy. The total energy content of the embryo (kcal.g^{-1} x embryonic mass) is presented as a percentage of the energy required for storage (the tissue energy plus the energy consumed to produce that amount of tissue energy). Early in incubation, the storage efficiency is low, about 20% at day 4-6, and about 45% at age 8-11 days. During early incubation, the metabolic cost of producing extra-embryonic membranes (eg. chorioallantois) is incurred, and lowers the efficiency for the production of embryonic tissue. Thereafter, the energetic storage efficiency rises to about 60% from age 30 days to hatching (Fig. 4). The overall storage efficiency of 56.3% for the hatchling reflects the cost of producing and maintaining extra-embryonic tissue, and the costs associated with hatching. Consumed energy by the Brown Noddy egg continues to increase exponentially throughout incubation (Fig. 4).

Ricklefs and Cullen (1973) determined that the efficiency of growth in the Green Iguana (_Iguana iguana_) embryo ranged from 44.7 - 51.6%. They defined growth efficiency as the caloric content of the embryo divided by the calories lost in yolk, ignoring the caloric contribution of albumen to the growth process. They speculated that slow growth of the iguana embryo contributed to the low growth efficiency. Other estimates of the embryonic efficiency of biosynthesis range from 59% for the sea urchin (Brody, 1945); 59.3% in the lizard _Gerrhonotus coeruleus_ (Vitt, 1974); and 61% in the slipper shell, _Crepidula fornicata_ (Pandian, 1969).

Lipid Content and Metabolic Water Production

The energy content of the avian egg may be calculated from the proportions of yolk and albumen in the egg, if one assumes caloric densities of 0.65 kcal per gram of albumen and 4.0 kcal per gram of yolk (Ricklefs, 1974). The calculated energy based upon the yolk: albumen ratio is remarkably similar to the observed mean values (Table 6). Thus, this rough estimate is a useful means of calculating the egg energetic content from the proportions of yolk and albumen.

The lipid content of the egg may be calculated by assuming protein contains 5.65 kcal per gram of combustible energy and fat contains 9.5 kcal.g^{-1} (Ricklefs, 1974). The caloric density of the dry egg (kcal/g) may be used to estimate the lipid content:

$$\% \text{ lipid} = \frac{(\text{kcal/g} - 5.65)}{9.5 - 5.65} \times 100$$

The calculated lipid content ranges from 13.1% in the altricial booby egg to 38% in the semi-precocial White Tern egg. Furthermore, by assuming 9.5 kcal per gram of fat are oxidized to meet the consumed energy needs of the developing embryo, and as lipids provide the energy for 94% of the total energy consumed during

Table 6. Yolk, lipid and energy content of the freshly-laid egg together with the amount of lipid consumed and the metabolic water production.

| | Yolk mass (g) | Albumen mass (g) | Calculated Energy[a] (kcal) | Observed Energy[b] (kcal) | Lipids | | | | Metabolic Water | |
					% of Yolk	Mass (g)	Consumed[c] (g)	% of Total[d]	(g)	% of Egg Water loss
PROCELLARIIFORMES										
Black-footed Albatross	92.9	188.7	494.3	477.9	29.1	27.0	15.0	55.6	16.1	32.7[e]
Laysan Albatross	91.8	172.0	479.0	456.4	27.1	24.9	15.0	60.2	16.1	35.2[e]
Wedge-tailed Shearwater	22.5	33.8	112.0	107.4	37.9	8.5	4.0	47.1	5.5	53.4[f]
Bonin Petrel	15.0	22.5	82.0	77.2	34.8	5.2	2.8	53.8	3.0	41.1[g]
PELECANIFORMES										
Great Frigatebird	21.1	48.2	124.1	121.5	24.1	5.1	----	----	----	----
Red-tailed Tropicbird	15.6	61.1	93.7	93.7	21.6	3.4	----	----	----	----
Red-footed Booby	8.3	44.4	62.1	53.1	13.1	1.1	----	----	----	----
CHARADRIIFORMES										
Brown Noddy	13.2	24.2	68.5	63.4	26.1	3.4	2.4	70.6	2.6	38.4[h]
Sooty Tern	13.0	21.5	66.0	66.1	29.3	3.8	----	----	----	----
Gray-backed Tern	9.8	17.0	50.3	45.5	26.5	2.6	----	----	----	----
Black Noddy	8.2	15.3	42.7	39.9	29.9	2.5	1.6	64.0	1.7	31.5[h]
White Tern	8.4	13.7	42.5	45.4	38.1	3.2	1.8	56.3	1.9	46.6[i]

[a]Calculated Energy derived from yolk and albumen masses where 1 g fresh yolk = 4.0 $kcal \cdot g^{-1}$ and 1 g fresh albumen = 0.65 $kcal \cdot g^{-1}$.

[b]Observed Energy from calorimetry (Table 1).

[c]Consumed Lipids derived from Consumed Energy (Table 5) where 1 g lipid = 9.5 kcal.

[d]% of Total Lipids Consumed

[e]Grant et al., 1982a.

[f]Whittow et al., 1982.

[g]Grant et al., 1982.

[h]Pettit and Whittow, unpubl. data.

[i]Pettit et al., 1981.

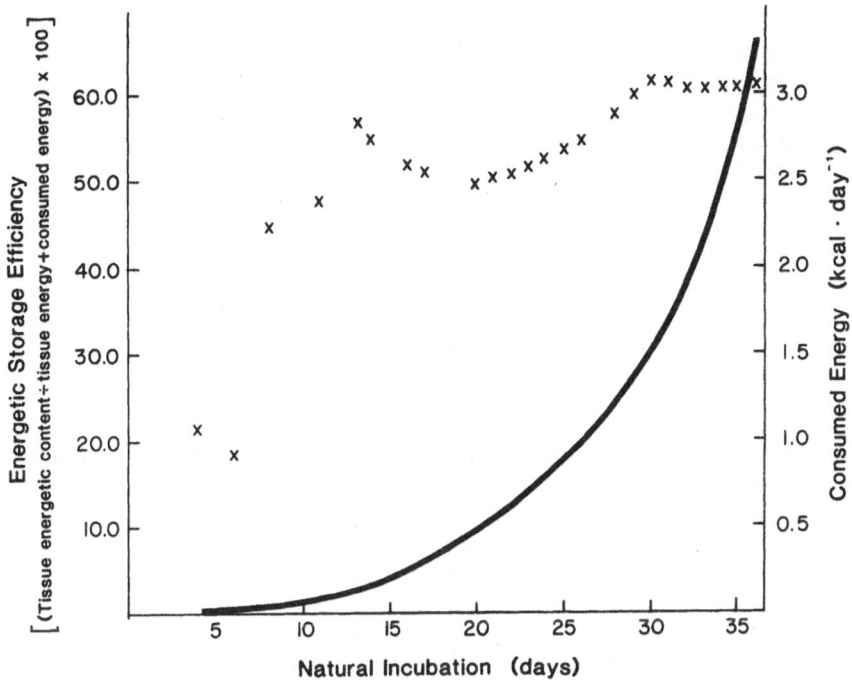

Fig. 4. The efficiency of energy storage (x) in the embryo of the Brown Noddy, and the energy consumed by the embryo (solid line) during incubation. See text for definitions of the ordinates.

incubation (Murray, 1925), it is possible to calculate the total amount of lipid and the fraction of the lipid content of the egg which is consumed as energy substrate. Accordingly, the seven species of seabirds utilize a mean of 60.1% $\pm$ 5.8 (SD) of the lipid content of the egg to meet the total metabolic requirements of the developing embryo.

Calculation of metabolic water production from lipid oxidation is based upon the total O_2 uptake during incubation multiplied by 0.54 mg H_2O produced for each ml of O_2 consumed (Schmidt-Nielsen, 1975). This analysis (Table 6) suggests that an average of 39% of the water lost from the egg during incubation is replaced by metabolic water. These data support the concept of "constant hydration" of egg contents (Ar and Rahn, 1980). This concept suggests that the relative water content of the egg is constant during incubation. This is true because the fraction of consumed energy is approximately equal to the metabolic water produced since lipid oxidation releases an almost equal amount of metabolic water. Thus, the mean consumed energy for seven seabird eggs is approximately 36%

(Table 5). A high fat (caloric) content means that relatively more water is produced during lipid oxidation. Thus the high caloric content of tropical seabird eggs with long incubation periods is also a compensation for extended egg water loss during incubation (Grant et al., 1982a): the greater water loss is balanced by greater metabolic water production.

The water budget for embryonic development in the Bonin Petrel egg (Fig. 5) shows that allocations of water to the hatchling, yolk reserve, extra-embryonic membranes, and daily water loss ($\dot{M}_{H_2O}$) are balanced by the water content of the fresh egg (26.1 g) and metabolic water produced by lipid oxidation (2.9 g). Thus the overall water budget of 29.0 g required for development can be accounted for at hatching.

In summary, several conclusions may be affirmed regarding the energetic budget of tropical seabird eggs. In general, the yolk content, yolk:albumen ratio and caloric density of the freshly-laid egg is associated with the maturity of the chick at hatching. In addition, prolonged incubation and semi-precocial status requires a higher yolk and caloric content than does prolonged incubation which produces an altricial chick. The caloric density of pro-cellariiform and charadriiform hatchlings is similar to that of the fowl (1.2 kcal.g^{-1}) and supports their inclusion in the category of semi-precocial chicks. The caloric content of yolk reserves in the hatchling provides enough energy to sustain resting metabolism without feeding for several days. The time required for depletion of yolk energy reserves may be associated with the timing of incubation spans and feeding intervals for each species.

Production efficiency, efficiency of biosynthesis and the energetic storage efficiency decrease as the incubation period becomes relatively longer than predicted on the basis of egg mass. Total consumed energy increases with increasing relative incubation period, presumably due to increased maintenance costs of the embryo. Caloric density of noddy embryos increased in a sigmoidal fashion throughout incubation, while water and ash content decreased. The energetic storage efficiency of embryonic tissue is low early in incubation (20-45%) and rises to about 60% near the end of incubation.

ACKNOWLEDGEMENTS

Our research was supported by a National Science Foundation Grant (PCM 76-12351-A01) and the Sea Grant College Program (N1/R-13, 14). We are grateful to Commander J.C. Barnes, Commanding Officer, for assistance during our stay at the U.S. Naval Air Facility Midway Islands. We are grateful also to the U.S. Fish and Wildlife Service in Honolulu for assistance during our stay on Tern Island,

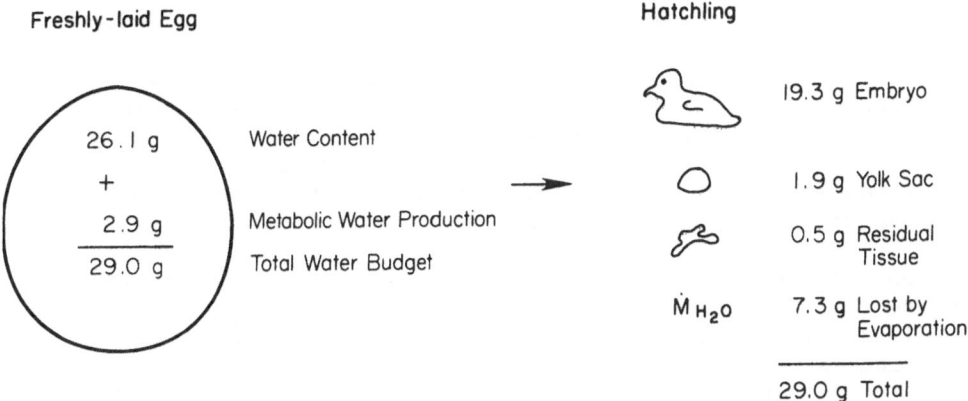

Fig. 5.   The allocation of the water content of the freshly-laid
egg and the metabolic water production, to the hatchling,
yolk sac, residual tissue and to loss by evaporation, in
the Bonin Petrel.

French Frigate Shoals, for the procurement of equipment, and for
granting permits to conduct this work. Colonel C.D. Robinson,
Lt. Col. A.L. Mize and Dr. Diane Drigot of the Kaneohe Marine Corps
Air Station, Oahu, Hawaii, kindly granted permission to study the
colony of Red-footed Boobies at Ulupau Crater. Our thanks are due
also to the State of Hawaii Division of Forestry and Wildlife for
permission to work at French Frigate Shoals. M. Naughton (U.S.
Fish and Wildlife Service - Honolulu) assisted in collection of
fresh eggs on Laysan Island. This chapter is dedicated to the
memory of Professor Samuel Brody.

REFERENCES

Ackerman, R. A., Whittow, G. C., Paganelli, C. V., and Pettit,
     T. N., 1980, Oxygen consumption, gas-exchange and growth of
     embryonic Wedge-tailed Shearwaters (Puffinus pacificus
     chlororhyncus), Physiol. Zool., 53:210.
Ar, A., and Rahn, H., 1980, Water in the avian egg: overall budget
     of incubation, Amer. Zool., 29:373.
Ar, A., and Tom-Tov, Y., 1978, The evolution of parental care in
     birds, Evolution, 32:655.
Brody, S., 1945, "Bioenergetics and Growth", Waverly Press,
     Baltimore.
Carey, C., Rahn, H., and Parisi, P., 1980, Calories, water, lipid
     and yolk in avian eggs, Condor, 82:335.
Evans, R. M., 1980, Development of behavior in seabirds: an ecolo-
     gical perspective, in "Behavior of Marine Animals, Vol. 4:

Marine Birds, J. Burger, B. L. Olla, and H. E. Winn, eds.,
    Plenum Press, New York.
Grant, G. S., Pettit, T. N., Rahn, H., Whittow, G. C., and
    Paganelli, C. V., 1982a, Water loss from Laysan and Black-
    footed Albatross eggs, Physiol. Zool., 55:405.
Grant, G. S., Pettit, T. N., Rahn, H., Whittow, G. C., and
    Paganelli, C. V., 1982b, Regulation of water loss from Bonin
    petrel (Pterodroma hypoleuca) eggs, Auk, 99:236.
Hazelwood, R. L., 1965, Carbohydrate metabolism, in "Avian
    Physiology", P. D. Sturkie, ed., Cornell Univ. Press, Ithaca.
Montevecchi, W. A., Kirkham, I. R., Roby, D. D., and Brink, K. L.,
    1983, Size, organic composition, and energy content of
    Leach's storm-petrel (Oceanodroma leucorhoa) eggs with refer-
    ence to position in the precocial-altricial spectrum and
    breeding ecology, Can. J. Zool., 61:1457.
Murray, H. A, 1925, Chemical changes in fertile eggs during incuba-
    tion, J. Gen. Physiol., 9:1.
Needham, J., 1931, "Chemical embryology", Vol. 2, Cambridge Univ.
    Press, Cambridge.
Nice, M. M., 1962, Development of behavior in precocial birds, Trans.
    Linn. Soc. N.Y., 8:1.
Pandian, T. J., 1969, Yolk utilization in the gastropod Crepidula
    fornicata, Marine Biol., 8:117.
Pettit, T. N., Grant, G. S., Whittow, G. C., Rahn, H., and
    Paganelli, C. V., 1981, Respiratory gas exchange and growth
    of White Tern embryos, Condor, 83:355.
Pettit, T. N., Grant, G. S., Whittow, G. C., Rahn, H., and
    Paganelli, C. V., 1982a, Embryonic oxygen consumption and
    growth of Laysan and black-footed albatross, Am. J. Physiol.,
    242:R121.
Pettit, T. N., Grant, G. S., Whittow, G. C., Rahn, H., and
    Paganelli, C. V., 1982b, Respiratory gas exchange and growth
    of Bonin petrel embryos, Physiol. Zool., 55:162.
Pettit, T.N., and Whittow, G. C., 1983, Embryonic respiration and
    growth in two species of noddy terns, Physiol. Zool., 56:455.
Ricklefs, R. E., White, S., and Cullen, J., 1980, Postnatal develop-
    ment of Leach's Storm Petrel, Auk, 97:768.
Ricklefs, R. E., 1974, Energetics of reproduction in birds, in
    "Avian Energetics", R. A. Paynter, Jr., ed., Publ. Nutall.
    Ornithol. Club No. 15.
Ricklefs, R. E., 1977, Composition of eggs of several bird species,
    Auk, 94:350.
Ricklefs, R. E., and Cullen, J., 1973, Embryonic growth of the
    Green Iguana, Iguana iguana, Copeia, 2:296.
Ricklefs, R. E., and Montevecci, W. A., 1979, Size, organic com-
    position and energy content of North Atlantic Gannet Morus
    bassanus eggs, Comp. Biochem. Physiol., 64A:161.
Romanoff, A. L., 1944, Avian space yolk and its assimilation, Auk,
    61:235.
Romanoff, A. L., and Romanoff, A. J., 1949, "The avian egg", Wiley,
    New York.

Romanoff, A. L., 1967, "Biochemistry of the Avian Embryo", Wiley, New York.

Schmidt-Nielsen, K., 1975, Animal Physiology, Cambridge Univ. Press, New York.

Vanheel, B., Vandeputte-Poma, J., and Desmeth, M., 1981, Resorption of yolk lipids by the pigeon embryo, Comp. Biochem. Physiol., 68:641.

Vitt, S., 1974, Reproductive effort and energy comparisons of adults, eggs and neonates of Gerrhonotus coeruleus principis, J. Herpetol, 8:165.

Vleck, C., and Kenagy, G. J., 1980, Embryonic metabolism of the Fork-tailed Storm-Petrel: physiological patterns during prolonged and interrupted incubation, Physiol. Zool., 53:32.

Vleck, C. M., Vleck, D., and Hoyt, D. F., 1980, Patterns of metabolism and growth in avian embryos, Amer. Zool., 20:405.

Warham, J., 1983, The composition of petrel eggs, Condor, 85:194.

Whittow, G. C., 1980, Physiological and ecological correlates of prolonged incubation in seabirds, Amer. Zool., 20:427.

Whittow, G. C., Ackerman, R. A., Paganelli, C. V., and Pettit, T. N., 1982, Pre-pipping water loss from the eggs of the Wedge-tailed Shearwater, Comp. Biochem. Physiol., 72:29.

Williams, A. J., Siegfried, W. R., and Cooper, J., 1982, Egg composition and hatchling precocity in seabirds, Ibis, 124:456.

METABOLIC RESPONSES OF EMBRYONIC SEA BIRDS TO TEMPERATURE

William R. Dawson

Museum of Zoology and Division of Biological Sciences
The University of Michigan
Ann Arbor, Michigan 48109

INTRODUCTION

Surprisingly little information is available concerning metabolic responses of embryonic sea birds to temperature, despite the challenging thermal conditions under which many species nest. Knowledge of such responses should be useful in tracing fully the ontogeny of thermoregulation in these animals and in defining the thermal limits for development. It should also allow identification of acute effects of variations in egg temperature and any compensatory mechanisms that serve to damp these effects. Development of a comprehensive understanding of these topics for sea birds is a challenging task, for these animals comprise a group defined by habitat preference rather than taxonomic affinity. It easily includes more than half a dozen orders if it is broadly defined to include littoral as well as near-shore and pelagic birds, and transients as well as species that are resident in marine situations. This diverse group includes altricial, semi-altricial, semi-precocial, and precocial species, with all the physiological diversity that this implies. Despite these difficulties it seems worthwhile, within the framework of this symposium on the energetics of sea birds, to assemble information on embryonic responses to temperature so that we may begin to formulate meaningful questions for further investigation.

THERMAL RELATIONS OF AVIAN EGGS

The colonial nesting habits of many sea birds have facilitated observations on the biology of incubation. Extensive reviews of this subject (see, for example, Drent, 1970, 1973, 1975; White and Kenney, 1974) document patterns of parental attentiveness that tend to

139

restrict egg temperature over the course of embryonic development to a narrow band of temperatures within the interval between 30° and 40° C. The parental behavior leading to control of egg temperature can be quite elaborate and vary with external circumstances (see, for example, Bartholomew and Dawson, 1979). The incubating adult serves as a heat source for the eggs under cooler circumstances and a growing body of information indicates that it may also serve as a heat sink in very hot environments (see, for example, Russell, 1969; Grant, 1982; Walsberg and Voss-Roberts, 1983). Where high temperature and intense insolation combine to produce a persistent danger of over-heating, shorebirds and terns may protect their developing embryos by applying water from moistened abdominal feathers directly to the eggs (Purdue, 1976; Grant, 1982) or to a thin layer of sand covering them (Howell, 1979).

Despite the impressive thermal protection that incubating birds provide their eggs, circumstances producing fluctuations of egg temperature do arise regularly. In the case of the Gray Gull (Larus modestus) these involve the substantial diurnal variation in thermal conditions characterizing its inland nesting grounds in the desert of northern Chile. During the breeding season, air temperature varies as much as 30°-35°C between day and night and solar radiation is intense at the upland (1800 m) site where Howell et al. (1974) carried out their study of nesting by this species. A diurnal cycle of incubation temperature is associated with this variation, with eggs generally maintained at 35°-37°C during the day and 33°C at night.

Behavioral factors also contribute to variation in egg temperature. In various gulls, continuous incubation does not commence until the clutch is completed (Drent, 1970) and it is delayed 3 or 4 days beyond this in Kittiwakes, Rissa tridactyla (Barrett, 1980). Drent (1973) also notes that the onset of nocturnal incubation may be delayed in colonies of Herring Gulls (Larus argentatus) with low densities of incubating adults. Such densities appear to reduce the readiness with which predators such as foxes can be detected and so increase the risk for the parents. Drent (1973) also notes that Herring Gulls with isolated nests have been found to abandon their eggs regularly at night for up to 16 days after clutch completion. Consequently, ample opportunities exist for the eggs to cool to ambient temperature, particularly at night when inattentiveness is most pronounced. However, observations on Black-tailed Godwits (Limosa limosa) indicate that nocturnal incubation can be initiated in the laying period, if risk of freezing the eggs exists (Lind, 1961).

The actual approach of predators to a colony can interfere with incubation behavior by stimulating nest owners either to attack intruders or flee. The thermal consequences of such responses in a hot environment are illustrated by Grant's (1982) observations on nesting Forster's Terns (<u>Sterna</u> <u>forsteri</u>) at the Salton Sea, a saline lake in the Colorado Desert of southeastern California. One of these birds left its nest during a summer midmorning to chase a Western Gull (<u>Larus</u> <u>occidentalis</u>) flying over the tern colony. The temperature of the tern's egg rose approximately 6°C to 45.7°C while exposed for the 10-min period required for the chase. Later on the same day, all of the incubating terns in the colony left their nests in apparent response to an alarm call of a nearby shorebird. In this case the nest referred to previously was untended for approximately 25 min. This time the egg became heated to a lethal level of 50°C.

Territorial behavior is a further activity that can disrupt incubation rhythms as parents defend their nesting or feeding territories against conspecific intruders. Such contests, which appear to play an important role in establishing spacing within breeding colonies of sea birds (Patterson, 1965), lead to exposure of the eggs to ambient conditions. These exposures lasted only a minute or two in Heermann's Gulls (<u>Larus</u> <u>heermanni</u>) nesting on an island in the Gulf of California, Mexico, where the diurnal thermal environment was dominated by solar radiation (Bartholomew and Dawson, 1979). Exposures of this duration produced only a small rise in egg temperature, but this could be physiologically significant, owing to the narrow interval existing between incubation temperature and upper lethal temperature for avian embryos (Bennett and Dawson, 1979). Grant's (1982) account of a territorial contest in Black-necked Stilts (<u>Himantopus</u> <u>mexicanus</u>) on the extremely hot shores of the Salton Sea further illustrates this point. Late in the summer afternoon, a pair of these birds left their nest uncovered as they attempted to defend their feeding territory from another pair of stilts accompanied by three chicks. The temperature of the eggs in the uncovered nest rose from 38.4° to 41.7°C over the 113-sec exposure. A substantially greater rise capable of injuring the embryos might well have occurred if the contest had taken place earlier in the day, when solar radiation was more intense (Grant, 1982).

Maintenance requirements of parents can also interfere with incubation. In species nesting in hot environments, these requirements can include bathing as a form of behavioral heat defense, and drinking as well as feeding (Grant, 1982). Terns in flight on hot days may touch the water with their ventral feathers and by such skimming not only cool themselves but also obtain fluid that can be used to cool their eggs (Mes et al., 1978; Grant, 1982). In the case of birds in which incubation is performed by only one sex, maintenance needs can lead to exposure of eggs to ambient conditions for substantial periods. For example, though female Northern Shovelers (<u>Anas</u> <u>clypeata</u>) absent themselves from their eggs only a little more than twice a day on the average, the duration of these feeding recesses exceeds 1.5 h (Afton, 1979). This was long enough in the Delta area, Manitoba, Canada, to allow egg (air cell) temperature to cool an average 7.6°C. In some of the species in which only one sex incubates, inattentive periods tend to be minimized by the non-incubating partner's feeding its mate at the nest (Skutch, 1957, 1962). In species in which incubation duties are shared, the process of nest relief can involve brief exposures of eggs to heating or cooling. Substantial changes in egg temperatures apparently can result, particularly in situations where intense insolation is a factor (Grant, 1982). Incubating birds that depend on distant and possibly unpredictable food resources have to expend substantial time in travel and foraging and their eggs may be left unattended for lengthy periods. Prominent among these are various procellariiform birds (Roberts, 1940; Matthews, 1954; Wilbur, 1969; Beck and Brown, 1972; Boersma and Wheelwright, 1979). For example, the Fork-tailed Storm Petrel (Oceanodroma furcata) lays a single egg for which incubation duties are shared by both parents (Wheelwright and Boersma, 1979). These birds must commute long distances between the burrows within rock crevices that serve as nests, and their feeding areas at sea. The time required to effect these movements and forage necessitates both parents' being away from the egg for periods ranging from a few hours up to 7 days at a time (Boersma et al., 1980). Egg temperatures during these absences can fall to below 10°C (Boersma and Wheelwright, 1979).

The circumstances described here do result in fluctuations of egg temperature, particularly in extreme environments. Consideration of both acute and long term effects of fluctuations in temperatures of embryos is thus of more than just theoretical interest.

Thermal Tolerances of Embryonic Sea Birds

Information on the temperature tolerances of embryonic sea birds affords some insight concerning the demands for attentive behavior that successful development places on their parents. The respective tolerances of these embryos to acute heating and cooling from incubation temperature are distinctly asymmetrical, with a far smaller change required for producing heat injury. For example, week-old embryos of the Heermann's Gull, in which incubation temperature approximates 37°C, suffer blockage of heart action when warmed to 41.1°C. Interestingly, this is actually below the diurnal level of body temperature for adults. The blockage in the embryos is reversible even after an hour's exposure to 42°-43°C, but temperatures above 43°C are immediately lethal. In contrast, cold blockage of heart action in embryonic Heermann's Gulls does not occur until they are cooled to 7.9°-13.1°C and this is reversible down to at least 6°C (Bennett and Dawson, 1979).

The above information pertains to acute exposure. Longer term displacements of egg temperature from normal incubation temperature appear to result in a somewhat different set of relationships, judging by information in the domestic fowl (Lundy, 1969). The optimal incubation temperature for its embryos is 37°-38°C, with the incidence of hatching declining precipitously with deviation from this interval. Lundy (1969) noted that no embryos of the domestic fowl survived continuous exposure to temperatures lying outside the range of 35°-40.5°C. Duration of exposure appears critical, for Kühn et al. (1982) reported successful hatching of chicks of the Rhode Island Red breed following incubation at 33.8°C from day 17 until emergence from the egg. Pipping and hatching occurred significantly later in these birds and the interval between these events was significantly longer than in their counterparts incubated continuously at 37.8°C. Growth pattern in the days after hatching also differed between the two groups. The effects of cooling on avian embryos are complex. If embryos of the domestic fowl are subjected to prolonged moderate cooling which slows but does not stop development, anomalies arise (Romanoff, 1972). However, cooling for moderate periods to non-freezing temperatures below 25°-27°C, where development stops, produces no lasting effects (Lundy, 1969), even in Fork-tailed Storm Petrels that were absent from their eggs for a total of 28 days over the incubation period (Boersma et al., 1980). Drent (1973) succinctly summarizes these relationships ".....the [temperature] range for optimal development is narrow: death through overheating is uncomfortably close at all times; a slight fall in internal temperature, if long maintained, will lead to abnormal development; and an internal temperature between physiological zero (no development) and freezing can be tolerated for prolonged periods, the most important limitation being that tolerance declines with age."

The thermal tolerances of the embryos of wild birds, including sea birds, will not necessarily mimic those of the domestic fowl. In this connection it is reasonable to ask whether the tolerances of these other species reflect any special adaptations to extreme environments. Curiously, the heart blockage in embryos of Heermann's Gull described above occurs with acute exposure to only 41.1°C, despite the fact that short daytime exposures of the eggs of this species to the intense insolation on their breeding grounds can easily heat them beyond this temperature (Bennett and Dawson, 1979). On the other hand, such blockage does not occur until 46°C in the Western Gull (Larus occidentalis wymani), which breeds in somewhat cooler situations. The significance of the difference between the two species is not clear. The attentive behavior of adult Heermann's Gulls normally serves to protect their embryos from being disadvantaged by a rather limited tolerance of hyperthermia.

At present, embryonic procellariiform birds provide some suggestion of interspecific variation in response to cooling, though the information is largely anecdotal. The development of some of these organisms can be interrupted by cooling for hours or days as their parents forage at sea (see, for example, Matthews, 1954). For example, embryonic Fork-tailed Storm Petrels apparently can survive cooling to 3°C for 48 h (Boersma and Wheelwright, 1979) and some individuals that had been cooled to 10 C every 4 days hatched successfully (Vleck and Kenagy, 1980). On the other hand, half-grown embryos of the Southern Giant Fulmar (Macronectes giganteus), another procellariiform, did not survive 48 h exposure to 3°C (Williams and Ricklefs, 1984).

The information now available on the thermal tolerances of embryonic sea birds is insufficient to identify any more than suggestions of special adaptations to extreme environments or unusual patterns of incubation. More information on a greater assortment of species is needed. Hopefully, future observations will be conducted in such a way as to explore the reciprocity between exposure temperature and length of exposure in producing injury at thermal extremes.

METABOLIC RESPONSES TO TEMPERATURE BY EMBRYONIC SEA BIRDS

Introduction

The preceding discussion identifies circumstances that can foster short-term departures of egg temperature from the narrow range to which parental attentive behavior normally restricts it. We may now move to assess the effects such departures can exert on embryonic metabolism. A necessary prelude to such as assessment is some consideration of the metabolic patterns evident in embryos at normal incubation temperatures, from the initiation of development to hatching. This consideration is particularly critical for sea birds with their heterogeneous patterns of development.

Metabolism and Developmental Pattern

Vleck et al. (1980a) provide a useful overview of the metabolic patterns evident in developing embryos of precocial and altricial birds: metabolic rates of embryos of species in the former group increase rapidly over the first 80% of incubation and then increase slowly, stabilize, or actually decline in the remainder of the period before hatching. Conspicuous among species in which embryonic metabolic rate declines over the last 20% of development are ratites: Rhea, Rhea americana, and Emu, Dromeceius novaehollandiae (Vleck et al., 1980b), and Ostrich, Struthio camelus (Hoyt et al., 1978). In altricial birds, on the other hand, embryonic metabolic rates (per embryo) increase at an accelerating rate throughout incubation (Vleck et al., 1979, 1980a). These metabolic patterns correlate with growth patterns manifested in the two modes of development. In precocial birds, embryonic mass approximates hatching mass as early as the 80th percentile of incubation and growth rate declines during the remainder of the period before hatching. In altricial species, on the other hand, embryonic mass and growth rates tend to increase continuously throughout incubation (Vleck et al., 1979). Vleck et al. (1980a) conclude that the total cost of pre-hatching development is higher in precocial than in altricial birds because a larger mass of tissue must be maintained for a longer period. Difficulties are to be anticipated in identifying truly comparable metabolic values for use in various allometric analyses pertaining to embryonic development in sea birds and in birds generally. Hoyt and Rahn (1980) have attempted to minimize many of these by utilizing values for embryos just before internal pipping, i.e., just before the developing chick's bill breaks into the air cell of the egg. Analysis of the available data for this stage (34 observations on 28 species) yields the following equation describing the relation of pre-internal pipping (pre-IP) metabolic rate to initial egg mass:

$$\dot{V}O_2 = 28.9m_e^{0.714} \tag{1};$$

where $\dot{V}O_2$ is oxygen consumption (ml. day^{-1}) and m_e is initial egg mass (g). They also detected an inverse relation between pre-IP metabolic rate and duration of incubation. The multiple regression equation describing the relation of pre-IP metabolic rate to initial egg mass and incubation period is:

$$\frac{\dot{V}O_2 = 139m_e^{0.85}}{I^{0.65}} \tag{2};$$

where $\dot{V}O_2$ and m_e are as in eq (1) and I is incubation period (days). It seems best to regard these equations as early approximations of embryonic metabolism-size relations, for internal pipping appears to be a less rigidly fixed reference point interspecifically than was initially assumed. Internal pipping follows star fracturing of the egg in some species (Pettit et al., 1982) and is absent in megapodes (Vleck et al., 1984). For future allometric analyses of the metabolism of avian embryos, Whittow (1984) advocates use of metabolic values obtained just before the initial event in the pipping process, whether this is internal pipping or star-fracturing of the egg shell.

Metabolic Responses to Temperature by Embryonic Seabirds during the Earlier Stages of Development

Several studies include data on the effects of temperature on energy metabolism and/or heart rate of embryonic seabirds in the ectothermic phase of their development. Difficulties exist in attempting to compare these, for specification of the developmental stages of the experimental subjects in some of the studies is either vague or lacking. Despite this limitation, it does appear feasible to assess the general thermal sensitivity of metabolism or heart rate of ectothermic embryos through consideration of the relevant temperature coefficients (Q_{10}) for various intervals. Information on these

145

coefficients (Table 1) indicates that the general form of the responses of rate processes of embryos to temperature resembles that in adult ectotherms (see, for example, Scholander et al., 1953; Dawson, 1967), in that Q_{10} tends to decline with increasing temperature. This coefficient ranges from 3.6 to 15 in the embryos between 10° and 15°C, but only from 1.1 to 2.0 between 35° and 40°C (Table 1). Over the intervals of 30°-35°C and 35°-40°C, in which the incubation temperatures of birds lie (Drent, 1973), embryos of only two species come close to complete independence in their metabolic and heart rates. These are the domestic fowl (Q_{10} for heart rate between 30° and 40°C and oxygen consumption between 34° and 40°C, approximately 1.2 and 1.0, respectively; Romanoff, 1960; Grieff, 1952) and Heermann's Gull (Q_{10}'s for heart rate between 35° and 40°C and oxygen consumption between 30° and 40°C, 1.1 and 1.0, respectively; Bennett and Dawson, 1979). However, the South Polar Skua's (Catharacta skua mccormicki) Q_{10} for oxygen consumption between 30° and 35°C is only 1.3 (Williams and Ricklefs, 1984).

Perhaps the most interesting feature of the information on Q_{10} concerns storm petrels, which may leave their eggs unincubated for relatively long periods (see discussion of fluctuations in egg temperature). This circumstance might make it advantageous for the metabolism of the embryos of these birds to be relatively insensitive to temperature. However, the Q_{10}'s for oxygen consumption and heart rate of these embryos are as high or higher than corresponding values for embryos of other birds (Table 1). While the available information is certainly not comprehensive, it provides no indication of any compensation of rate processes of embryonic storm petrels for temperature.

The relatively high Q_{10}'s characterizing metabolism and heart rate of avian embryos below 25°C dramatize the sensitivity of these processes to moderate and cool temperatures. This sensitivity presumably contributes to the cessation of development that occurs in eggs undergoing substantial cooling. Such cessation appears advantageous, for, as noted above, the development of avian embryos appears more adversely affected by slowing than by stopping temporarily (Drent, 1973).

The Acquisition of Thermogenic Capacity in Developing Chicks

The establishment of thermoregulatory capabilities at moderate and cool ambient temperatures in developing birds is intimately linked with a rise in the level of metabolism and with the acquisition of abilities to augment heat production beyond this level (Dawson and Hudson, 1970). The timing of these changes is linked with the mode of development, altricial or precocial, characterizing the particular species. Their effect upon the heat budget of embryos of the semi-precocial Herring Gull has been estimated by Drent (1970). He calculates that these embryos contribute about one-fifth

Table 1. Temperature Coefficients (Q_{10}) for Rate Processes in Embryos of Sea Birds and Other Species

Age[b] (day)	Body Mass (g)	Process[c]	Q_{10} Temperature Range (°C)[a]						Reference[d]
			10–15	15–20	20–25	25–30	30–35	35–40	
Southern Giant Fulmar (_Macronectes giganteus_)									
	110	O_2	6.1	2.2	2.0	3.0	2.3		1
Wilson's Storm Petrel (_Oceanites oceanicus_)									
2–3	6	HR	5.0	7.1	4.0	2.6	2.4	1.4	1
		O_2	6.9	4.1	2.9	2.8	2.3		1
Leach's Storm Petrel (_Oceanodroma leucorhoa_)									
2–3	4	HR	15.0	7.5	4.8	3.1	2.7	1.7	1
		O_2		5.1	3.5	2.4	2.0		1
Fork-tailed Storm Petrel (_Oceanodroma furcata_)									
		O_2	3.4(10–34)				2.4(30–34)		2
Heermann's Gull (_Larus heermanni_)									
2–9		HR		3.0	2.6	1.9	1.6	1.1	3
		O_2			2.1(20–30)		1.0(30–34)		3
Western Gull (_L. occidentalis_)									
5–9		HR		6.9	5.4	3.0	1.9	1.9	4
South Polar Skua (_Catharacta skua mccormicki_)									
	35	O_2	3.6	2.8	3.7	2.2	1.3		1
Duck									
3		HR					1.9		5
Domestic Fowl (_Gallus gallus_)									
2–8		HR					1.7(30–40)		6
3		HR					1.6		7
6		HR					1.8(33–39)		8
7		HR					1.4(31.5–40.5)		9
7		HR					1.5		10
		HR	3.3	3.0	2.1	1.3	1.2	1.1	11
		O_2					1.0(34–40)		12
		O_2		1.6(15–38)					13

Table 1 (cont.)

[a] In some cases, Q_{10}'s were determined for temperature intervals other than the 5°C ones indicated in the table heading. In such cases, the actual interval (in °C) is given in parentheses after the value for the temperature coefficient to which it pertains.

[b] Age refers to the approximate number of days of incubation.

[c] O_2 and HR refer to rate of oxygen consumption and frequency of heart beat, respectively.

[d] The references for the data presented are as follows: 1, Williams and Ricklefs (1984); 2, Vleck and Kenagy (1980); 3, Bennett and Dawson (1979); 4, Bennett et al. (1981); 5, Inukai, 1925; 6, Cesana (1911-12); 7, Paff et al. (1963); 8, Cohn (1928); 9, Romanoff and Sochen (1936); 10, Parpart and Glaser, 1930; 11, calculated from Romanoff (1960) by Williams and Ricklefs (1984); 12, Grieff (1952); 13, Hasselbalch (1900).

of the energy required for their incubation, with the daily contribution rising from a negligible proportion in the first week of development to over half just before hatching. This leads to a rise in embryo temperature above incubation temperature as development proceeds (Drent, 1972). Comparison of the metabolic rate of embryonic Herring Gulls just before internal pipping, with rates obtained at thermal neutrality in hatchlings of various species of gulls (Table 2) marks the last few days before hatching as a time of conspicuous intensification of metabolism in these semi-precocial birds. The relatively low level of embryonic mass-specific metabolism evident late in incubation appears to be a broadly shared characteristic of precocial and semi-precocial birds. Whittow (1984) notes that oxygen uptake by eggs of seven tropical sea birds just before pipping represents 32-41% of the rates of oxygen consumption noted in hatchlings. Low metabolic levels are also apparent in Hoyt and Rahn's (1980) analysis of energetic relations of avian eggs (but see comment in "Metabolism and Developmental Pattern" on complications introduced by interspecific variation in pipping patterns). The proportionality constant for eq. (1), which they obtained for the relation of metabolic rate just before internal pipping to initial egg mass, is only about a quarter that for one of the equations (Lasiewski and Dawson, 1967) describing metabolism-size relations for adult non-passerine birds. Hoyt and Rahn (1980) have approached this situation in an additional way by considering the body mass-specific metabolic rates of parent birds as a function of the mass of their eggs. The equation for this is:

$$\dot{V}O_2 = 70.6 m_e^{-0.36} \tag{3};$$

where $\dot{V}O_2$ is adult metabolic rate in ml O_2 (g.day)$^{-1}$ and m_e is as in eq. (1) and (2). Rates predicted from eq. (3) approximate the

Table 2. Metabolic Levels of Late Embryonic and Hatchling Gulls
(Larus)[a]

Body Mass (g)	Metabolic Level[b] (ml $O_2 \cdot g^{-1} \cdot h^{-1}$)	Metabolic Level as Percentage of Predicted Level[c]	Reference
		Herring Gull (L. argentatus) Embryo	
50.1[d]	0.68	44	Rahn et al. (1974)
53.6	0.79	52	Drent (1967)
		Herring Gull Hatchling	
57.4	1.38 (day)	92	Drent (1967)
59.1	1.06 (night)	71	Drent (1967)
		Glaucous-winged Gull (L. glaucescens)	
60.3	1.22	82	Drent (1967)
		Black-headed Gull (L. ridibundus)	
26.8	2.27	123	Palokangas and Hissa (1971)
		Laughing Gull (L. atricilla)	
28.4	1.99	109	Dawson et al. (1972)
		Ring-billed Gull (L. delawarensis)	
34.6	1.62	94	Dawson et al. (1976)
		Western Gull (L. occidentalis livens)	
65.4	1.35	93	Dawson and Bennett (1980)
		Western Gull (L. o. wymani)	
58.0	1.48	99	Dawson and Bennett (1981)

[a] Data on gulls other than L. argentatus pertain to hatchlings.
[b] The metabolic values presented represent rates for hatchlings resting in their zone of thermal neutrality or for late embryos at incubation temperature just before internal pipping.
[c] Predicted values refer in each case to the basal metabolic rate of an adult nonpasserine of similar body mass to that of the embryo or hatchling. Predictions were made using the appropriate equation from Lasiewski and Dawson (1967).
[d] The body masses for embryos of L. argentatus were estimated to comprise 57% of egg mass (Rahn et al., 1974).

pre-internal pipping mass-specific rates for embryos developing in eggs of comparable size, even though the adults, in fact, have body masses that are 5 and nearly 10 times those of their eggs and late embryos, respectively. Mass-specific metabolism scales interspec-ifically with the -0.28 power of body mass in adult non-passerine birds (Lasiewski and Dawson, 1967). Consequently, size ratios of 5 or 10 between adult birds would involve mass-specific rates for the smaller individuals that are 1.6 and 1.9 times, respectively, those for the larger ones. This comparison, though based on inter- rather than intra-specific observations on the size dependence of metabol-ism, provides a further suggestion that the thermogenic capacities of embryos are sufficiently small, even in the later stages of develop-ment, so that these animals would have only limited abilities to withstand a significant cold challenge; parental attentiveness remains critical for thermostasis.

As noted above, the data on gulls summarized in Table 2 suggest that a substantial intensification of metabolism occurs during the interval between internal pipping and the hours just after hatching in these semi-precocial birds. This also may be the case for ducks, judging by Dawson et al.'s (1976: Table 2) analysis of Koskimies and Lahti's (1964) data on metabolic rates of hatchlings. Part of this intensification may involve nothing more than the developing chicks' gaining access first to the air cell of the egg and then, with external pipping, to the open air. However, it appears to reflect some physiological changes as well, if observations on the domestic fowl are any guide. We shall return to these following a consider-ation of the capacities of embryonic birds for acutely increasing their heat production in the interest of cold defense.

Evidence that embryonic birds can increase heat production in response to a cold challenge comes primarily from studies of the domestic fowl. Freeman (1964) has shown that cooling the egg of this bird three days before hatching produces a progressive decline in the oxygen consumption of the embryo. However, a similar treatment on the succeeding day produces a transient rise in oxygen uptake. Ther-mogenic capacity improves further over the remainder of incubation and chicks at the moment of hatching can sustain the increased level of metabolism elicited by cold. The transient increase in metabolism occurring in embryos during the initial phase of cold exposure is paralleled by an increase in ventilation, internal pipping having occurred. This ventilatory increase is achieved through a greater amplitude of breathing that more than compensates for an associated decline in respiratory frequency (Dawes, 1981). A similar pattern has been observed in Japanese Quail (Coturnix coturnix japonica) in the period just before hatching (Nair and Dawes, 1980). Unfortunately, observations on metabolic responses to cooling by advanced embryos of sea birds appear unavailable. It is likely that responses similar to those described for embryonic domestic fowl exist in gulls and more precocial species. This prediction is based on field observations of

pipped eggs of the Western Gull that had not been incubated for
approximately 45 min (Bartholomew and Dawson, 1952). The chicks in
these eggs were approximately 5°-7°C warmer than their surroundings
(32.3° and 33.8°C vs 27°C). It is also consistent with laboratory
observations of hatchling gulls at ambient temperatures of 20°-25°C,
which show that these chicks can establish and sustain metabolic
rates that are approximately twice their level of metabolism in the
zone of thermal neutrality (see Dawson et al., 1976, and Dawson and
Bennett, 1980, for details). It would be of interest to determine the
metabolic responses to cooling of embryos of gulls and other sea
birds in the days immediately preceding hatching.

The changes in metabolic capacities of precocial and semi-precocial birds in the interval between the initiation of pipping (where
this can be precisely defined) and the first few hours after hatching
undoubtedly have a complex basis. As noted above, establishment of
pulmonary ventilation and the direct access to atmospheric oxygen
that external pipping provides probably are involved (see Whittow,
1984). Certainly a plateau exists in the metabolic rate of embryonic
domestic fowl between the 17th day of incubation and the onset of
pipping, which Freeman and Vince (1974) postulate is a reflection of
the maximum theoretical flux of oxygen and carbon dioxide through the
egg shell. The ability to augment muscular thermogenesis in the face
of cold challenges should also be beneficially affected by the emergence of hatchlings from their constrained circumstances in the egg.
However, more fundamental factors are probably also involved in the
changes in metabolic capacity. For example, maturation of the mechanisms controlling thermoregulation probably contributes. Endocrin-
ological changes also appear to be involved, for Decuypere et al.
(1978) report that serum levels of the thyroid hormones T_3 and T_4
undergo substantial increases in developing domestic fowl commencing
on the 19th and 17th days of incubation, respectively. These levels
reach maxima at external pipping on the 21st day of incubation, when
heat production undergoes a sharp rise from the plateau evident over
the preceding couple of days. The serum concentrations of T_3 and T_4
decline between the onset of pipping and hatching approximately 24 h
later, but are still more than twice their respective levels on the
17th day of incubation. These observations are of interest in view
of evidence that thyroid hormones are calorigenic in neonatal domestic fowl (Freeman, 1970, 1971). Perhaps development of non-shivering
thermogenesis contributes to the ability of late embryos and hatchlings of precocial and semi-precocial birds to increase heat production in cool surroundings. This process is poorly characterized in
birds and evaluation of observations pertaining to it is beyond the
scope of this review. However, a couple of observations concerning
young sea birds should be mentioned. First, the augmentation of
metabolic rate occurring in young Black-headed Gulls (Larus
ridibundus) with exposure to cold principally involves an increment
in muscular activity (Palokangas and Hissa, 1971). Second, the claim
of the existence of non-shivering thermogenesis in young South Polar

Skuas is based on the impairment of their temperature control produced by administration of the beta-receptor blockading agent propranolol (Murrish and Guard, 1973). Freeman (1977) suggests that such impairment is linked with the side effects this agent produces in young birds, rather than with a specific curtailment of any non-shivering thermogenesis.

Of the remaining circumstances that could account for the increased metabolic capacities appearing in late embryos and hatchlings of precocial and semi-precocial birds, changes in enzyme profiles of skeletal muscle appear potentially prominent. Barnes and Hasson (1983) trace the rise in cytochrome oxidase and succinoxidae activity in leg and breast muscle of embryonic domestic fowl from 11 to 19 days of age. The results obtained suggest at least a partial differentiation of these skeletal muscles over this interval. An example of the type of measurement it would be desirable to carry out on chicks and embryos of sea birds, with the wealth of experimental material their colonial nesting habits can provide, is supplied by Marsh and Wickler's (1982) study of Bank Swallows (Riparia riparia). During the first 65% (up to 12g) of their growth, the altricial nestlings of this species do not increase their metabolism in response to cold. Beyond this size, cold-stimulated peak metabolic rate rises substantially. This appears to reflect in part changes in the mass and aerobic capacity (indicated by citrate synthase activity) of skeletal muscle, which increase steadily over posthatching development. However, the relatively abrupt rise in peak metabolic rate after the 65th percentile of growth is attained appears most intimately linked with an abrupt increase in myofibrillar ATPase activity in the pectoralis muscles. The availability of sensitive radio-immunoassays for hormones and test reagents for determinations of enzyme activity should make it quite feasible to analyze the chemical correlates of the changes in thermoregulatory capacity occurring in the advanced embryos and hatchlings of sea birds and other wild species.

CONCLUDING STATEMENT

In this presentation, we have considered three facets of the thermal responses of embryonic birds to temperature. Two of these, thermal tolerances and sensitivity of metabolic responses to temperature, pertain to embryos in the ectothermic phase of development. The third emphasizes changes in metabolic level and capacity that are involved in the establishment of thermoregulation. Most of the data assembled pertain to precocial and semi-precocial birds in which a portion of these changes tends to occur in the days just before hatching. In keeping with the theme of this symposium an effort has been made to feature findings on sea birds, supplemented by data for other species, notably the domestic fowl. Information on thermal responses of avian embryos is far from complete and particularly rich opportunities appear to exist for investigating these in sea birds. The colonial nesting habits of many of these provide particularly convenient access to a wealth of experimental material.

In this presentation, the possibility of special adaptations to adverse climatic conditions in the temperature tolerances and thermal sensitivity of metabolic processes of sea birds has been examined. There are indications of such adjustments, but much more information is required before these possibilities can be evaluated conclusively. With respect to the metabolic development of advanced embryos, it is clear that events occur prior to hatching that directly affect the thermoregulatory capacities of these animals. Again, a great deal more information is needed to characterize these events fully. Although understanding of all these topics is at a very preliminary stage for sea birds as well as for other species, one cannot but be heartened by the significant progress that has occurred over the past decade.

ACKNOWLEDGEMENTS

Work of W. R. Dawson and associates and preparation of this paper were supported in part by grants from the National Science Foundation (most recently, BSR 80-21389).

REFERENCES

Afton, A. D., 1979, Incubation temperatures of the Northern Shoveler, Can. J. Zool., 57:1052.
Barnes, W. S., and Hasson, S. M., 1983, Respiratory capacity of chick red and white muscle, Comp. Biochem. Physiol., 75A: 491.
Barrett, R. T., 1980, Temperature of Kittiwake Rissa tridactyla eggs and nests during incubation, Ornis Scand., 11:50.
Bartholomew, G. A. Jr., and Dawson, W. R., 1952, Body temperatures in nestling Western Gulls, Condor, 54:58.
Bartholomew, G. A., and Dawson, W. R., 1979, Thermoregulatory behavior during incubation in Heermann's Gulls, Physiol. Zool. 52:422.
Beck, J. R., and Brown, D. W., 1972, The biology of Wilson's Storm Petrel, Oceanites oceanicus (Kuhl), at Signy Island, South Orkney Islands, Br. Antarctic Survey Sci. Rep, 6:911.
Bennett, A. F., and Dawson, W. R., 1979, Physiological responses of embryonic Heermann's Gulls to temperature, Physiol. Zool., 52:413.
Bennett, A. F., Dawson, W. R., and Putnam, R. W., 1981, Thermal environment and tolerance of embryonic Western Gulls, Physiol. Zool. 54:146.

Boersma, P. D., and Wheelwright, N. T., 1979, Egg neglect in the Procellariiformes: reproductive adaptations in the Fork-tailed Storm Petrel, Condor, 81:57.

Boersma, P. D., Wheelwright, N. T., Nerini, M. K., and Wheelwright, E. S., 1980, The breeding biology of the Fork-tailed Storm-petrel (Oceanodroma furcata), Auk, 97:268.

Cesana, G., 1911-12, Interno al coefficiente termico del cuore embrionale di pollo nei primi giorni dello sviluppo, Arch. Fisiol., 10:193.

Cohn, A. E., 1928, Physiology ontogeny. A. Chicken embryos. XIII. The temperature characteristic for the contraction rate of the whole heart, J. Gen. Physiol., 11:369.

Dawes, C. M., 1981, The effects of cooling the egg on the respiratory movements of the hatching fowl, Gallus g. domesticus), with a note on vocalization, Comp. Biochem. Physiol., 68A:399.

Dawson, W. R., 1967, Interspecific variation in physiological responses of lizards to temperature, in: "Lizard Ecology," W. W. Milstead, ed., University of Missouri Press, Columbia.

Dawson, W. R., and Bennett, A. F.,, 1980, Metabolism and thermoregulation in hatchling Western Gulls, Condor, 82:103.

Dawson, W. R., and Bennett, A. F., 1981, Field and laboratory studies of the thermal relations of hatchling Western Gulls, Physiol. Zool., 54:155.

Dawson, W. R., Bennett, A. F., and Hudson, J. W., 1976, Metabolism and thermoregulation in hatchling Ring-billed Gulls, Condor, 78:49.

Dawson, W. R., and Hudson, J. W., 1970, Birds, in: "Comparative Physiology of Thermoregulation," G. C. Whittow, ed., Academic Press, New York.

Dawson, W. R., Hudson, J. W., and Hill, R. W., 1972, Temperature regulation in newly hatched Laughing Gulls (Larus atricilla), Condor, 74:177.

Decuypere, E., Nouwen, E. J., Kühn, E. R., Geers, R., and Michels, H., 1978, Heat production and serum concentration of thyroid hormones in the chick embryo at the end of the incubation period, I.R.C.S. Med. Sci.: Dev. Biol. Med.; Endocr. Syst.; Met. Nutr.; Physiol; Rep. Obst. Gyn., 6:336.

Drent, R., 1967, "Functional Aspects of Incubation in the Herring Gull," E. J. Brill, Leiden.

Drent, R., 1970, Functional aspects of incubation in the Herring Gull, Behav. Suppl., 17:1.

Drent, R., 1972, Adaptive aspects of the physiology of incubation, in: "Proceedings of the XVth International Ornithological Congress," K. H. Voous, ed., E. J. Brill, Leiden.

Drent, R., 1973, The natural history of incubation, in: "Breeding Biology of Birds," D. S. Farner, ed., National Academy of Sciences, Washington, D. C.

Drent, R., 1975, Incubation, in: "Avian Biology, Vol. 5," D. S. Farner and J. R. King, eds., Academic Press, New York.

Freeman, B. M., 1964, The emergence of the homeothermic metabolic response in the fowl (Gallus domesticus), Comp. Biochem. Physiol., 13:413.

Freeman, B. M., 1970, Thermoregulatory mechanisms of the neonate fowl, Comp. Biochem. Physiol., 33:219.

Freeman, B. M., 1971, Non-shivering thermogenesis in birds, in: "Nonshivering Thermogenesis," L. Jansky, ed., Academia, Prague.

Freeman, B. M., 1977, Lipolysis and its significance in the response to cold of the neonatal fowl, Gallus domesticus, J. Therm. Biol., 2:145.

Freeman, B. M., and Vince, M. A., 1974, "Development of the Avian Embryo," Wiley and Sons, New York.

Grant, G. S., 1982, Avian incubation, egg temperature, nest humidity and behavioral thermoregulation in a hot environment, Ornithol. Monogr., 30.

Grieff, D., 1952, The metabolic interactions of intracellular parasites and embryonate eggs, Ann. New York Acad. Sci., 55:254.

Hasselbalch, K. A., 1900, Ueber den respiratorischen Stoffwechsel des Huhnerembryos, Skand. Arch. Physiol., 10:353.

Howell, T. R., Araya, B., and Millie, W. R., 1974, Breeding biology of the Gray Gull, Larus modestus, Univ. Calif. Publ. Zool., 104:1.

Howell, T. R., 1979, Breeding biology of the Egyptian Plover, Pluvianus egyptius (Aves: Glareolidae), Univ. Calif. Publ. Zool., 113:1.

Hoyt, D. F., and Rahn, H., 1980, Respiration of avian embryos--a comparative analysis, Resp. Physiol., 39:255.

Hoyt, D. F., Vleck, D., and Vleck, C. M., 1978, Metabolism of avian embryos: ontogeny and temperature effects in the Ostrich, Condor, 80:265.

Inukai, T., 1925, Über den Einfluss der Temperature auf die Pulsationzahl bei den Amphibienlarven und Vogelembryonen, Jap. J. Zool., 1:67.

Koskimies, J., and Lahti, L., 1964, Cold-hardiness of the newly hatched young in relation to ecology and distribution in ten species of European ducks, Auk, 81:281.

Kühn, E. R., Decuypere, E., Colen, L. M., and Michels, H., 1982, Posthatch growth and development of a circadian rhythm for thyroid hormones in chicks incubated at different temperatures, Poultry Sci., 61:540.

Lasiewski, R. C., and Dawson, W. R., 1967, A re-examination of the relation between standard metabolic rate and body weight in birds, Condor, 69:13.

Lind, H., 1961, "Studies on the Behaviour of the Black-tailed Godwit (Limosa limosa) (L.), Meddelelse fra Naturfredningsradets reservatudvalg Nr. 66, Munksgaard, Copenhagen.

Lundy, 1969, A review of the effects of temperature, humidity,, turning and gaseous environment in the incubator on the hatchability of the hen's egg, in: "The Fertility and Hatchability of the Hen's Egg," T. C. Carter and B. M. Freeman, eds., Oliver and Boyd, Edinburgh.

Marsh, R. L., and Wickler, S. J., 1982, The role of muscle development in the transition to endothermy in nestling Bank Swallows Riparia riparia, J. Comp. Physiol., 149:99.

Matthews, G. V. T., 1954, Some aspects of incubation in the Manx Shearwater Procellaria puffinus, with particular reference to chilling resistance in the embryo, Ibis, 96:432.

Mes, R., Schuckard, R., and Wattel, J., 1978, Visdieven Sterna hirundo zocken koelte, Limosa, 51:64.

Murrish, D. E., and Guard, C. L., 1973, Sympathetic control of nonshivering thermogenesis in South Polar Skua chicks, Antarctic J. U. S., 8:197.

Nair, G., and Dawes, C. M., 1980, The effects of cooling the egg on the respiratory movements of the hatching quail (Coturnix c. japonica), Comp. Biochem. Physiol.,67A:587.

Paff, G. H., Boucek, R. J., Nieman, R. E., and Deichmann, W. B., 1963 The embryonic heart subjected to radar, Anat. Rec., 147:379.

Palokangas, R., and Hissa, R., 1971, Thermoregulation in young Black-headed Gull (Larus ridibundus), Comp. Biochem. Physiol. 38A:743.

Parpart, E. R., and Glaser, O., 1930, Temperature and heart rate in chick embryos, J. Exp. Biol., 7:143.

Patterson, I. J., 1965, Timing and spacing of broods in the Black-headed Gull Larus ridibundus, Ibis, 107:433.

Pettit, T. N., Grant, G. S., Whittow, G. C., Rahn, H., and Paganelli, C. V., 1982, Respiratory gas exchange and growth of Bonin Petrel embryos, Physiol. Zool., 55:162.

Purdue, J. R. 1976, Thermal environment of the nest and related parental behavior in Snowy Plovers, Charadrius alexandrinus, Condor, 78:180.

Rahn, H., Paganelli, C. V., and Ar, A., 1974, The avian egg: air cell gas tension, metabolism and incubation time, Respir. Physiol., 22:297.

Roberts, B., 1940, The life cycle of Wilson's Petrel Oceanites oceanicus (Kuhl), British Graham Land Expedition:1934-37, Sci. Rep.,1:141.

Romanoff, A. L., 1960, "The Avian Embryo," Macmillan, New York.

Romanoff, A. L., 1972, "Pathogenesis of the Avian Embryo," Wiley-Interscience, New York.

Romanoff, A. L., and Sochen, M., 1936, Thermal effect on the rate and duration of the embryonic heart beat of Gallus domesticus, Anat. Rec., 65:59.

Russell, S. M., 1969, Regulation of egg temperatures by incubating White-winged Doves, in: "Physiological Systems in Semiarid Environments," C. C. Hoff and M. L. Riedesel, eds., University of New Mexico Press, Albuquerque.

Scholander, P. F., Flagg, W., Walters, V., and Irving, L., 1953, Climatic adaptation in arctic and tropical poikilotherms, Physiol. Zool., 26:67.

Skutch, A. F., 1957, The incubation period of birds, Ibis, 99:69.

Skutch, A. F., 1962, The constancy of incubation, Wilson Bull., 74:115.

Vleck, C. M., Hoyt, D. F., and Vleck, D., 1979, Metabolism of avian embryos: patterns in altricial and precocial birds, Physiol. Zool., 52:363.

Vleck, C. M., and Kenagy, G. J., 1980, Embryonic metabolism of the Fork-tailed Storm Petrel: physiological patterns during prolonged and interrupted incubation, Physiol. Zool., 53:32.

Vleck, C. M., Vleck, D., and Hoyt, D. F., 1980a, Patterns of metabolism and growth in avian embryos, Am. Zool., 20:405.

Vleck, D. , and Vleck, C. M., and Hoyt, D. F., 1980b, Metabolism of avian embryos: ontogeny of oxygen consumption in the Rhea and Emu, Physiol. Zool, 53:125.

Vleck, D., Vleck, C. M., and Seymour, R. S., 1984, Energetics of embryonic development in the megapode birds, Mallee Fowl (Leipoa ocellata) and Brush Turkey (Alectura lathami), Physiol. Zool., in press.

Walsberg, G. E., and Voss-Roberts, K. A., 1983, Incubation in desert-nesting doves: mechanisms for egg cooling, Physiol. Zool., 56:85.

White, F. N., and Kenney, J. L., 1974, Avian incubation, Science, 186:107.

Wheelwright, N. T., and Boersma, P. D., 1979, Egg chilling and the thermal environment of the Fork-tailed Storm Petrel (Oceanodroma furcata) nest. Physiol. Zool., 52:231.

Whittow, G. C., 1984, Physiological ecology of incubation in tropical seabirds, Stud. Avian Biol., No. 8, in press.

Wilbur, H. M., 1969, The breeding biology of Leach's Petrel, Oceanodroma leucorhoa, Auk, 86:433.

Williams, J. B., and Ricklefs, R. E., 1984, Egg temperature and embryo metabolism in some high-latitude procellariiform birds, Physiol. Zool., in press.

ENERGETICS OF BREEDING DARK-RUMPED PETRELS

T. R. Simons[1] and G. C. Whittow[2]

[1]Wildlife Science Group
College of Forest Resources
University of Washington
Seattle, Washington
[2]Department of Physiology
John A. Burns School of Medicine and
P.B.R.C. Kewalo Marine Laboratory
University of Hawaii
Honolulu, Hawaii

INTRODUCTION

The Dark-rumped Petrel (Pterodroma phaeopygia sandwichensis) is an endangered gadfly petrel that nests in the Hawaiian Islands and ranges throughout the central Pacific. The species was once common in Hawaii with large colonies on all of the main islands, but its numbers have recently been reduced to several small relict populations. Over 85% of the estimated 450 breeding pairs known today nest in and around Haleakala National Park on the island of Maui, the site of a three-year study begun in 1979 (Simons 1983). Like most Procellariiformes, the Dark-rumped Petrel exploits what is generally assumed to be a widely dispersed and unpredictable food resource (Lack, 1967; 1968). This food resource is thought to place important energetic constraints on these birds, and we wanted to examine how those constraints might have shaped the petrel's breeding biology. In addition these birds breed at an elevation of almost 3000 m in one of the highest seabird nesting colonies in the world. We have described the adaptations of the Dark-rumped Petrel's egg to high altitude nesting elsewhere (Whittow et al., 1983). In this paper we shall examine several aspects of the energetics of reproduction. Apart from a study of the much smaller Leach's Storm-Petrel (Oceanodroma leucorhoa)

(Ricklefs et al., 1980a), the energetics of reproduction in these birds has received little attention.

MATERIALS AND METHODS

The metabolic rates of nestling and adult Dark-rumped Petrels were estimated by measuring their rates of oxygen uptake. A portable system using an air-tight chamber and a manometer was used, and most measurements were made at the nest sites. The device consisted of a chamber containing soda lime (to absorb carbon dioxide) a calibrated manometer, and a source of oxygen. Measurements were made on 72 occasions on seven nestlings and five adult birds. A minimum of five measurements were made on each occasion, and these values were averaged to obtain a final estimate. A small chamber of approximately 4 liters in size was used for small nestlings and a larger 10 liter chamber was used for larger nestlings and adults. The system was calibrated in the laboratory by measuring the oxygen uptake of a Rock Dove (<u>Columba livia</u>), and the estimates obtained were within 10% of those reported by other authors (Dawson and Hudson, 1970). It was assumed that the air within the chamber was saturated with water vapor due to the respiration of the bird, and based on this assumption, measurements were converted to standard temperature and pressure dry. Birds were weighed with 500 or 1000 g Pesola scales to the nearest 1.0 g. Nestling heart rates were measured with a small ECG transmitter (E M Telemetry Transmitter #FM-1100-E1). Proventricular temperatures of birds were measured with a Wescor model TH-65 digital thermocouple thermometer calibrated against a laboratory mercury thermometer in a water bath. Burrow attendance patterns were monitored using specially designed event recorders at 10 - 12 nests each season (Simons, 1981a; 1981b; 1983). We collected six Dark-rumped Petrel food samples for caloric analysis. All of the samples were obtained from adult birds returning to feed their chicks. The samples were frozen soon after weighing, and at a later date, they were thawed and dried to a constant mass in an oven at 45°C. The water content of the samples was determined by subtracting the dry mass fraction from the initial weight. Dried samples were homogenized by blending with a mortar and pestle, and the caloric content of approximately 10 mg samples was determined using a Phillipson micro-bomb calorimeter with a benzoic acid standard. Duplicates of each sample were assayed. The ash content was determined by burning 6 aliquots of each dried sample in a muffle furnace at 600°C for 6 hours. The caloric content of stomach oil was determined by adding a known amount of oil (approximately 3 mg) to 10 mg pellets of enriched baking flour with a predetermined caloric content (16.73 kJ/g dry, S.D. 0.02). The total caloric content of the stomach oil and flour pellet was measured with the

micro-bomb calorimeter, and the caloric content of the oil was
determined by subtracting the caloric content of the flour.

RESULTS AND DISCUSSION

Metabolic Rate

The rates of oxygen uptake have been summarized by age class
and are presented as total oxygen uptake and mass specific rates
(Table 1). The high variance in the metabolic estimates for
nestlings is probably related to the variance in chick mass
(Simons, 1983). Nestlings of the same age can vary in their body
mass by a factor of three. Nestlings reach their highest total
energy expenditure near the middle of the nestling period, after
which energy expenditures decline until fledging. This reduction
reflects the sharp decline in body mass towards the end of the
nestling period which is characteristic of the species. The
mass-specific metabolic rate is highest in very young chicks and
then declines. This pattern is common in precocial or
semi-precocial birds (Ricklefs, 1974). The average basal
metabolic rate of adult birds is approximately twice the predicted
value based on the equations of Aschoff and Pohl (1970). This is
similar to the findings of Ricklefs et al. (1980a) for the Leach's
Storm Petrel.

Energy Conservation on the Nest

The energy demands of Dark-rumped Petrel nestlings and
incubating adults are probably somewhat less than indicated by the
oxygen uptake data. Behavioral observations (Simons, 1983)
indicate that chicks are inactive in the burrow, and spend
approximately 65% of their time resting quietly and 25% of their
time sleeping on the nest. This behavior has obvious energetic
benefits. Because chicks measured in the field never rested as
quietly as they did in their burrows, two chicks were monitored
regularly for 24 hour periods to obtain some indication of the
relationship between a chick's behavior and its energetic demands.
The chicks were each about three months old and weighed
approximately 550 g. They were placed in a large ventilated
chamber in a darkened room and their oxygen uptake was measured
every two to three hours. The oxygen consumption of these chicks
varied considerably depending on the bird's activity (Table 2).

The resting metabolic rates of these chicks were similar to
those obtained in the field for other nestlings of about the same
age (Table 1), but the level of oxygen consumption in sleeping
chicks was substantially reduced. These data suggest that
nestlings can reduce their metabolic rates by 20 to 30% below

Table 1. Oxygen consumption of individual Dark-rumped Petrel nestlings at ambient temperatures 5–25°C

	1–15	16–30	31–45	46–60	61–75	76–90	91–105	106+	Adult
					Age (days)				
Total									
n	4	13	8	6	10	11	11	4	5
x	419.1	576.8	1061.7	1021.4	1046.9	965.8	854.2	668.4	940.6
S.D.	53.5	306.7	201.7	133.3	183.1	193.6	204.6	128.3	170.8
Mass-specific[b]									
n	4	13	8	6	10	11	11	4	5
x	2.6	2.2	2.3	1.9	1.8	1.6	1.8	1.6	2.3
S.D.	0.79	0.82	0.59	0.47	0.43	0.25	0.44	0.13	0.43

[a]Units are ml O_2/hr.

[b]Units are ml O_2/g hr.

Table 2. Nestling behavior, oxygen uptake, and heart rate in the Dark-rumped Petrel.

| NESTLING | BEHAVIOR | OXYGEN UPTAKE[1] | | AVERAGE HEART RATE[2] |
		ml/g hr	ml/hr	(Beats/min.)
	Active	2.8	1540.0	251
A	Resting	1.8	990.0	201
	Sleeping	1.4	770.0	195
	Active	3.0	1677.0	249
B	Resting	1.8	1006.2	203
	Sleeping	1.2	670.8	171

[1]Nestling A age 99 days, Nestling B age 77 days

[2]Nestling A age 107 days, Nestling B age 105 days

their alert resting rate by sleeping. Walker et al. (1983) concluded that sleep in many birds lies on a metabolic continuum between activity and torpor. They found that sleeping doves reduced their metabolic output substantially and that fasting birds frequently entered the early stages of torpor during sleep. Dark-rumped Petrel burrows are cold, averaging 9.59 ± 2.39°C (Simons, 1983). Because chicks can fast for periods of up to two weeks (Simons, 1983) and spend much of their time sleeping in the burrow, it is probable that the energy budget calculated from field measurements may overestimate the acutal energy demands of a developing Dark-rumped Petrel nestling.

A second estimate of the relationship between a chick's behavior and its metabolic rate was made by monitoring the heart rates of two nestlings over a 24 hour period. The chicks were fitted with an ECG transmitter, returned to their burrows, and their heart rates were monitored regularly for at least 24 hours. Observations of the chick's activity were made in conjunction with the heart rate counts and the results (Table 2) show a relationship between behavior and metabolic rate similar to that derived from the measurement of oxygen uptake. Resting heart rates (f_h) were slightly higher than the predicted value of 192 beats per minute obtained from Calder's (1968) predictive equation

$$(f_h = 763 \times W^{-0.23})$$

based on bird mass (W).

A similar relationship probably holds for incubating adults. Direct observations indicate that adult petrels spend approximately 95% of their incubation shifts sleeping (Simons, 1983), and this behavior may permit the extremely long incubation shifts characteristic of the species. Grant and Whittow (1983) found that the rate of oxygen uptake in incubating adult Bonin Petrels (Pterodroma hypoleuca) was 18% less than the resting rate recorded in the laboratory. They also found that the respiratory quotient (RQ) of incubating Bonin Petrels was approximately 0.7 and they concluded from this that the birds were metabolizing fat. They reasoned that because the body mass change due to the exchange of carbon dioxide and oxygen is very small when fat is oxidized (Schmidt-Nielsen, 1974) and defecation is rare, the weight loss of incubating birds was due entirely to water loss. Birds do not drink in the fasting state, therefore their only source of water is metabolic water production. Because metabolic water production is known to be equal to 1.07 times the quantity of fat consumed (Schmidt-Nielsen, 1974), the amount of fat burned by an incubating bird equals metabolic water loss divided by 1.07 (that is, body mass loss divided by 1.07). Assuming that fat produces approximately 39.7 kJ/g (Petrusewicz and Macfayden, 1970), we can then estimate the metabolic rate of the incubating bird from its weight loss.

This is a useful method for estimating the metabolic rate of an incubating Procellariiform and it can be compared to estimates obtained by other methods. For example, it provides a comparison with the estimates of metabolic rate in the Dark-rumped Petrel made from measurements of oxgyen uptake. Dark-rumped Petrel incubation shifts can extend for up to 23 days, perhaps the longest of any Procellariid, and it is likely that they also metabolize fat over those intervals. Long incubation shifts are characteristic of Procellariiformes and longer shifts appear to be associated with poorer feeding conditions (Harris, 1969; Simons, 1981a). We weighed one adult bird over a twelve day incubation shift and it lost 85 g or an average of 7.08 g/day. This value is equivalent to an estimated loss of 1.7% of body weight per day which is slightly higher than the average daily loss of 1.2% reported for two species of albatross (Prince et al., 1981). Given the assumptions just discussed, this bird would have used 7.08/1.07 or 6.62 g of fat per day. This is equivalent to an energetic expenditure of 6.62 g/day x 39.7 kJ/g or about 263 kJ/day. Based on a conservative assumption that alert resting adult petrels use about 760 ml oxygen/hr we then find that they burn about .760 1/hr x 24 hr/day x 20.1 kJ/1 = 366.6 kJ/day. Thus, it appears that incubating Dark-rumped Petrels metabolize about 28% (263/366) less energy than they do in an alert resting state and that like nestling petrels they are capable of substantially reducing their energetic needs while they are on the nest.

Energy Budget

We calculated an energy budget for Dark-rumped Petrel nestlings following the methods outlined by Ricklefs et al. (1980a). We applied Ricklefs' values for the rate of accumulation of lipid and nonlipid dry matter in Leach's Petrel to the Dark-rumped Petrel. We assumed that these components accumulated, as a percentage of body weight, at approximately the same rates in both species, and used these estimates in conjunction with the data for oxygen consumption to estimate the total energy expenditure of growth and maintenance in a developing nestling. Clearly an extrapolation of this sort should be viewed as a first approximation requiring validation from other field studies, but it is the best estimate possible from the data currently available. Like Ricklefs et al. (1980a), we assumed that the energy requirement of in-burrow activity and the energy excreted in feces were negligible and ignored them in these calculations. The completed energy budget is shown in Table 3 and Fig. 1. The cost of biosynthesis is assumed to be about one-third the energy equivalent of tissue accumulation (Ricklefs, 1974). Therefore, total energy for growth is 1.33 times the energy equivalent of tissue accumulation. Maintenance energy is the energy equivalent of oxygen consumption minus the energy cost of biosynthesis and total energy is the sum of growth and maintenance. Energy demands increased during the early nestling phase, reaching a peak of about 600 kJ/day by day 30. They remained at about that level during development until the last third of the nestling period when they decreased to less than 500 kJ/day. The energy requirement for growth decreased from almost 50% of total energy demands early in the nestling period to less than 25% in older nestlings. It is likely that a substantial percentage of the energy expenditure of young nestlings is devoted to thermoregulation given their small size, limited insulation, and low burrow temperatures. Based on these calculations, the total energy requirement of a developing Dark-rumped Petrel nestling is estimated to be approximately 54,000 kJ.

Body Temperature and Thermoregulation

Adult Dark-rumped Petrels have an average core temperature of 38.6°C (Table 4). This is very close to the average of 38.78°C reported by Warham (1971) for 31 species of petrels. Chicks may be capable of thermoregulation from shortly after hatching, but their temperatures are quite variable for the first two weeks (Table 4). They do not appear to require much brooding by their parents, and although some adults remain in the burrow for up to six days after their chicks hatch, most depart within two days. Ricklefs (1983) has pointed out that the energetic demands on adult pelagic seabirds are greatest during the chick brooding period and this may favor minimizing the brooding period in some

Table 3. An energy budget for Dark-rumped Petrel nestlings

Age interval (days)	(1)	(2)	(3)	(4)	(5)	(6)	(7)	(8)	(9)	(10)
0-15	419.1	1.90	1.10	72.2	22.0	94.2	202.2	125.3	171.1	296.4
16-30	576.8	1.90	1.00	72.2	20.0	92.2	278.2	122.6	247.8	370.4
31-45	1061.7	1.90	1.20	72.2	24.0	96.2	512.2	127.9	480.5	608.4
46-60	1021.4	1.90	0.93	72.2	18.6	90.8	492.7	120.8	462.7	583.5
61-75	1046.9	1.90	0.67	72.2	13.4	85.6	505.0	113.8	476.8	590.6
76-90	965.8	1.90	0.40	72.2	8.0	80.2	461.6	106.7	435.1	541.8
91-105	854.2	1.90	0.38	72.2	7.6	79.8	412.1	106.1	385.8	491.9
105-110	668.4	1.90	0.36	72.2	7.2	79.4	322.4	105.6	296.2	401.8

(1) Oxygen consumption (cc O_2/hr) from Table 1.
(2) Rate of accumulation of lipid (g/day) estimated from Ricklefs et al. (1980a).
(3) Rate of accumulation of nonlipid dry matter (g/day) estimated from Ricklefs et al. (1980a).
(4) Energy equivalent of lipid accumulation (kJ/day) = (2) x 38 kJ/g lipid.
(5) Energy equivalent of accumulation of nonlipid dry matter (kJ/day) = (3) x 20 kJ/g.
(6) Energy equivalent of tissue accumulation (kJ/day) = (4) + (5).
(7) Energy equivalent of oxygen consumption (kJ/day) = ((1) x 24 hr/day x 20.1 kJ/1 02) / 1000 ml/1.
(8) Total energy requirement for growth (kJ/day), assuming production efficiency of 75% = (6) x 1.33.
(9) Total energy expenditure for maintenance (kJ/day) = (7) - 0.33 x (6).
(10) Total energy expenditure for growth and maintenance (kJ/day) = (8) + (9).

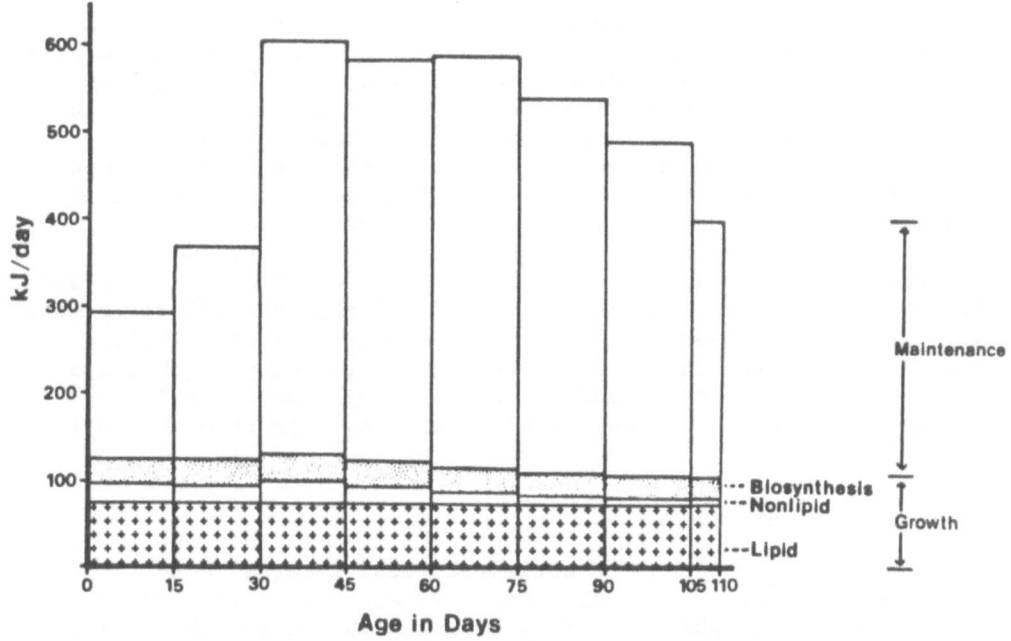

Fig. 1. Energy budget of Dark-rumped Petrel nestlings. Rates
were estimated from the accumulation of lipid and
nonlipid matter (Ricklefs et al., 1980a) and oxygen
consumption.

Table 4. Body temperatures of Dark-rumped Petrels in degrees
centigrade.

AGE	n	AVERAGE TEMPERATURE	S.D.	RANGE
ADULT	9	38.6	1.1	36.8 – 40.5
LESS THAN TWO WEEKS	13	35.2	2.1	31.7 – 38.4
MORE THAN TWO WEEKS	14	37.5	0.96	36.4 – 39.7

species. Adults that remain with their chicks often do not brood them directly, but rest next to them on the nest and feed them at regular intervals.

It appears that the ability of an adult to feed its chick determines the length of the brooding period and adults may depart as soon as their food reserves are depleted. Chicks that were only brooded for short periods developed normally as long as they were fed regularly. Well fed chicks maintained near adult internal temperatures shortly after hatching, but changing food reserves caused their temperatures to fluctuate over a range of several degrees. Chicks with low body temperatures were characteristically light in weight, often shivering, and in extreme cases showed signs of hypothermia such as closed eyes and lethargic semi-conscious behavior. As in other Procellariiformes, (Boersma et al., 1980; Ricklefs et al., 1980a; Simons, 1981a), Dark-rumped Petrel chicks were capable of remaining in this state of partial torpor for periods of several days, occasionally longer, and they recovered quickly when they were brooded and fed by their parents.

Nestling Feeding Rates

Table 5 summarizes the estimated feeding rates in the Dark-rumped Petrel. The table was derived from several sources including, the weight loss of fasting chicks, weight changes of nestlings, and data from event recorder monitored burrows. Because the only accurate way to determine food load size, i.e. weighing chicks before and after each feeding, was not possible, we used average chick weight loss during periods of fasting, determined from 112 measurements of 31 nestlings, and changes in chick weight to estimate the size of the food loads delivered by adult birds. The weight loss of chicks between feedings is variable, and depends on the initial size and age of the chick, the size of the most recent feed, and the duration of the fast. The rate of weight loss is high following a large feed and declines throughout the fasting period. The following values were used as average rates of weight loss: 15 g/day in chicks age 0 - 15, 20 g/day in chicks age 16 - 60, 25 g/day in chicks age 61 - 90, and 13 g/day in chicks older than 90 days. By combining these estimates with event recorder information on the rate of burrow visitation by adult birds and chick weights, we were able to es- timate the feeding rates of nestlings. We did this for 12 chicks that were monitored over the three years of the study. An example of the calculations for a single chick is shown in Table 6. The chick feeding patterns of adult petrels were irregular and they presumably reflected the variability of the food supply. Nestlings were capable of consuming large quantities of food when it was available and they frequently fasted for extended periods between feedings (Simons, 1983). Feeding rates were maximum

168

Table 5. Estimated feeding rates in Dark-rumped Petrel nestlings [1].

Chick Age (days)	n [2]	Est. food delivered to chick (g)	% Total food	Avg. number chick feeding visits	Est. Avg. Food load size (g)
0 - 30	9	791.7 (53.0)	32	12.5 (1.8)	63.3
31 - 60	10	901.5 (173.77)	36	14.3 (4.03)	62.9
61 - 90	9	682.6 (79.70)	27	12.3 (3.28)	55.4
91 - 120	9	114.7 (67.5)	5	3.2 (1.9)	35.6
0 - Fledgling (Total)	7	2501.7 (94.6)	100	44.5 (7.6)	56.2

[1] Values in parentheses are one standard deviation.

[2] Event recorder-monitored chicks.

Table 6. Chick weight changes, burrow visitation, and estimated chick feeding rates at one Dark-rumped Petrel nest.

Nestling Age (days)	Date	Nestling weight (g)	Interval length (days)	# Feeding visits	Nestling weight gain/ loss (g)	Estimated metabolic weight loss (g)	Estimated total food delivered in interval (g)
0	6/30	60					
2	7/2	94	2	1	34	30	64
3	7/3	85	1	0	-9	15	-
4	7/4	80	1	0	-5	15	-
7	7/7	148	3	1	68	45	113
9	7/9	234	2	1	86	30	116
11	7/11	286	2	1	52	30	82
15	7/15	323	4	2	37	60	97
17	7/17	359	2	1	36	40	76
20	7/20	373	3	1	14	60	74
23	7/23	395	3	1	22	60	82
24	7/24	385	1	0	-10	20	-
25	7/25	386	1	1	1	20	21
26	7/26	375	1	0	-11	20	-
27	7/27	405	1	1	30	20	50
28	7/28	379	1	0	-26	20	-
29	7/29	363	1	0	-16	20	-
30	7/30	352	1	0	-11	20	-
33	8/2	413	3	1	61	60	121
37	8/6	572	4	3	159	80	239
44	8/13	538	7	2	-34	140	106
49	8/18	470	5	0	-68	100	-
52	8/21	654	3	2	184	60	244

58	8/27	687	6	2	33	120	153
61	8/30	630	3	0	−57	75	–
64	9/2	580	3	0	−50	75	–
68	9/6	800	4	4	220	100	320
71	9/9	712	3	0	−88	75	–
74	9/12	662	3	0	−50	75	–
77	9/15	792	3	1	130	75	205
80	9/18	846	3	2	54	75	129
83	9/21	820	3	1	−26	75	49
86	9/24	798	3	1	−22	75	53
91	9/29	700	5	0	−98	85	–
98	10/6	662	7	1	−38	91	53
103	10/11	594	5	0	−68	65	–
105	10/13	562	2	0	−32	26	–
109	10/17	537	4	0	−25	52	–
112	10/20	502	3	0	−35	39	–
113	10/21	490	1	0	−12	13	–
114	10/22	482	1	0	−8	13	–
117	10/25	452	3	0	−30	39	–
118	10/26	442	1	0	−10	13	–
119	10/27	432	1	0	−10	13	–
120	10/28	422	1	0	−10	13	–
121	10/29	Fledged					

Total 2447 g

between 30 and 60 days which coincides with the period of maximum chick growth, the inflection point of the logistic chick growth curve (Ricklefs, 1967; 1968; Simons, 1983). Maximum estimated food load sizes ranged from less than 10 g to over 110 g. The latter figure is about 26% of mean adult weight and it is probably close to the maximum adult seabirds can carry (Ashmole, 1971). This is especially likely in the Dark-rumped Petrel in light of the altitude and distance from the sea of their nesting colonies. Nestlings were fed almost 70% of their total food during the first half of the nesting period and about 95% of their total by the time they were 90 days old. We estimated that chicks were fed a total of approximately 2500 g of raw food during their development period on the nest. Chicks on average were only fed a few times during their last month on the nest which suggests that, like the Leach's Storm Petrel (Ricklefs et al., 1980a), they store and metabolize a large amount of fat prior to fledging. However, there is much variability in the fledging pattern because some chicks were deserted entirely for up to three weeks before fledging, whereas others were visited and fed immediately prior to fledging (Simons, 1983).

Caloric Analysis of Nestling Food

A caloric analysis of Dark-rumped Petrel food samples is summarized in Table 7. Combining this information with the physiological and behavioral data discussed previously provides several additional insights into the feeding ecology of these birds. We estimate that adult petrels need to deliver about 54,000 kJ of energy to raise their chicks. The fish and squid that petrels feed on provide between 4.5 and 5.0 kJ of energy per gram. These values are similar to those reported by other authors. Clarke and Prince (1980) reported wet weight caloric values of 4.51 kJ/g for krill, 4.64 kJ/g for squid, and 5.61 kJ/g for fish from food samples of two species of albatross. Assuming an assimilation efficiency of 80% (Ricklefs, 1979), the caloric analysis leads us to the conclusion that if adult petrels were only feeding their chicks fish and squid, they would have to deliver over 14,000 grams of food during the nestling period. This estimate conflicts with the feeding data obtained from event recorder monitored chicks (Table 5), and in fact it is more than five times the amount of food adults birds are estimated to bring to their chicks. The missing factor appears to be stomach oil.

The Ecological Significance of Petrel Stomach Oil

The stomach oil of petrels has intrigued researchers for decades. The oil, which is peculiar to birds in the order Procellariiformes, is a fine, light, transparent oil varying in color from light yellow to a deep red, and is stored by the birds, often in large quantities, in the fore-gut or proventriculus

Table 7. Analysis of Dark-rumped Petrel food samples.[1]

Sample #	Description	Fresh Weight (g)	%H$_2$O	%Ash[2]	kJ/g[3] dry	kJ/g ash-free	kJ/g wet
1	Oil and digested food	49.38	78.7	6.21	24.38	25.90	5.19
2	Digested fish-squid	49.96	79.4	7.71	22.01	23.70	4.53
3	Digested squid	34.99	78.0	8.23	21.82	23.62	4.80
4	Digested squid	23.00	76.2	8.41	20.95	22.72	4.99
5	Squid pieces	40.40	78.1	8.78	21.63	23.53	4.74
6	Oil	-----	----	----	41.74	-----	----

[1] This analysis was kindly performed by Dr. T. N. Pettit at the University of Hawaii

[2] Six determinations per sample

[3] Two determinations per sample

(Warham et al., 1976; Warham, 1977; Jacob, 1982). Many species
are capable of regurgitating their oil for defensive purposes, and
there are numerous accounts of the effectiveness of this
anti-predator behavior (Warham, 1977; Jacob, 1982).

The nutritive value of stomach oil was suspected by early
researchers because it is fed to nestlings, and it is found in
many species that do not use it extensively for defensive
purposes. Early workers concluded that the oil was secreted by
special glands in the proventriculi of adult birds (Mathews,
1949). Subsequent work has shown that this is not the case, and
it is now generally agreed that stomach oil is strictly of dietary
origin (Lewis, 1966; 1969; Cheah and Hanson, 1970a, b; Clarke and
Prince, 1976; Imber, 1976; Warham et al., 1976; Warham, 1977).
The oil is composed mostly of wax esters and/or triglycerides
which are thought to be concentrated from the fish, squid and
crustaceans that constitute the principal prey of most
Procellariiformes (Warham et al., 1976; Warham, 1977). The
caloric value of the oil is high (slightly less than that of
commercial diesel oil), and in eleven species studied by Warham et
al. (1976), it ranged in value from 39 - 41 kJ/g and averaged 40.3
kJ/g.

Procellariiformes, in general, rely on a food resource that
is unpredictable in time and space and often dispersed at
considerable distances from their breeding grounds (Lack, 1966;
1968; 1976). Ashmole and Ashmole were some of the first to
speculate on the role of stomach oil in allowing these birds to
utilize such a food resource (Ashmole and Ashmole, 1967; Ashmole,
1971). They postulated that stomach oil would provide adult birds
with an efficient vehicle for the transportation of a highly
concentrated, high-energy food supply to their nestlings. These
ideas have been further developed by Ricklefs (1983) in a model of
the reproductive energetics of pelagic seabirds. Warham (1977)
and others have also speculated that stomach oil is important in
allowing adult birds to maintain the long incubation shifts that
are characteristic of Procellariiformes, and in allowing nestlings
to survive extended fasting periods between feedings.

Some indirect evidence can be found to reinforce these views.
We might expect that oil would be most important to chicks early
in the nestling period when their relative energetic demands are
highest. Rice and Kenyon (1962) found that oil was an important
component in the diet of young Laysan and Black-footed Albatrosses
and Boersma et al. (1980) found that Fork-tailed Storm Petrel
adults delivered more oil to their chicks during the first half of
the nestling period. Warham (1962) reported that Giant Petrel
(Macronectes giganteus) chicks contained stomach oil up to 14 days
after their last meal, suggesting that it may be acting as an
energetic reserve, although it is not known if nestling
Procellariiformes are capable of concentrating stomach oil from

the raw food fed to them by their parents. Dark-rumped Petrel chicks have been observed to regurgitate oil more than ten days after their last feeding. Finally, it is interesting to note that the only Procellariiform that does not appear to utilize stomach oil is the Common Diving Petrel (Pelecanoides urinatrix; Warham, 1977; Roby and Ricklefs, 1983). This species is ecologically very different from the rest of the Procellariiformes. It is a near-shore feeder that feeds its chicks daily, and it apparently does not require a concentrated food supply like the rest of the petrels.

The stomach oil of the Dark-rumped Petrel is over nine times richer in calories than the fish, squid, and crustaceans from which it is derived (Table 7). Because it is clear that adult birds are not feeding their chicks anything near the 14,000 or so grams of raw food that the chicks would require during their development, they may be making up the caloric deficit with stomach oil. Our data indicate that if 40 - 50% of the food adult petrels delivered to their chicks was oil, they would be able to meet their chicks' energetic demands and still deliver the estimated 2500 g of food during the nestling period. If, as behavioral observations suggest, we have overestimated the actual energetic needs of the nestling by perhaps 30%, then a diet of 25 - 35% stomach oil would be sufficient. Thus, stomach oil in the Dark-rumped Petrels may be viewed as an adaptation to the unpredictable and widely dispersed food supply upon which the birds rely. The production of this concentrated food by adult birds is ecologically similar to the production of milk by pigeons, or mammals for that matter, or the storage of high energy oils by crustaceans for use when food supplies dwindle.

These findings raise a number of questions that remain to be answered. We still know nothing about how stomach oil is metabolized by the birds, nor do we know how or with what efficiency it is extracted from the raw food that the petrels eat. We don't know if oil is stored by adult birds prior to the breeding season or if it is concentrated on a continual basis? Nor do we know if adult birds would be capable of raising a chick entirely on raw food or if that would require too many and too frequent feeding trips to the colony? The importance of stomach oil to adult birds maintaining long incubation shifts is also not known. These questions remain to be answered, but it is clear that stomach oil is an important adaptation to the energetic contraints of breeding in the Dark-rumped Petrel and this is probably true of most other Procellariiformes.

Reproductive Energetics of Four Species of Seabirds

In Table 8 we have compared the results of this study with recent work on Leach's Storm Petrels (Ricklefs et al., 1980a, b)

Table 8. Reproductive energetics of four species of seabirds[1].

	Leach's Petrel	Dark-rumped Petrel	Common Tern	Sooty Tern
Brood Size	1	1	3	1
Mass of newly hatched chick (g)	7.3	60.0	14.9	20.5
Adult mass (g)	45.0	425.0	111.5	176.7
Adult mass - neonate mass (fledgling mass) (g)	37.7	365.9	(289.8)	156.2
Fledgling period/ fledgling mass (days/g)	1.59	0.30	(0.01)	0.32
Fledgling period (days)	60	111	30	50
Total energy required (kJ)	4811.97	54231.50	(13228.68)	5864.06
Energy/g fledgling (kJ/g)	127.63	148.20	45.65	37.54
Energy delivery rate (kJ/day)	80.20	488.57	(440.96)	117.28

[1]Data for Leach's Petrel from Ricklefs et al. (1980a, b), data for Common and Sooty Terns from Ricklefs and White (1981), numbers in parentheses refer to an estimated brood of three.

and Common (<u>Sterna</u> <u>hirundo</u>) and Sooty Terns (<u>Sterna</u> <u>fuscata</u>)
(Ricklefs and White, 1981). The purpose of the comparison is to
determine if any patterns exist that distinguish the two
Procellariiformes with their relatively long development periods
from the more rapidly developing terns. In addition, it allows
the comparison of the Common Tern, with relatively rapid growth,
and the Sooty Tern with relatively slow growth, to the other
species studied to date.

Both the Leach's Storm Petrel and the Dark-rumped Petrel are
slow growing Procellariiformes that raise at best a single chick
to fledging each season. Both are thought to utilize a variable
offshore food resource. The two species of terns develop more
rapidly but the breeding biology of the Sooty Tern resembles that
of the petrels. Sooty Terns lay a single egg, they are offshore
feeders, and they develop slowly compared to other terns. Common
Terns, on the other hand, raise up to three chicks each season and
the chicks develop rapidly. They are nearshore feeders and their
food supply appears to be more abundant and predictable. For the
purpose of emphasis it was assumed in Table 8 that Common Terns
fledged three chicks each season although in some seasons fledging
success is lower (Langham, 1972).

Several trends are suggested. When the fledgling period is
expressed as fledgling period/fledgling mass or days/gram the
Leach's Storm Petrel has the highest ratio by a factor of at least
five. The Dark-rumped Petrel and the slow growing Sooty Tern are
similar in this respect with a ratio of about 0.30 while the
Common Tern has a ratio of only about 0.10. Thus the species with
the shortest fledgling period also accumulates fledgling tissue
most rapidly. It has been demonstrated that Procellariiformes
produce eggs that are high in yolk and therefore energetically
expensive (Ackerman et al., 1980; Grant et al., 1982). It is also
clear that the petrels require considerably more energy to produce
a gram of fledgling than other seabirds such as terns. This may
be due to the fact that the petrels deposit more fat than the
terns or to a lower requirement to produce heat for thermoregulation in the terns. Thermoregulation costs may also explain
the values for energetic cost/gram of fledgling produced in the
two tern species which are opposite to what we might predict on
the basis of fledgling period. The Dark-rumped Petrel has the
highest total energy delivery rate followed by the Common Tern,
the Sooty Tern and Leach's Storm Petrel.

It might be argued that the assumptions made for the
accumulation of tissue in the Dark-rumped Petrel, invalidate any
comparisons with the other species. However, the total volume of
oxygen consumed by the chicks during their nestling period, which

was measured in both species, was clearly greater (2,237 1) in the Dark-rumped Petrel than in the Leach's Storm-Petrel (194 1), paralleling the calculated total energy requirements for the two species.

Without additional information, it is difficult to interpret these figures any further. It is important to understand that the breeding biology of these birds is molded to a large extent by their feeding ecology. Until we understand where and how these birds feed, what they eat and its caloric value, and how efficient they are at converting that food into tissue, the patterns in these data will continue to be obscured.

ACKNOWLEDGMENTS

We would like to express our thanks to D. A. Manuwal, R. E. Ricklefs, P. D. Boersma, and S. West, for their helpful comments on the manuscript and T. N. Pettit for performing the caloric analysis. We would also like to thank K. Belt, B. Hann, and especially Pam Simons and Jitsumi Kunioki for help in the field. Major funding and material support for the study was provided by the U.S. National Park Service, contract # CX800090017. Additional funding was provided by the Cooperative Park Studies Unit at the University of Hawaii, the U.S. Fish and Wildlife Patuxent Wildlife Research Center, the Frank M. Chapman Memorial fund for Ornithology, NSF grant PCM76-12351-A01, and the Sea Grant College Program N1/R-13.

REFERENCES

Ackermann, R. A., Whittow, G. C., Paganelli, C. V., and Pettit, T. N., 1980, Oxygen consumption, gas exchange, and growth of embryonic Wedge-tailed Shearwaters (Puffinus pacificus chlororhynchus), Physiol. Zool., 53:210.

Aschoff, J., and Pohl, H., 1970, Rhythmic variations in energy metabolism. Proc. Fedn. Am. Socs. Exp. Biol., 29:1541.

Ashmole, N. P., and Ashmole, M. J., 1967, Comparative feeding ecology of sea birds of a tropical oceanic island. Peabody Mus. Nat. Hist., Yale Univ., Bull., 24:1.

Ashmole, N. P., 1971, Seabird ecology and the marine environment, in: "Avian Biology", Vol. 1., D. S. Farner and J. R. King, eds., Academic Press, N.Y.

Boersma, P. D., Nerini, M. K., Wheelwright, N. T., and Wheelwright, E. S., 1980, The breeding biology of the Fork-tailed Storm Petrel (Oceanodroma furcata), Auk, 97:268.

Calder, W. A., 1968, Respiratory and heart rates of birds at rest, Condor, 70:358.

Cheah, C. C. and Hansen, I. A., 1970a, Wax esters in the stomach oils of petrels, Int. J. Biochem, 1:198.

Cheah, C. C., and Hansen, I. A., 1970b, Stomach oil and tissue lipids of the petrels Puffinus pacificus and Pterodroma macroptera, Int. J. Biochem, 1:203.

Clarke, A., and Prince, P. A., 1976, The origin of stomach oil in marine birds: analysis of the stomach oil from six species of subantarctic procellariiform birds, J. Exp. Mar. Biol. Ecol., 23:15.

Clarke, A. and Prince, P. A., 1980, Chemical composition and calorific value of food fed to Mollymauk chicks Diomedea melanophris and D. chrysostoma at Bird Island, South Georgia, Ibis, 122:488.

Dawson, W. R., and Hudson, J. W., 1970, Birds, in: "Comparative Physiology of Thermoregulation", Vol. 1., G. C. Whittow, ed., Academic Press, N.Y.

Grant, G. S., Pettit, T. N., Rahn, H., Whittow, G. C., and Paganelli, C.V., 1982, Regulation of water loss from Bonin Petrel (Pterodroma hypoleuca) eggs, Auk, 99:236.

Grant, G. S., and Whittow, G. C., 1983, Metabolic cost of incubation in the Laysan Albatross and Bonin Petrel, Comp. Biochem. Physiol., 74:77.

Harris, M. P., 1969, Food as a factor controlling breeding of Puffinus lherminieri, Ibis, 11:139.

Imber, M. J., 1976, The origin of petrel stomach oils - a review, Condor, 78:366.

Jacob, J., 1982, Stomach Oils, in: "Avian Biology", Vol. 6., D. S. Farner, J. R. King, and K. C. Parkes, eds., Academic Press, N.Y.

Lack, D., 1966, "Population studies of birds", Clarendon Press, Oxford.

Lack, D., 1968, Interrelationships in breeding adaptations as shown by marine birds, Proc. 14th Int. Ornithol. Congr., 1966, pp. 3-42.

Lack, D., 1968, "Ecological Adaptations for Breeding in Birds", Methuen, London.

Langham, N. P. E., 1972, Chick survival in terns (Sterna spp.) with particular reference to the Common Tern, J. Anim. Ecol., 43:385.

Lewis, R. W., 1966, Studies on the glycerylethers of the stomach oils of Leach's Petrel Oceanodroma leucorphoa (Viellot), Comp. Biochem. Physiol., 19:363.

Lewis, R. W., 1969, Studies on the stomach oils of marine animals - II. Oils of some procellariiform birds, Comp. Biochem. Physiol., 31:725.

Mathews, L. H., 1949, The origin of the stomach oil in the petrels, with comparative observations on the avian proventriculus, Ibis, 91:373.

Petrusewicz, K., and Macfayden, A., 1970, "Productivity of terrestrial animals", I.B.P. Handbook No. 13, Blackwell, London.

Prince, P. A., Ricketts, C., and Thomas, G., 1981, Weight loss in incubating albatrosses and its implications for their energy and food requirements, Condor, 83:238.

Rice, D. W., and Kenyon, K. W., 1962, Breeding cycles and behavior of Laysan and Black-footed Albatrosses, Auk, 79:517.

Ricklefs, R. E., 1967, A graphical method of fitting equations to growth curves, Ecology, 48:978.

Ricklefs, R. E., 1968, Patterns of growth in birds, Ibis, 110:419.

Ricklefs, R. E., 1974, Energetics of reproduction in birds, in: "Avian Energetics", R. A. Paynter Jr., ed., Publ. Nuttall Ornithol. Club, No. 5, Cambridge.

Ricklefs, R. E., 1979, "Ecology," Chiron, New York.

Ricklefs, R. E., 1983, Some considerations on the reproductive energetics of pelagic seabirds, Studies in Avian Biology, in press.

Ricklefs, R. E., White, S. C., and Cullen, J., 1980a, Energetics of postnatal growth in Leach's Storm Petrel, Auk, 97:566.

Ricklefs, R. E., White, S. C., and Cullen, J., 1980b, Postnatal development of Leach's Storm Petrel, Auk, 97:768.

Ricklefs, R. E., and White, S. C., 1981, Growth and energetics of the Sooty Tern (Sterna fuscata) and Common Tern (S. hirundo), Auk, 98:361.

Roby, D. D., and Ricklefs, R. E., 1983, Life-history observations on the diving petrels Pelecanoides georgicus and P. urinatrix exsul at Bird Island, South Georgia, Bull. Brit. Ant. Surv.

Schmidt-Nielsen, K., 1974, "Animal Physiology," Prentice Hall, Englewood Cliffs, New Jersey.

Simons, T. R., 1981a, Behavior and attendance patterns of the Fork-tailed Storm Petrel, Auk, 98:145.

Simons, T. R., 1981b, A simple event recorder for monitoring cavity-dwelling animals, Murrelet, 62:27.

Simons, T. R., 1983, "Biology and conservation of the endangered Hawaiian Dark-rumped Petrel (Pterodroma phaeopygia sandwichensis)," National Park Service, Cooperative Park Studies Unit, University of Washington, CPSU/UW83-2, Seattle.

Walker, L. E., Walker, J. M., Palca, J. W., and Berger, R. J., 1983, A continuum of sleep and shallow torpor in fasting doves, Science, 221:194.

Warham, J., 1962, The biology of the Giant Petrel Macronectes giganteus, Auk, 79:139.

Warham, J., 1971, Body temperatures of petrels, Condor, 73:214.

Warham, J., Watts, R., and Dainty, R. J., 1976, The composition, energy content and function of the stomach oils of petrels (order Procellariiformes), J. Exp. Mar. Biol. Ecol., 23:1.

Warham, J., 1977, The incidence, functions and ecological
    significance of petrel stomach oils, Proc. N.Z. Ecol. Soc.,
    24:84.
Whittow, C. G., Simons, T. R., and Pettit, T. N., 1983, Water loss
    from the eggs of a tropical seabird (Pterodroma phaeopygia)
    at high altitude, Comp. Biochem. Physiol., In press.

THERMOREGULATION IN ADULT SEABIRDS

Sheldon Lustick

Zoology Department
The Ohio State University
Columbus, Ohio 43210

INTRODUCTION

In simple terms, an animal is fit in an evolutionary sense, if
it can reproduce, obtain food, and avoid becoming someone else's
food. Those animals leaving the most viable offspring at the least
possible cost can be considered the most fit. In birds, a major
portion of the cost of survival is spent in maintaining a relatively
high (38-42°C) body temperature (homeothermy). The maintenance of
a relatively constant body temperature is based on a balance between
heat production (metabolism), heat transfer from the environment,
and heat transfer to the environment. The influence of the environment on homeothermy can best be expressed by the following heat
balance equation:

$$\Delta T_b = R_s \pm R \pm C \pm K + MR - E + S \qquad (1)$$

where

ΔT_b = change in body temperature
R_s = heat gain from direct solar radiation
R = heat gain or loss from thermal radiation
C = heat gain or loss from thermal convection
K = heat gain or loss from conduction
MR = heat gain from metabolism
E = heat loss from evaporation
S = heat storage

Marine birds, though studied little when compared to other

183

species, are an excellent group of organisms to study thermoregulation in for the following reasons. Except for the penguins, Ratites, and Procellariiformes (Boyd and Slader, 1971; Warham, 1971) they show little adaptation in body temperature being good homeotherms (T_b between 38-40°C). They vary in size from British Storm Petrel (Hydrobates pelagicus)(0.028 Kg)(Lockley, 1983) to the Emperor Penguin (Aptenodytes forsteri)(26-38 Kg, Pinshow et al., 1974; Le Maho et al., 1976). Their distribution is worldwide, thus they are exposed to the extremes of temperature on earth (Maher, 1962; Bartholomew, 1966; Stonehouse, 1967; Howell et al., 1974).

Even more important to understanding the mechanism of thermoregulation is that within a single genus or group they are exposed to wide environmental variations. Though most of the species within the genus Larus are temperate or boreal, there are a few species breeding in deserts where the daily air temperatures range from 0-40°C (Howell et al., 1974; Bartholomew and Dawson, 1979). Also, penguins breed from the Antarctic to the tropics (Stonehouse, 1967). That seabirds spend a good deal of time in the water stresses the thermoregulatory mechanism even more, water being such a good conductor of heat (24 times that of air). Microclimate conditions on the nesting sites when the birds are tied to land are often extremely demanding, with extreme substratum temperatures, intense solar radiation, and high wind speeds (Stonehouse, 1967; Howell et al., 1974; MacMillen et al., 1977; Lustick et al., 1978; Bartholomew and Dawson, 1979).

In general, seabirds are relatively well-insulated (feather and fat) and seem to tolerate cold better than heat. All birds can tolerate some hypothermia (Dawson and Hudson, 1970), but heat defenses are important because birds regulate their body temperatures (T_b) close to the upper lethal body temperature. Herring Gulls (Larus argentatus) can maintain body temperatures at air temperatures (Ta's) as low as minus 50°C in the absence of wind, whereas they were heat stressed at 3-5°C when exposed to intense solar radiation at wind speeds under 3 m/sec on the nesting grounds (Lustick et al., 1978). The Adelie penguins on Cape Royds pant at air temperatures of -4°C when exposed to full sun during incubation (Stonehouse, 1967).

The Sooty Tern (Sterna fuscata), a small bird (0.15 Kg) nesting in the tropics, has a lower critical temperature of 30°C (MacMillen et al., 1977). Yet at air temperatures of 10°C its metabolism is only twice that of what it was at 30°C. Assuming that birds can increase their metabolic rate 3 to 5 times what it is in thermal neutrality by shivering thermogenesis (Dawson and Hudson, 1970), the Sooty Tern could maintain T_b to at least minus 30°C.

Birds respond to their thermal environment in three ways. There is a morphological response, a physiological response, and a behavioral response. Together they lead to thermal homeostasis.

MORPHOLOGY AND ITS EFFECT ON THERMOREGULATION

As Bartholomew (1964) pointed out, the morphological adaptations to the thermal environment have been selected for throughout the evolution of the species. At present these adaptations are in a genetic trap and cannot respond to a rapid change in environmental conditions. The morphological factors affecting thermal homeostasis in seabirds are similar to those of other homeotherms and are summarized in Table 1.

Surface Area/Volume Ratio (SA/V).

The SA/V ratio is influenced mainly by the size and shape of the bird and its extremities. The larger the organism and the smaller its extremities, the smaller its surface area is relative to its volume. One cannot forget the influence of shape and SA/V. Two birds could weigh the same but have different shapes (sphere vs elongate) and thus have different surface areas. The assumption with regard to thermoregulation is that large size and small extremities facilitate heat conservation and animals within a species or group evolved larger size and shorter appendages in colder climates. Thus, one would expect to find the larger seabirds with shorter appendages in the colder latitudes and the smaller birds in the tropics. The best-studied group of marine birds with regard to body size and extremity size are the penguins (Stonehouse, 1967). Eudyptids form two distinct groups with small temperate species (3 kg) in warm climates and larger species (4-4.5 kg) in cooler climates. Yet the Pygoscelids and the Gentoos defy Bergmann's rule. In general, penguins nesting in cooler climates

Table 1. Morphological Factors Affecting Thermoregulation

Factor	Influenced By	Effects
Surface area/ volume ratio	Size and shape of body and extremities,	Heat loss and gain
Insulation	SA/V ratio, plumage, fat, wetability piloerection	Heat loss and gain
Color	–	Radiative heat load
Peripheral heterothermy	Anatomy of circulation to legs and feet	Heat loss and gain

do have smaller extremities. The Peruvian, Blackfooted, Erect-crested, Northern Blue, and the Chatham Island subspecies of the Southern Blue, have flippers over 10% larger in size than the theoretical value for their weight. All nest in exposed areas and are unable to avoid strong insolation and must shed heat. Four other species, Magellanic, Yellow-eyed, King, and Emperor, have flippers over 10% smaller than their theoretical value. The first two are from cool climates and select a microhabitat to avoid insolation. Both the King (12% smaller extremities) and the Emperor (23% smaller extremities) are subjected to extreme cold (Stonehouse, 1967). Not only are flippers smaller but head and feet are also smaller in the Emperor and Adélie Penguins than would be expected. Penguins due to their spherical shape have a 15% smaller surface area than predicted from weight (Stonehouse, 1967), an adaptation to cold waters and climates.

Though within the penguins you do find some correlation between size and distribution, in general seabirds do not demonstrate Bergmann's rule very well. You find big seabirds in the tropics and small seabirds in the Arctic and Antarctic.

As Scholander (1956) and Irving (1956) state, animals have a number of other ways of adapting to extreme temperatures (insulation change, metabolic adjustments, and behavioral adjustments) thus negating the selective pressure to evolve larger size as they moved into colder climates. Though we discussed large size as an adaptation to cold stress, it could also help with heat stress. Large mass means a higher heat capacity (mass x specific heat) which means larger birds can store more heat (hyperthermia), an advantage when tied to land and exposed to insolation in hot climates.

Insulation

Seabirds, because of their aquatic habitat, are notorious for having thick feather insulation, especially on the ventral surfaces, and large deposits of subcutaneous fat. The plumage covering the breast of the Herring Gull has twice the insulating capacity of the dorsal plumage. Also plumage with a natural fluff has twice the insulating capacity of compressed plumage (Lustick et al., 1978), stressing the importance of piloerection. Stonehouse (1967) demonstrates a relationship between plumage length and distribution in penguins. In cooler climates penguins have longer feathers than those in warmer climates. Plumage insulation is reinforced by sub-dermal fat which is found in all marine birds. Subdermal fat in Emperor and Adélie penguins is almost 2.3 centimeters thick. Fat has a low coefficient of thermal conductivity 0.0003-0.00049 kcal/cm. sec.$^{\circ}$C (Lipkin and Hardy, 1954). In Emperor Penguins, feathers account for 84% of the insulation, important because they lose fat during the long incubation period.

186

Another important aspect of the plumage is its wetability. Wetting of the plumage is extremely detrimental to the birds' ability to maintain T_b (Lustick and Adams, 1977). The ability of feathers to resist wetting is extremely important in marine birds. Recent studies by Sheila Mahoney (personal communication) have suggested that diving birds do not get as wet as non-diving birds, implying the feathers are more resistant to water penetration.

Penguins in the Antarctic are so well-insulated that the surface temperature approximates the ambient thus reducing radiative and convective heat loss. The stiff feathers of a penguin, arranged like tiles on a roof, are pressed more closely together in wind and maintain a boundary layer of air close to the skin, reducing heat loss.

Color

Factors affecting animal coloration are precision of microhabitat selection, body size, predator-prey interaction, and individual recognition, among others. Although concealment and social interactions are the most easily recognized selective advantages of coloration pattern, researchers have long been calling attention to the importance of coloration to thermoregulation (Cartwright and Harrold, 1925; Cole, 1943; Hamilton and Heppner, 1967; Stonehouse, 1967; Lustick, 1969). Yet, with regard to feathers, there is still a question as to whether light or dark coloration is a better absorber of radiant energy. Monteith (1973) points out that "reflectivity is an important discriminant in the heat balance of animals but the relationship between coat color and the radiative heat load is complex." Marder (1973) found that though black plumage heats more than white, skin temperatures were similar. Walsberg et al. (1978) stated that convective cooling differentially affects the radiative heating of dark and light plumages, so that at very low wind speeds, black plumages acquire a greater radiative heat load than do white plumages, but the heat loads of black and white plumages rapidly converge as wind speed is increased. This effect was observed best when the feathers were erected and not when they were smooth. It has also been demonstrated that darker plumage colors absorb more solar radiation (DeJong, 1976; Hamilton and Heppner, 1967; Lustick, 1969; Heppner, 1970; Lustick et al., 1970). Lustick (1969) and Heppner (1970) show that dark birds had a significantly greater reduction in metabolism at air temperatures below the lower critical temperature than did white birds when receiving radiation. More recently Chappell (1980) working with white and dark pelage of winter and summer acclimatized ermine and mink, and Finch et al. (1980), working with light and dark desert goats, showed dark pigmentation to be a better absorber of solar radiation than light pigmentation (Table 2). One could conclude that barring the erection of the plumage, dark pigmentation is a better absorber of solar energy. Why, then, do we find dark birds

Table 2. Theoretical Effect of Color on the Radiative Energy
Exchange of a 1000 g Bird With its Feathers Smooth[*]

Color	Absorptivity	Air Temperature	Net Radiation kcal/Bird
Gray	0.50	10	125.1
Brown	0.75	10	255.8
White	0.15	10	-37.0
White	0.30	10	42.0

[*]No wind

in the desert (Howell et al., 1974; Bartholomew and Dawson, 1979) and white birds in the Arctic and Antarctic (Maher, 1962)? From our work (Lustick et al., 1978) with Herring Gulls we know that subtle adjustments in posture and orientation can minimize the exposed surface area and thus the absorptivity of solar energy. Dark birds approach the radiative heat exchange of light birds if they assume the proper orientation and posture with regard to the incident radiation (Lustick et al., 1980). As Walsberg et al. (1978) have pointed out, feather erection can modify the effect of color on the absorption of solar energy. We can now ask, "Is thermoregulation important in the selection for color?" The arguments supporting the hypothesis that thermoregulation is an important determinant of color are similar to those supporting Bergmann's climatic rule, that increased size is selected for in colder latitudes. Large size like dark color, is advantageous in a cold climate but not necessary because of the many physiological and behavioral adjustments birds can make.

Circulation to Legs and Feet

Since the work of Irving and Krog (1955) on Herring Gulls, it has been known that the legs and feet of aquatic birds play an important role in thermoregulation. Feet and legs are the major avenue of heat loss in gulls, their importance increasing with T_a (Steen and Steen, 1965). At a T_a of 30°C the heat loss from the legs and feet of a gull was 4 times that at 20°C and accounted for 40% of the total heat loss. This was substantiated by Lustick et al. (1979) when they placed insulated boots on Herring Gulls' legs and feet at high T_a's (25°C) and increased evaporative water loss 5 fold. Obviously the ability to control the heat loss from the feet and legs is due to the ability to control blood flow to them and the presence of countercurrent heat exchangers in them. As

Guard and Murrish (1974) point out, blood circulation through these extremities must balance the demands of (1) tissue nutrition, (2) prevention of tissue freezing, and (3) core body temperature regulation. There seems to be a conflict between nourishing the tissue, keeping them from freezing, and reducing heat loss. To date one of the best studies on the control of peripheral circulation in birds has been that of Guard and Murrish (1974) on the Giant Petrel (Macronectes giganteus). Unlike the gull, the Giant Petrel has a dual system of venous return from the web and lower leg. One set of veins returns to the body core as an anastomosing plexus surrounding the supplying artery. Another larger venous drainage returns in parallel with the countercurrent veins. The countercurrent return in the giant petrel functions primarily for tissue nutrition and heat conservation while the more peripheral return functions for heat loss. An overheated Giant Petrel (T_b = 39.2°C) cooled to 37°C within five minutes when its foot was placed in ice water, a remarkable feat for a 3.4 Kg bird (Guard and Murrish, 1974).

PHYSIOLOGY AND ITS EFFECT ON THERMOREGULATION

The physiological adaptations for thermoregulation in marine birds are similar to those of birds in general and have been summarized extensively by Calder and King (1974). Like the morphological adaptations, the physiological adaptations (Table 3) are in a genetic trap and change slowly.

Metabolic Rate

Though there have been a number of studies dealing with thermoregulation in marine birds, only a few have actually determined the metabolic rate directly (Scholander, 1950; Tucker, 1972; Pinshow et al., 1974; MacMillen, 1977; Lustick et al., 1978) while others have calculated the metabolic rate from weight loss (Le Maho, 1976). In most of these studies the standard metabolic rate (SMR) in thermal neutrality approximated those predicted by equations for nonpasserine birds (Lasiewski and Dawson, 1967; Aschoff and Pohl, 1970). The exception was the Sooty Tern which had a metabolic rate 26% lower than predicted for a nonpasserine bird of similar weight (MacMillen et al., 1977). This was considered an adaptation to a tropical environment especially since the Sooty Tern's powers of evaporative cooling were not very good. As pointed out in the introduction, birds can increase their metabolic rate 3-5 times that in thermal neutrality (Dawson and Hudson, 1970) by shivering thermogenesis. In the Sooty Tern (0.15 Kg tropical bird) this would be sufficient to tolerate much lower (-30°C) temperatures than it encounters on land and becomes important when in the water. Though shivering is the main means of increasing heat production in birds, both the heat increment of feeding (specific dynamic action, SDA) and that of exercise can influence the ability of a

Table 3. Physiological Factors Affecting Thermoregulation

Factor	Influenced By	Effects
Metabolic heat production	Shivering, exercise specific dynamic action	Increase heat production
Evaporative water loss	Panting, gular flutter, cutaneous	Heat loss and conservation
Blood flow to extremities	Autonomic nervous system	Heat loss and conservation
Hypothermia	Interaction between rate of heat loss and rate of heat production	Conserves energy
Hyperthermia	Interaction between rate of heat loss and rate of heat production	Conserves energy and water

bird to regulate its body temperature. Since the SDA is greater for animals digesting protein and marine birds feed mainly on protein diets, it might be important to thermoregulation in marine birds in both hot and cold climates. Further study is needed in this area. Calder and King (1974) suggest that if flight metabolism is 25% efficient, 75% is converted to heat and is potentially available for thermoregulation in the cold or could be a disposal problem during flight in a warm climate. Body temperatures of birds returning to the nest after flight were elevated as much as 2°C over that of incubating birds, implying an elevated MR and heat disposal problem (Howell and Bartholomew, 1962). Obviously size of the bird, type of flight (soaring vs flapping), and environmental conditions influence whether the increased heat from exercise can be used for thermoregulation or is a disposal problem in birds. Small birds have a much higher thermal conductance and thus could lose more heat than they produce during flight.

Evaporative Heat Loss

Lasiewski and Seymour (1972) have demonstrated clearly that the thermoregulatory effectiveness of evaporative heat dissipation is determined by the relationship between evaporation and heat production or gain. That is to say that a bird with a high metabolic rate

when heat stressed must evaporate more water. Evaporative heat loss
mechanisms in marine birds once again are similar to those of other
birds and occur through panting, gular flutter, and insensible
cutaneous water loss (for a detailed review of evaporative water
loss in birds, see Calder and King, 1974). Gular flutter is the
more efficient means of evaporative cooling since heat production
is less during gular flutter (Lasiewski et al., 1969). The Poor-will,
a bird which gular flutters, lost 352% of the heat produced and
gained from the environment (at T_a 47°C), by evaporation. Among the
marine birds are those that pant and those that possess the ability
to gular flutter. In general those exposed to intense heat and/or
solar radiation in hot environments can gular flutter. Howell and
Bartholomew (1962) suggest that the lack of a well-developed gular
pouch and gular flutter in the Red-tailed Tropic Bird (Phaethon
rubicauda) which is characteristic of other Pelecaniforms, may keep
the effectiveness of evaporative cooling in these birds below the
level achieved by other members of this order. The shade nesting
habitat of the tropic birds as opposed to the full sun nesting of
other marine Pelecaniforms may explain their lack of the capacity
for gular flutter. The importance of cutaneous water loss has not
been investigated in marine birds, though it accounts for as much
as 62% of the total in those birds investigated (Bernstein, 1971;
Lasiewski et al., 1971). It should be pointed out that the ability
of the nasal gland to concentrate salt permits marine birds to
drink seawater and is important to their ability to lose heat by evaporation. Marine birds use evaporative heat loss somewhat during
flight (Howell and Bartholomew, 1962) and during the nesting season
when tied to land. At other times, they can enter the water and
dump heat through their legs and feet (Steen and Steen, 1965;
Lustick et al., 1979).

Control of Blood Flow to the Legs and Feet

Though the arrangement of the blood vessels in the legs and
feet of marine birds may be considered a morphological adaptation
to heat conservation and heat loss, the ability to control blood
flow to this region is a physiological adaption within the auto-
nomic nervous system (Guard and Murrish, 1974). Heat stress in
Giant Petrels caused an increased blood flow to the arteriovenous
anastomoses (near periphery) in an effort to dissipate heat, with
little blood flow through the countercurrent heat exchangers; cold
stress reversed these responses. Beta-blockade reduced blood flow
to the arteriovenous anastomoses. Alpha-blockaded birds with feet
in cold water showed high blood flow to the arteriovenous anastomoses but the countercurrent veins were acting to cool arterial
blood which reduced heat loss. Drugs working on the postganglionic
parasympathetic nervous system had little effect on leg blood flow
(Guard and Murrish, 1974). The importance of the leg and feet to
thermoregulation in marine birds cannot be overemphasized.

191

Body Temperature (hypothermia and hyperthermia)

Though marine birds are good homeotherms maintaining a fairly constant T_b, they can withstand, as do other birds, various degrees of hypothermia and hyperthermia. Howell and Bartholomew (1962) found a 2°C lower nocturnal T_b (37 vs 39) in Red-tailed Tropic Birds even though air temperatures were similar and the nests were in the shade. The differences in day-night T_b may be due to the differences in the metabolic rates during resting phase (night) and activity phase (day) of the daily cycle (Aschoff and Pohl, 1970). Hypothermia with a concomitant decrease in metabolism can save a bird a great deal of energy (Irving, 1955; Budd, 1972; Mayer et al., 1982). Huddling Emperor Penguins (Le Maho et al., 1976) have a 2 to 4° lower T_b than isolated Emperor Penguins and it is estimated from the rate of weight loss that this hypothermia and huddling amounts to a 13-37% savings in energy expenditure.

Most birds can withstand hyperthermia of a few degrees, body temperatures reaching 44°-45°C. Hyperthermia usually occurs during activity (flight) exposure to high air temperature, exposure to intense solar radiation, or any combination of these factors. Hyperthermia can be considered a physiological adaptation to heat stress. The need for evaporative cooling is minimized by having a high T_b; it either reduces the thermal gradient into the bird, reducing rate of heat gain, or maintains a thermal gradient favorable for heat loss from bird to environment. As pointed out, large birds can store more heat. This is important for marine birds nesting in hot environments because it increases the time they can spend incubating without seeking H_2O and thus reduces conspecific predation.

BEHAVIORAL THERMOREGULATION

As pointed out previously, birds in general are in a sort of genetic trap with regards to the kinds of physiological and morphological adjustments that they can make to environmental conditions. In contrast, most birds have evolved very complex behavioral patterns and behavioral adjustments to the environment can be rapid, precise, and highly flexible. More important is the fact that many of the behavioral adjustments are less costly than the physiological or morphological adjustments. With this in mind, let's look at the types of behavioral adjustments displayed by marine birds (Table 4).

Microclimate Selection

The ability to select a favorable thermal environment is an important aspect in the survival of free-living birds. This ability is especially important to marine birds during the nesting period when they are tied to land. In fact, the water may be considered a microclimate, tropical birds entering the water to cool. The importance of the microclimate to the survival of marine birds

Table 4. Behavioral Patterns Affecting Thermoregulation

Pattern	Effect
Microhabitat selection	Modify the climate space with regard to conduction, convection, radiation and evaporation
Linear nesting adjacent to water	Can get water for cooling without risking predation of egg and chicks
Posture and orientation adjustments	Reduce or maximize the effects of wind and insolation
Wing drooping	Increase convective heat loss, shades feet and sides
Tucking beak and sitting on feet	Decreases exposed surface area and conserves heat
Huddling	Decreases exposed surface area and conserves heat
Move chicks closer to water	Decreases predation while adults dump heat in water
Cease conspecific predation	Allows bird to leave nest for water enhancing their ability to cool
Soaring	Increase convective heat loss

has been noted by Maher (1962); Snow Petrels (_Pagodroma nivea_) nesting in cavities in the Antarctic to reduce the effects of wind and snow (Farner and Serventy, 1959); burrow temperatures of the Slender-billed Shearwater are much more stable than air temperatures. Red-tailed Tropic Birds nesting in vegetation were more successful than those exposed to direct sunlight (Howell and Bartholomew, 1962). As one goes from the Antarctic to the tropics, one finds that penguins seek a microclimate to reduce the effect of intense solar radiation (Stonehouse, 1967). Though the importance of a micro-climate to energy savings has been quantified for many birds (Calder, 1974; Mayer et al., 1982; Mugaas and King, 1981), it has not been quantified for adult marine birds, though it is obvious from nesting success that it is a significant behavioral response to extreme environments.

Linear Arrangement of Nests Just Above High Tide Line

Hand et al. (1981) state that the Western Gull (Larus occidentalis livens) nests linearly along the beach as a response to conspecific predation and thermoregulation. They hypothesized that L. o. livens nests in close proximity to water so that at high Ta's (30°C and blackbulb temp exceeding 40°C) they could enter the water to cool and still protect their eggs and chicks from their neighbor. Panting always ceased when they returned from the water and nasal drip increased, implying that they cooled and drank. This is further supported by the fact that species like the Sooty Terns (S. fuscata) which are not conspecific predators, fly to drink water, leaving the nest exposed (Dinsmore, 1972). Gulls which are con-specific predators either decrease the intervals between nest exchanges (Grant, 1979) or stay on territory until night when all the adults leave (Herring Gulls and Heermann's Gulls, Lustick et al., 1978, and Bartholomew as cited by Hand et al., 1981).

Another behavior pattern to decrease the possibility of predation and enhance thermoregulation is to lure the chicks closer to water so the adult can be in the water and keep an eye on the chicks.

Posture and Orientation

Behavioral patterns prevalent in marine birds not using a microhabitat to alleviate thermal stress are orientation and postural adjustments to wind and solar radiation. That orientation and postural adjustments can make a significant difference in the net energy exchange between the bird and its thermal environment has been demonstrated by Lustick et al. (1978,1980) and Mugaas and King (1981). By assuming various postures and orientations with regard to solar radiation and wind, a bird can vary its surface area exposed to direct radiation, vary its characteristic dimension to wind, or if bicolored, expose different-colored plumage to solar radiation. All of these variables will affect the net energy balance (Fig. 1). Herring Gulls on days with low wind velocities, high solar intensities and air temperatures between 5 and 15°C rotated 180° during the day, always facing the sun. When facing the sun their posture is such that the darker dorsal surface is exposed to only atmospheric and diffuse radiation while only a small portion of the white ventral surface is exposed to direct solar radiation (Lustick et al., 1978). Thus, the gulls reduced

194

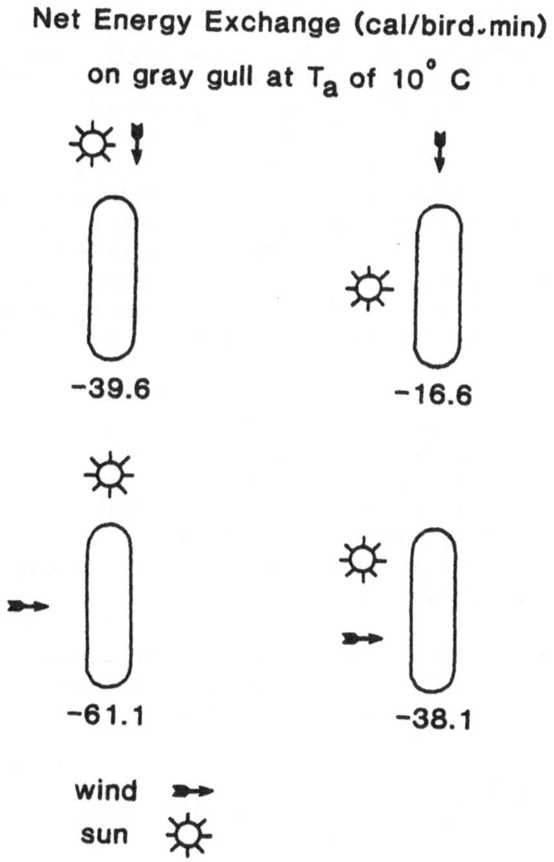

Net Energy Exchange (cal/bird.min)

on gray gull at T_a of $10°$ C

Fig. 1. Effects of orientation with regard to wind and sun on the energetics of a 1000 g gray gull, at 10°C. The metabolic rate of a 1000 g gull at 10°C is 53.8 cal/bird.min.

heat stress by (1) reducing the area exposed to direct solar radiation to approximately one-fourth the total surface area; (2) exposing the less absorptive white plumage to direct radiation; (3) exposing the better insulated plumage (breast plumage) to solar radiation. Other investigators have now demonstrated orientation toward the sun in Brown Pelicans (Pelecanus occidentalis) (James Keith, personal communication). Bartholomew (1966) found the Masked Booby (Sula dactylatra) faced away from the sun, shading feet and gular area, thus increasing their ability to lose heat. Howell et al. (1974) and Bartholomew and Dawson (1979) found the gray gull (Larus modestus) and Heermann's Gull (Larus heermanni) respectively to show little orientation though both nested in the desert with intense solar radiation and high Ta's. Why the paradox? Why do some seabirds orient toward the sun while others show no orientation or orient away from the sun? It should be pointed out that the Masked Booby, Gray Gull, and Heermann's Gull all nest where the sun is more directly overhead than in the study by Lustick et al. (1978) on Herring gulls. When the sun is overhead, orientation is of little value in reducing surface area exposed to direct solar radiation. Birds nesting in a hot environment and exposed to intense solar radiation overhead have evolved another behavioral mechanism to retard heat gain from solar radiation and that is the erection of the scapular feathers. This increases convective heat loss while at the same time shielding the body from direct solar radiation. Regal (1975) sets forth the hypothesis that feathers evolved not for flight or the maintenance of endothermy, but as heat shields. As suggested by Walsberg et al. (1978), feather erection can alter the radiative properties of the plumage; erected white feathers absorb more solar radiation than erected dark feathers when winds are greater than 3 m/sec.

The effects of wind on the energetics of a bird is best expressed by an equation approximating the heat loss from a hypothetical cylindrical bird (Calder and King, 1974).

$$\frac{H_c}{A} = 6.17 \times 10^{-3} \frac{v^{0.5}}{D^{0.65}} (Ts-Ta) \qquad (2)$$

Where

H_c = convection heat energy
A = surface area
v = velocity in centimeters (cm) per second
D = the characteristic dimension in cm
Ts = temperature of plumage surface
Ta = air temperature

The value of D in equation 2 changes with the orientation of
the bird with respect to wind direction. For a bird facing parallel
to the wind direction, D becomes the length of the bird from beak
to base of tail, while for a bird whose long axis is perpendicular
to the direction of the wind, D is the diameter at its widest point.
As the characteristic dimension increases, the boundary layer
increases thus reducing the effect of wind on convective heat loss
(Fig. 1). Herring gulls at low air temperature (5-10°C) faced into
the wind (ignoring orientation toward the sun) when wind velocities
exceeded 3 meters/sec (Lustick et al., 1978). In fact, it is
common to see whole colonies of seabirds facing into the wind,
probably to avoid convective heat loss. By subtle postural and
orientation, adjustments with respect to sun and wind direction,
birds can regulate the exchange of thermal energy between themselves
and their environment (Fig. 1) minimizing the cost of regulating
their T_b.

Tucking Beak, Sitting on Legs and Feet, Huddling, and Wing Drooping

Since in most birds the beak, legs, and undersides of wings are
unfeathered, they are potential heat sinks. Most birds when cold-
stressed, tuck the beak under the scapular wing feathers while
sitting on their feet, thus reducing exposed surface area and con-
ductive, convective, and radiative heat loss. Huddling in a sense
reduces exposed surface area so that thermal radiation and convection
heat loss is reduced. As pointed out earlier, the metabolic rate of
huddling Emperor penguins is 13-37% below the metabolic rate of
isolated winter birds (Le Maho et al., 1976). The investigators
concluded that the thermoregulatory behavior of huddling was
essential for the survival of the male Emperor penguin while fasting
during the long periods of incubation at winter temperatures. Under
heat stress, wing drooping increases convective heat loss by
increasing the circulation of air along the sides and under the wings
of the bird (Bartholomew, 1966). Though never mentioned in the
literature, wing drooping might reduce the net radiative heat gain
in the same way that elevated scapular feathers reduce radiative
heat gain, shading the sides.

Soaring

Though I do not know of any studies dealing with the effect of
soaring on body temperature, seabirds spend a great deal of time
soaring. We know from the Tucker (1972) study on the energetics of
flight in the Laughing Gull (<u>Larus</u> <u>atricilla</u>) that the thermal
conductance is 5.8 times greater during flight than at rest at 30°C.
We also know that the cost of soaring flight is only 2-3 times basal,
thus, it is possible that soaring at altitude (lower Ta's) may be
a way to dump excessive heat. Further research is needed on the
body temperatures of soaring birds over a range of ambient
temperatures.

Seabirds through morphological, physiological, and behavioral adaptations for thermoregulation have been successful in an evolutionary sense in some of the harshest environments known to man. One cannot overemphasize the importance of the behavioral adjustments since they are rapid, precise, and highly flexible while morphological and physiological adaptation are in a genetic trap.

There have been many qualitative studies dealing with behavioral thermoregulation in seabirds; what is needed at this time is to quantify the energy-saving afforded by the various behavior patterns observed. We need studies on the annual variation of the daily energy expenditure in adult seabirds. This is not an easy task but a necessary one.

REFERENCES

Aschoff, J., and Pohl, H., 1970, Rhythmic variations in energy metabolism, Fed. Proc., 29:1541.
Bartholomew, G. A., 1964, The roles of physiology and behaviour in the maintenance of homeostasis in the desert environment, in: "Homeostasis and Feedback Mechanisms," 18th symposium of The Soc. Exp. Biol., Cambridge Univ. Press, London.
Bartholomew, G. A., 1966, The role of behavior in temperature regulation of the Masked Booby, Condor, 68:523.
Bartholomew, G. A., and Dawson, W. R., 1979, Thermoregulatory behavior during incubation in Heermann's Gulls, Physiol. Zool., 52:422.
Bernstein, M. H., 1971, Cutaneous and respiratory evaporation in Painted Quail Excalfactoria chinensis during ontogeny of thermoregulation, Comp. Biochem. Physiol., 38(A):611.
Boyd, J. C., and Sladen, J. L., 1971, Telemetry studies of the internal body temperatures of Adélie and emperor penguins at Cape Crozier, Ross Island, Antarctica, Auk, 88:366.
Budd, S. M., 1972, Thermoregulation in Black-capped Chickadees (Parus aticapillus), Amer. Zool., 12, Abstr. No. 402.
Calder, W., 1974, Factors in the energy budget of mountain humming-birds, in: "Perspectives in Biophysical Ecology," D. Gates and R. Schmerl, eds., Ecological Studies, Analysis and Synthesis, Springer-Verlag, New York.
Calder, W. A., and King, J. R., 1974, Thermal and caloric relations in birds, in: "Avian Biology," D. S. Farner and J. R. King, eds., Academic Press, New York.
Cartwright, B. W., and Harrold, C. G., 1925, An outline of the principles of the natural selective absorption of radiant energy, Auk, 42:94.
Chappell, M. A., 1980, Insulation, radiation, and convection in small arctic mammals, J. Mammal., 61:268.
Cole, L. C., 1943, Experiments on the toleration of high tempera-

tures in lizards with reference to adaptive coloration, Ecology, 24:94.

Dawson, W. R., and Hudson, J. W., 1970, Birds, in: "Comparative Physiology of Thermoregulation," G. C. Whittow, ed., Academic Press, New York.

DeJong, A. A., 1976, The influence of simulated solar radiation on the metabolic rate of white-crowned sparrows, Condor, 78:174.

Dinsmore, J. J., 1972, Sooty tern behavior, Bull. Fla. State Mus. Biol. Sci., 16:129.

Farner, D. S., and Serventy, D. L., 1959, Body temperature and ontogeny of thermoregulation in the slender-billed shearwater, Condor, 61:426.

Finch, V. A., Omi'el, R., Boxman, R., Shkolnik, A., and Taylor, C. R., 1980, Why black goats in hot deserts? Effects of coat color on heat exchanges of wild and domestic goats, Physiol. Zool., 53:19.

Grant, G., 1979, Avian incubation: egg temperature, nest humidity and behavioral thermoregulation in a hot environment, Ph.D. Diss., Univ. Calif., Los Angeles.

Guard, C. L., and Murrish, D. E., 1974, Blood flow in the giant petrel, Antarctic J. U.S., 9:101.

Hamilton, W. J. II, and Heppner, F., 1967, Radiant solar energy and the function of black homeotherm pigmentation: an hypothesis, Science, 155:196.

Hand, J. L., Hunt, G. L., and Warner, M., 1981, Thermal stress and predation: Influences on the structure of a gull colony and possibly on breeding distributions, Condor, 83:193.

Heppner, F., 1970, The metabolic significance of different absorption of radiant energy by black and white birds, Condor, 75:50.

Howell, T. R., Araya, B., and Millie, W. R., 1974, Breeding biology of the Gray Gull, Larus modestus, Univ. Calif. Publ. Zool. 104.

Howell, T. R., and Bartholomew, G. A., 1962, Temperature regulation in the Red-tailed Tropic Bird and the Red-footed Booby Bird, Condor, 64:6.

Irving, L., 1955, Nocturnal decline in the temperature of birds in cold weather, Condor, 57:362.

Irving, L., 1956, The usefulness of Scholander's views on adaptive insulation of animals, Evolution, 10:257.

Irving, L., and Krog, J., 1955, Temperature of skin in the Arctic as a regulator of heat, J. Appl. Physiol., 7:355.

Lasiewski, R. C., 1969, Physiological responses to heat stress in the Poorwill, Amer. J. Physiol., 217:1504.

Lasiewski, R. C., Bernstein, M. H., and Ohmart, R. D., 1971, Cutaneous water loss in the roadrunner and poor-will, Condor, 73:470.

Lasiewski, R. C., and Dawson, W. R., 1967, A re-examination of the relation between standard metabolic rate and body weight in birds, Condor, 69:13.

Lasiewski, R. C., and Seymour, R. S., 1972, Thermoregulatory responses to heat stress in four species of birds weighing approximately 40 grams, Physiol. Zool., 45:106.

LeMaho, Y., Delclitte, P., and Chatonnet, J., 1976, Thermoregulation in fasting emperor penguins under natural conditions, Amer. J. Physiol., 231:913.

Lipkin, M., and Hardy, D. D., 1954, Measurements of some thermal properties of human tissues, J. Appl., Physiol., 7:212.

Lockley, R. M., 1983, "Flight of the Storm Petrel," Paul S. Ericsson, Middlebury.

Lustick, S. I., 1969, Bird energetics: effects of artificial radiation, Science, 163:387.

Lustick, S., Adam, M., and Hinko, A., 1980, Interaction between posture, color, and the radiative heatload in birds, Science, 208:1052.

Lustick, S., and Adams, J., 1977, Seasonal variation in the effects of wetting on the energetics and survival of starlings (Sturnus vulgaris), Comp. Biochem. Physiol., 56(A):173.

Lustick, S., Battersby, B., and Kelty, M., 1978, Behavioral thermoregulation: orientation toward the sun in herring gulls, Science, 200:81.

Lustick, S., Battersby, B., and Kelty, M., 1979, Effects of insolation on juvenile herring gull energetics and behavior, Ecology, 60(4):673.

Lustick, S., Talbot, S., and Fox, E. L., 1970, Absorption of radiant energy in redwinged blackbirds (Agelaius phoeniceus), Condor, 72:471.

MacMillen, R. E., Whittow, G. C., Christopher, E. A., and Ebisu, R. J., 1977, Oxygen consumption, evaporative water loss and body temperature in the sooty tern, Auk, 94:72.

Maher, W. J., 1962, Breeding biology of the snow petrel near Cape Hallett Antarctica, Condor, 64:488.

Marder, J., 1973, Body temperature regulation in the brown-necked raven (Corvus corax ruficollis). II. Thermal changes in the plumage of ravens exposed to solar radiations, Comp. Biochem. Physiol., 45(A):431.

Mayer, L., Lustick, S., and Battersby, B., 1982, The importance of cavity roosting and hypothermia to the energy balance of the winter acclimatized Carolina Chickadee, Int. Biometeor, 23:231.

Monteith, J. L., 1973, "Principles of environmental physics," American Elsevier Publishing Co., New York.

Mugaas, J. N., and King, J. R., 1981, Annual variation of daily energy expenditure by the black-billed magpie: A study of thermal and behavioral energetics, in: "Studies in Avian Biology," No. 5, Cooper Ornithological Soc.

Pinshow, B., Fedak, M. A., and Schmidt-Nielsen, K., 1974, Metabolic response of starving emperor penguins at low temperatures, Antarctic J. of U.S., 9:96.

Regal, P. J., 1975, The evolutionary origin of feathers, Quart. Rev. Biol., 50:35.

Scholander, P. F., 1956, Climatic rules, Evolution, 10:339.

Scholander, P. F., Hock, R., Walters, V., Johnson, F., and Irving, L., 1950, Heat regulation in some arctic and tropical mammals and birds, Biol. Bull., 99:237.

Steen, J., and Steen, I. B., 1965, The importance of the legs in the thermoregulation of birds, Acta. Physiol. Scand., 63:285.

Stonehouse, B., 1967, The general biology and thermal balance of penguins, in: "Advances in Ecological Research," J. B. Cragg, ed., Vol. 4, Academic Press, London.

Tucker, V. A., 1972, Metabolism during flight in the Laughing Gull, Larus atricilla, Amer. J. Physiol., 222:237.

Walsberg, G. E., Campbell, G. S., and King, J. R., 1978, Animal coat color and radiative heat gain: a re-evaluation, J. Comp. Physiol., 126:211.

Warham, J., 1971, Body temperature of petrels, Condor, 73:214.

ENERGETICS OF FREE-RANGING SEABIRDS

Hugh I. Ellis

Department of Biology
University of San Diego
San Diego, CA 92110

INTRODUCTION

I have chosen to address three major areas in the energetics
of free-ranging adult seabirds: basal metabolism, locomotion, and
energy budgets. More information is available on basal metabolism
than on the other major areas of this chapter, a reflection both of
the type of information most often collected in the past and of the
difficulties inherent in working with pelagic birds. Basal
metabolism not only provides a baseline and starting point for a
discussion on free-ranging energetics, but it also has certain interesting ecological correlates that may help elucidate the
energetics of seabirds. One such correlate involves plumage color
and has thermoregulatory implications (treated by Lustick in a
separate chapter), especially for birds of low (warm) latitudes.
Such other correlates of basal metabolism as climate and flight
and foraging behaviors are also discussed; these correlates may
suggest to readers possible avenues for future research.

The energetics of activity is of particular importance to a
general study of free-ranging energetics. Probably the activity
most amenable to measurement in adult seabirds is the cost of
incubation, which is discussed by Grant elsewhere in this book.
The only activity I treat here is locomotion. Although not numerous,
there are sufficient experimental and analytical studies to allow
comparisons of flapping and gliding flight as well as surface and
underwater swimming. Walking is treated briefly, but LeMaho provides a more detailed analysis for penguins in another chapter.

Finally, energy budgets are considered. Energy budgets may be
developed indirectly using time-activity information converted with

estimated energy costs or directly using isotopically labeled water. I have cited all the published and available but unpublished literature on seabird energy budgets measured directly using isotopes of water (see also the chapter on penguin energy budgets by Davis and Kooyman); these works shall be augmented soon by several studies currently in progress by different investigators. As more information becomes available, a general analysis of seabird energy budgets will be possible. For now, I have attempted to summarize the available information for seabirds and evaluate the different methods used to develop energy budgets. Regardless of whether energy budgets are developed directly or indirectly, they have profound ecological implications. Wiens, in his chapter, develops some of these implications for the population and community ecology of seabirds.

The survey of free-ranging energetics is confined largely to four orders of seabirds: Sphenisciformes (penguins), Procellariiformes (albatrosses, shearwaters, petrels), Pelecaniformes (pelicans, boobies, frigatebirds), and Charadriiformes (terns, gulls, auks). Other orders also include seabirds: Gaviiformes (loons), Podicipediformes (grebes), and Anseriformes (sea ducks), but these are not included in this survey due to a lack of data and because these groups include no entirely pelagic birds. The scientific names of most birds treated in this chapter are provided in Table 1 if not in the text.

BASAL METABOLISM

The most common unit used for energy expenditure is basal metabolic rate (BMR). Multiples of basal metabolism have been used to describe the cost of flight (Raveling and LeFebvre, 1967; Utter and LeFebvre, 1970; Tucker, 1972; Baudinette and Schmidt-Nielsen, 1974); energy budgets of individuals (Utter and LeFebvre, 1973, Pennycuick and Bartholomew, 1973; MacMillen and Carpenter, 1977; Ettinger and King, 1980; and reviews by Calder, 1974; King, 1974; and Walsberg, 1983; and population energetics (Furness, 1978). The use of basal metabolism as a base line is not unreasonable: BMR is measured on postabsorptive, resting endotherms in thermoneutrality and so provides a repeatable measure of the low end on the normal metabolic spectrum, notwithstanding shifts with circadian (Aschoff and Pohl, 1970) or seasonal (Kendeigh et al., 1977) cycles. King (1974:83) suggests BMR as a highly reliable (i.e., precise) measure.

Energy budgets are sometimes presented in terms of "predicted" metabolic rates. Lasiewski and Dawson (1967) provided an allometric equation to predict BMR for nonpasserine birds (including seabirds) on the basis of mass:

$$BMR = 327.8 \ m^{0.723} \tag{1}$$

where BMR is in kJ/da and m is in kg. Aschoff and Pohl (1970)
provided allometric equations which distinguish circadian differences for nonpasserines:

$$BMR_\alpha = 381.0 \; m^{0.729} \tag{2}$$

$$BMR_\rho = 307.7 \; m^{0.734} \tag{3}$$

where α represents the active phase and ρ the resting phase. The
units are as in equation (1). When multiples of BMR are not based
on measured values but on predicted ones, energy budgets may be
distorted, sometimes by appreciable amounts. Nevertheless, although
BMR is one of the most commonly measured physiological variables
among birds, energy budgets are occasionally constructed for birds
whose BMR is unknown and must be predicted (e.g., Withers, 1977;
Ashkenazie and Safriel, 1979; Burger, 1981; Ricklefs and White, 1981).

Measured basal metabolic rates of four orders and 12 families
of seabirds are presented in Table 1. The BMRs are also presented
as percentages of values predicted by equation (1). The Lasiewski-
Dawson equation is preferred here to those of Aschoff and Pohl
because 1) the time of BMR measurement was not reported for all
species cited, and Lasiewski-Dawson integrates both rest and activity
periods; 2) activity is not always associated clearly with day or
night (especially in Procelliiformes); and 3) it is unclear that
all polar birds show the circadian differences seen in birds of lower
latitudes (Ricklefs and Matthew, 1983). Table 1 shows that some seabirds deviate greatly from predicted values (i.e., 100%). Clearly,
energy budgets for seabirds derived from predicted BMRs should be
interpreted with care. However, the differences in BMRs among these
birds may also be instructive in attempting to analyze their free-
ranging energetics.

Figure 1 graphically represents the data of Table 1. For
comparison, the Lawieski-Dawson predictive line of equation (1) is
drawn. With the exception of two penguin species measured at night
and one species of shearwater, most seabirds fall above the line unless they are 1) largely tropical/subtropical in their nesting habits
and 2) dark in color. The Sooty Shearwater which is dark but nests
at higher latitudes (King, 1967) also falls below the line, as does
the Grey-backed Tern which is tropical though relatively light in
color; the latter bird will be discussed separately below. Seabirds
show an allometric relationship parallel to that of all nonpasserines as predicted by equation (1) but slightly elevated. Basal
metabolism for all seabirds in Table 1 is related to mass by the
equation:

$$BMR = 381.8 \; m^{0.721} \tag{4}$$

with the same units as in equations (1-3).

TABLE 1

Comparative Metabolism of Seabirds

Species	N	m^a	BMR^b	$\%BMR^c$	Source
Sphenisciformes					
Spheniscidae					
Aptenodytes forsteri (Emperor Penguin)	5	23.37	3704.1	116 (98α)	Pinshow, Fedak, Battles, and Schmidt-Nielsen 1976
Aptenodytes forsteri (Emperor Penguin)	11	24.80	4238.7	127 (107α)	Le Maho, Delclitte, and Chatonnet 1976
Pygoscelis adeliae (Adelie Penguin)	14	3.97	1060.1	119 (102α)	Kooyman et al. 1976 (as cited by Stahel and Nicol 1982)
Eudyptes pachyrhynchus (Fjordland Penguin)	4	2.60	599	92 (97ρ)	Drent and Stonehouse 1971
Megadyptes antipodes (Yellow-eyed Penguin)	1	4.80	996	98 (103ρ)	Drent and Stonehouse 1971
Eudyptula minor (Little Penguin)	6	0.90	383.5	126 (109α)	Stahel and Nicol 1982
Spheniscus humboldti (Peruvian Penguin)	3	3.87	821	94 (99ρ)	Drent and Stonehouse 1971

Procellariiformes

Diomedeidae

Diomedea immutabilis (Laysan Albatross)	5	3.103	637.2[d]	82.5	Grant and Whittow 1983

Procellariidae

Macronectes giganteus (Giant Petrel)	8	3.46	1465.8[e]	182	Ricklefs and Matthew 1983
Pterodroma hypoleuca (Bonin Petrel)	2	0.180	88.8[d]	94	Grant and Whittow 1983
Pterodroma hypoleuca (Bonin Petrel)	7	0.167	72.4	80.4 (69.9α/87.1ρ)	Pettit, Ellis and Whittow unpublished data
Bulweria bulwerii (Bulwer's Petrel)	6	0.087	44.0	78.4 (68.4α/85.4ρ)	Pettit, Ellis and Whittow unpublished data
Puffinus pacificus (Wedge-tailed Shearwater)	4	0.3378	128.5	85.9 (74.4α/92.3ρ)	Pettit, Ellis and Whittow unpublished data
Puffinus nativitatis (Christmas Island Shearwater)	6	0.3076	127.3	91.0 (79.0α/98.0ρ)	Pettit, Ellis, and Whittow unpublished data
Puffinus griseus (Sooty Shearwater)	3	0.740	249.5	95	Krasnow 1979

(continued)

TABLE 1 (continued)

Species	N	m^a	BMR^b	$\%BMR^c$	Source
Hybrobatidae					
Oceanodroma leucorhoa beali (Beal's [Leach's] Storm Petrel)	1	0.042	55.3	167[f]	Iverson and Krog 1972
Oceanodroma furcata (Fork-tailed Storm Petrel)	1	0.049	56.1	151[f]	Iverson and Krog 1972
Pelecaniformes					
Phaethontidae					
Phaethon rubricauda (Red-tailed Tropicbird)	5	0.5932	287.6	127.9 (110.5α)	Pettit, Ellis, and Whittow unpublished data
Pelecanidae					
Pelecanus conspicillatus (Australian Pelican)	1	5.09	1565.9	148	Benedict and Fox 1927
Pelecanus occidentalis (Brown Pelican)	1	3.510	1105.3	137	Benedict and Fox 1927
Pelecanus occidentalis (Brown Pelican)	3	3.038	896.0	123	Ellis and Hennemann unpublished data
Sulidae					
Sula dactylatra (Masked Booby)	1	1.289	475.6	122	Ellis unpublished data

Species					Reference
Sula sula (Red-footed Booby)	8	1.017	376.0	114	Ellis, Maskrey, Pettit, and Whittow 1982a
Phalacrocoracidae					
Phalacrocorax auritus (Double-crested Cormorant)	5	1.33	537.2 / 467.7	133 (155α) 116 (129ρ)	Hennemann 1983
Phalacrocorax atriceps (Blue-eyed Shag)	6	2.66	1307.5[e]	197	Ricklefs and Matthew 1983
Fregatidae					
Fregata magnifiscens (Magnificent Frigatebird)	4	1.078	239.9	69	Enger 1957
Charadriiformes					
Stercorariidae					
Catharcta skua (Great Skua)	1	0.97	410	128	Benedict and Fox 1927
Catharcta maccormicki (Southern Polar Skua)	9	1.13	701.3[e]	196	Ricklefs and Matthew 1983

(continued)

TABLE 1 (continued)

Species	N	m^a	BMR^b	$\%BMR^c$	Source
Laridae					
Gabianus pacificus (Pacific Gull)	1	1.210	531.7	141	Benedict and Fox 1927
Larus delawarensis (Ring-billed Gull)	3	0.439	249.5	139	Ellis 1980a
Larus canus (S)[g] (Common [= Mew] Gull) (W)	? ?	0.428 0.431	201.0 194.3	113 109	Gavrilov (in Kendeigh et al. 1977)
Larus argentatus (Herring Gull)	6	1.000	414.9	127	Lustick, Battersby, and Kelty 1978
Larus occidentalis (Western Gull)	7	0.761	293.9	109	Obst unpublished data
Larus hyperboreus (Glaucous Gull)	2	1.210	753.6	200	Scholander, Hock, Walters, and Irving 1950
Larus atricilla (Laughing Gull)	4	0.2756	162.0	126	Ellis 1980a
Larus ridibundus (S) (Black-headed Gull) (W)	? ?	0.252 0.289	187.6[f] 179.2[f]	155 134	Davydov 1972
Larus ridibundus (S) (Black-headed Gull) (W)	? ?	0.285 0.306	173.3 160.8	131 115	Gavrilov (in Kendeigh et al. 1977)

210

Species					Reference
Sterna lunata (Grey-backed Tern)	2	0.1317	60.7	80 (70α)	Pettit, Ellis, and Whittow unpublished data
Sterna fuscata (Sooty Tern)	6	0.148	68.7	84	MacMillen, Whittow, Christopher, and Ebisu 1977
Sterna maxima (Royal Tern)	3	0.373	217.3	135	Ellis 1980a
Anous stolidus (Brown Noddy)	16	0.1387	67.4	86	Ellis, Maskrey, Pettit, and Whittow 1982b
Anous tenuirostris (Black Noddy)	4	0.0902	54.8	95 (83α)	Pettit, Ellis, and Whittow unpublished data
Gygis alba (White [= Fairy] Tern)	6	0.0981	70.3	115 (100α)	Pettit, Ellis, and Whittow unpublished data
Alcidae					
Uria lomvia (Thick-billed Murre)	5	0.989	587.8	181	Johnson and West 1975
Uria aalge (Common Murre)	5	0.956	587.8	185	Johnson and West 1975

(continued)

^{a}mass in kg

^{b}basal metabolic rate in kJ/da

^{c}%of predicted BMR based on Lasiewski and Dawson (1967); values in parentheses based on Aschoff and Pohl (1970) as suggested in source (α=active phase, ρ=resting phase)

^{d}value in quiescent, fasting bird while incubating

^{e}values represent possibly non-postabsorptive birds measured at 2-14°C

^{f}measurement made at 20°C air temperature

^{g}S = summer, W = winter measurements

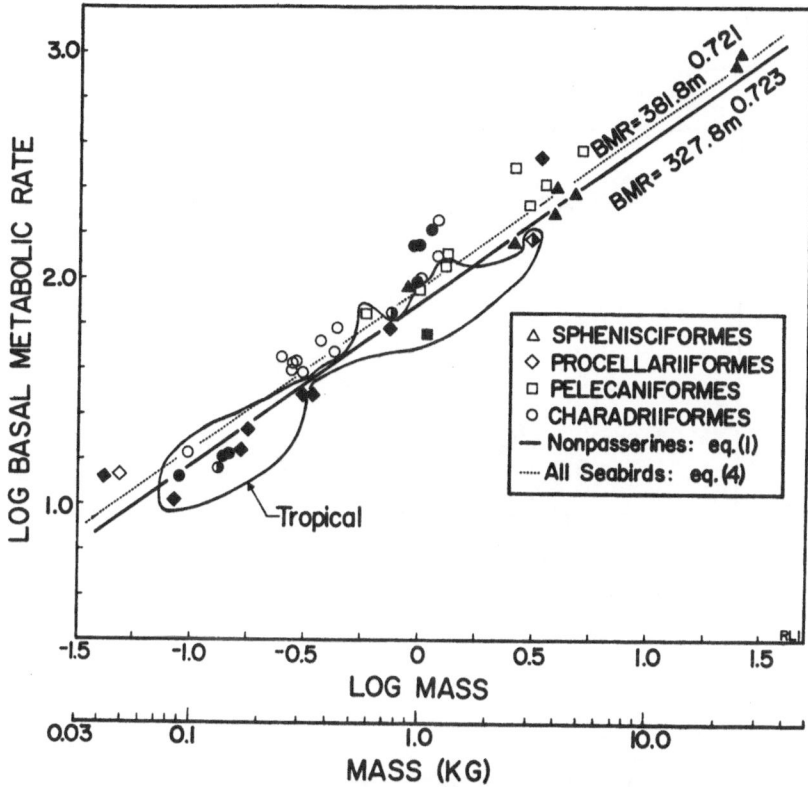

Fig. 1. Basal metabolism as a function of mass; data are taken from Table 1. Closed symbols are dark birds and open symbols are white birds. Symbols only half closed indicate birds that are black and white or grey.

Perhaps the most important reason why seabirds deviate from expected values of BMR is climate. This can be examined to the extent that climate is reflected in latitude. The order with the largest number of species (mostly in Laridae) and the widest geographic distribution in Table 1 is the Charadriiformes. In Figure 2, I have plotted % BMR as a function of breeding latitude for those charadriiform species not measured in zoos. The high latitudinal correlation ($r = 0.931$, $P \ll 0.001$, d.f. = 14) is not unexpected, having been suggested for terrestrial birds by Weathers (1979) and substantiated recently by Hails (1983), but it is perhaps stronger

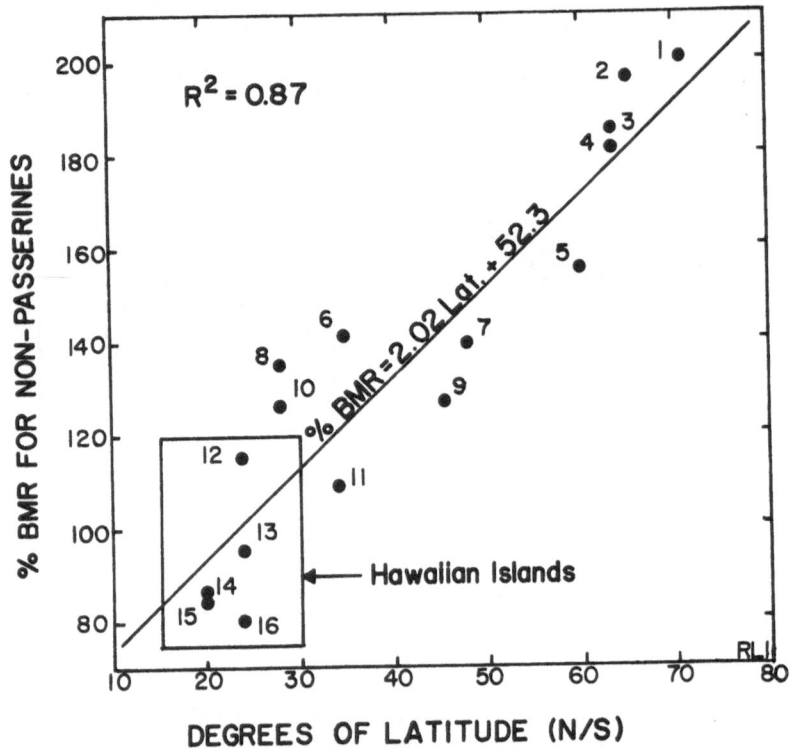

Fig. 2. Percent of basal metabolic rate (Lasiewski and Dawson, 1967) as a function of breeding (in species 5, capture) latitude. Species from zoos were not used. Data are from sources listed in Table 1 (species 5 from Davydov, 1972; summer measurement). 1 = Glaucous Gull, 2 = Southern Polar Skua, 3 = Common Murre, 4 = Thick-billed Murre, 5 = Black-headed Gull (S), 6 = Pacific Gull, 7 = Ring-billed Gull, 8 = Royal Tern, 9 = Herring Gull, 10 = Laughing Gull, 11 = Western Gull, 12 = White Tern, 13 = Black Noddy, 14 = Brown Noddy, 15 = Sooty Tern, 16 = Grey-backed Tern.

(slope = 2) than might be anticipated. There is some suggested latitudinal gradients in the other orders of seabirds listed in Table 1, but sample sizes and/or geographic distributions are too limited to analyze definitively. Weathers (1979) and Hails (1983) indicated that at low latitudes (terrestrial) birds foraging in the open have lower BMRs than those foraging in the shade. This may be particularly true for dark birds. Ellis (1980b) suggested that most dark birds living at low latitudes and exposed to high incident radiation have low metabolic rates. Figure 1 shows this relationship to continue when extended to seabirds. It is difficult not to draw thermoregulatory inferences from this correlation, but no causality has been experimentally demonstrated yet.

By isolating seabirds of one latitudinal region, non-climatic correlates of BMR can be more readily observed. I have chosen the latitudes of 24°S–24°N which Ashmole (1971) and Ainley and Boekelheide (1984) consider to include tropical and subtropical waters. Charadriiform, pelecaniform, and procellariiform seabirds of the Hawaiian Islands provided most of the following data. Dark seabirds at low latitudes not only show lowered BMRs by comparison to light ones, but also have different flight patterns and feeding methods. Table 2 shows that almost all those birds having low metabolic rates glide, flap/glide, or soar, although Sooty Terns may fly predominatly by flapping at certain speeds (see below) and Brown and Black Noddies were seen only to flap when observed by me on Tern Island, French Frigate Shoals (in the northeastern Hawaiian chain). Among pelecaniform birds of low latitudes, dark frigate-birds, which have quite low metabolic rates (Table 1: Enger, 1957), are capable of soaring for long periods (Pennycuick, 1983). By contrast, all light colored birds, except the Grey-backed Tern, have relatively high metabolic rates and are strong flapping fliers. At least one of these birds, the White Tern, is capable of hovering without wind assistance, a particularly energy-demanding task.

Table 2 lists the primary feeding method(s) of these low latitude seabirds. Most of the data are drawn from Ainley and Boekelheide (1984); the methods are defined elsewhere (Ashmole, 1971; Ainley, 1977; Ainley and Boekelheide, 1984). In general, seabirds that feed by shallow or deep plunging have high metabolic rates, whereas seabirds that feed by surface seizing or dipping usually have low BMRs. An interesting terrestrial analogy may be seen among falconiform birds: although Ospreys soar a good deal, they also plunge when feeding; the BMR of Ospreys is noticeably higher than those of hawks that soar but do not plunge (Wasser, 1979). Birds that plunge usually leave the water rapidly, requiring a large power output.

Ainley (1977) correlated feeding method to buoyancy index, a measure presented and described in Table 2. However, I was not able to correlate buoyancy index, wing-loading, or aspect ratio with BMR for any birds listed in Table 1 where wing characteristics could be found (N = 16, P > 0.05). For all those nonpasserine birds (including the 16 seabirds above) with known BMRs and wing characteristics (N = 46), a low inverse correlation (r = -0.42, P < 0.01) was found between BMR and buoyancy index (Ellis unpublished data). Such

TABLE 2
Activity and Metabolism of Tropical Seabirds

Species	Color[a]	%BMR[b]	BI[c]	Feeding[d]	Flight[e]
Procellariiformes					
Laysan Albatross	D/W	82.5	3.58[1]	surface seize	glide, flap/glide
Bonin Petrel	D	80.4	3.94[1,2]	surface seize	flap/glide
Bulwer's Petrel	D	78.4	4.2[3]	dip	flap/glide
Wedge-tailed Shearwater	D	85.9	3.3[3]	dip	flap/glide
Christmas Island Shearwater	D	91.0	3.8[3]	dip	flap/glide
Pelecaniformes					
Red-tailed Tropicbird	W	127.9	3.3[3]	deep plunge	flap, wind assisted hovering
Masked Booby	W	122	3.4[3]	deep plunge	flap
Red-footed Booby	W	114	3.9[3]	deep plunge, aerial pursuit	flap, flap/glide
Magnificent Frigatebird	D	69	5.48[4]	dip	glide, soar
Charadriiformes					

	[a]		[c]	[d]	[e]
Laughing Gull[f]	G	126	4.78[4]	scavenge, seize	flap
Royal Tern[f]	G/W	135	3.95[4]	shallow plunge	flap
Grey-backed Tern	G	80	4.2[3]	dip, shallow plunge	flap/glide, glide?
Sooty Tern	D	84	4.1[3]	dip	flap, soar, flap/glide
Brown Noddy	D	86	4.3[3]	dip	flap
Black Noddy	D	95	3.9[3]	dip	flap
White Tern	W	115	3.9[3]	dip	flap, hover

[a] colors of dorsum: D=dark, W=white, G=grey; D/W signifies a dark and white pattern

[b] based on equation (1) as in Table 1

[c] buoyancy index (Hartman, 1961 and as suggested by Ainley, 1977) = $\dfrac{\text{surface area}^{-2}}{\text{mass}^{-3}}$; data from 1=Warham (1977),

2=Ellis and Pettit (unpublished data), 3=E.Knudtson (pers. comm.), 4=Hartman (1961)

[d] as suggested by Ainley (1977), and Ainley and Boekelheide (1983)

[e] based on observation; also see text

[f] primarily subtropical to temperate in distribution

217

correlations deserve further analysis as more information becomes available.

In general, a case can be made that at low latitudes birds that primarily glide, soar, or flap/glide in flight and dip or seize in feeding have low metabolic rates. They are also usually dark in coloration; however, the presence of the Grey-backed Tern in this group suggests that color is not primary. In contrast, seabirds that are primarily flappers, hoverers, and plunge feeders have high metabolic rates. All the white seabirds of Table 2 fit into that category, but the inclusion of dark Laughing Gulls suggests again that color is not primary, although it may have thermoregulatory significance. The apparent correlation of high BMR with a more energetically expensive mode(s) of life suggests a linkage between BMR and maximal power output. This has been demonstrated for some mammals (Lechner, 1978) but not for birds and deserves future attention.

LOCOMOTION

One of the most important costs in free-ranging seabirds involves locomotion. With the exception of penguins and some isolated groups, all seabirds fly; most swim; and all, save the tropicbirds and storm petrels, walk. The cost of locomotion in animals has been described by Taylor et al. (1970) and Schmidt-Nielsen (1972) among others. They note that the total cost of movement includes a nonlocomotory component in addition to BMR; this may involve postural costs. In the following discussion, all values for metabolic expenditure during locomotion (measured as energy per unit time) are total metabolic costs of birds during movement; no attempt is made to provide the net cost of locomotion itself. Cost of locomotion is a common measure in the literature and allows a comparison of power output among species; it should not be confused with cost-of-transport, a term utilized below, which refers to the mass-specific energy expended per unit distance.

Flight

The metabolic cost during flight has been approached in several ways. Berger et al. (1970) measured oxygen consumption in Ring-billed Gulls during short flights (7-15 sec). They found the average metabolic rate of these birds to be 65.5 ml O_2/min. Because BMR in this species averages 8.74 ml O_2/min (Ellis, unpublished data), metabolism reached about 7.5 × BMR under those conditions. Tucker (1972) recorded the metabolism of Laughing Gulls flying in a wind tunnel. Like the Ring-billed Gulls studied by Berger et al. (1970), these birds wore a mask from which expired air could be analyzed. However, Tucker estimated the power input during flight in Laughing Gulls without a mask by applying a correction determined in an earlier study of Budgerigars (Tucker, 1968). He concluded that the metabolic cost during flight in Laughing Gulls unfettered by a mask and associated tube could be predicted by the equation:

$$P_i = (0.130V^2 - 1.86V + 34.5) \; m^{0.325} + 31 \; m \cdot V \cdot \sin\theta \qquad (5)$$

where P_i is the power input in W, V is the flight velocity in m/sec, m is mass in kg and θ is the angle of flight deviation from the horizontal. For a 0.322 kg Laughing Gull flying between 6 and 13 m/sec, he estimated metabolism to be 12-14 × BMR. However, his BMR was based on the prediction of the Lasiewski-Dawson equation (1), whereas Ellis (1980a) reported BMR for Laughing Gulls to be substantially higher. Using the higher value (126% of that expected from the Lasiewski-Dawson prediction) from Table 1, metabolism during flight is actually 9.5-11.1 × BMR. However, Tucker (1973) also estimated the power input in flying birds based on an aerodynamic model. This model predicts a metabolic rate 9% lower than his Laughing Gull measurements for flight velocites where P_i/V (cost-of-transport) is minimal. Greenewalt (1975), using aerodynamic theory, wing measurements, and power equations derived for pigeons (*Columba livia*) for which reliable metabolic date were available (LeFebvre, 1964), analyzed Tucker's data and argued that Tucker's values for power input were overestimated by at least 30%. Greenwalt (1975:35) suggested that the measurement of metabolism of birds wearing masks and flying in wind tunnels is highly artificial; the implication is that such measurements are bound to be inflated.

Also using wind tunnels and masks, Baudinette and Schmidt-Nielsen (1974) measured the metabolic rates of two Herring Gulls during gliding. These birds glided for 20 min periods and were assumed to be in steady state. Baudinette and Schmidt-Nielsen reported that gliding flight represents a doubling of resting metabolic values. However, using the measured BMR of the Herring Gulls found in Table 1 (based on Lustick et al., 1978), gliding flight represents 3.1 × BMR in this species.

Pennycuick (1982) has warned that there is an inherent problem when the energetics of flapping flight is compared (at minimal P_i/V) in birds of different mass. Because the mechanical component of P_i scales with mass more rapidly than metabolism does, P_i cannot be viewed as a constant multiple of BMR. Presumably, large birds would show lower multiples than small ones. This situation has not yet been shown for gliding (Pennycuick, 1982). The BMR of the three flying seabirds in Table 3 (less than one order of magnitude) probably differ more due to method of measurement than to inherent differences because the species are similar in size.

One study of flight in a free-ranging seabird has been done. Flint and Nagy (1984) studied the energetics of Sooty Terns. Because

Sooty Terns almost certainly do not land on water (Watson and Lashley, 1915; Gould, 1974; and Johnston, 1979) and apparently do not loaf on other islands when away from their nests during the breeding season (Flint, personal communication), they are ideal subjects for applying doubly labeled water studies to flight. Flint and Nagy found the metabolic rate of these birds during flight at low wind velocities (0-5.14 m/sec) to be 4.77 × BMR. At these speeds, Sooty Terns flap 94% of the time. Metabolic estimates based on the wind tunnel work of Tucker (1972) for flapping and Baudinette and Schmidt-Nielsen (1974) for gliding would predict a much higher multiple of BMR than Flint and Nagy found. The relatively low metabolic rate of Sooty Terns in flight suggests that birds flying free and able to take advantage of wind currents may fly at less cost than birds in wind tunnels, as Greenewalt (1975) implied.

The utilization of wind currents is only one way in which the energetic cost of flight may be reduced. Pennycuick (1975) discussed various types of soaring (slope soaring, wave soaring, wind-gradient soaring and thermal soaring) that may be used by seabirds and that reduce energy expenditure. He has also analyzed thermal soaring and slope soaring in Magnificent Frigatebirds and Brown Pelicans specifically (Pennycuick, 1983) as well as soaring and other flight modes of procellariiforms (Pennycuick, 1982). Withers and Timko (1977) showed for Black Skimmers (*Rynchops nigra*) how flying close to the water could reduce the cost of flight due to the ground effect which reduces drag and increases lift when an airfoil operates to the ground (here, water). They implied that ground effect might operate for a number of seabirds that fly close to the water's surface. In addition, Withers (1979) found "hovering" in the Wilson's Storm Petrel (*Oceanites oceanicus*) not to be energetically expensive normal hovering, but an inexpensive type of flight dependent in part on soaring and ground effect. Lissaman and Shollenberger (1970) demonstrated that birds flying in vee formation could increase their aerodynamic efficiency by reducing drag. By flying in formation, a group can substantially increase its (migratory) range over that of an individual. A few seabird species (e.g., Brown Pelicans, cormorants) do occasionally fly in formation. Brown Pelicans flying in true vee or (more often) offset vee formations often flap sequentially from the point to the end of the formation (personal observation); this may allow birds behind to take advantage of the vortex generated by the flapping of the preceding birds.

Finally, Schnell and Hellack (1979) analyzed flight speeds in 12 species of larids flying at a variety of speeds near their colonies. They hypothesized that the birds could minimize 1) metabolic rate in flight, 2) cost-of-transport through the air, or 3) cost-of-transport over the ground. Their computations indicated birds pick speeds that are a compromise between 1) and 3) but favor minimizing the cost-of-transport over the ground.

Swimming

Many seabirds swim on the surface of the water, paddling with their feet (e.g., albatrosses, gulls, grebes). There are currently no published energetics data on surface paddling in seabirds. However, Prange and Schmidt-Nielsen (1970) studied that mode in Mallard ducks (*Anas platyrhynchos*) and found them to have a minimum cost-of-transport (24.2 kJ/kg·km) at a speed of 0.5 m/sec. Metabolism in the Mallards appeared constant over speeds of 0.35-0.50 m/sec, equalling 2.2 × resting metabolic rate. Calculating BMR by using the value of 0.93 1 O_2/kg·hr (Prange and Schmidt-Nielsen, 1970) and the five values given for Mallards in Appendix 5.1 of Kengeigh et al. (1977), surface swimming costs 3.46 × BMR for speeds up to 0.5 m/sec. At higher speeds, metabolic rate increases rapidly to 6.45 × BMR at a maximum sustainable speed of 0.7 m/sec. Maximum speed for paddlers is dictated by hydrodynamic factors (e.g., the square root of "hull" length) as well as the power output of leg muscles (Prange and Schmidt-Nielsen, 1970). Because of these constraints surface swimming does not appear to be as metabolically demanding a form of locomotion as flight which can require in excess of 10 × BMR as discussed above.

Many seabirds swim underwater in pursuit of prey, including those described by Ainley (1977) as "divers" and "pursuit plungers". Seabirds that swim under the water use their feet (cormorants) or wings (penguins and alcids) for propulsion. Because submerged swimming, unlike surface swimming, generates no wave formation (an energy demanding process), birds swimming underwater can achieve greater speeds for the same metabolic output as they could by swimming on the surface. Hui (1983) has measured the total cost of swimming in Peruvian Penguins. He found that metabolic rate in two swimming penguins (average mass = 3.75 kg) could be described as:

$$P_i = 31.7 + 114.4V \qquad (6)$$

where P_i is metabolic rate in ml O_2/min and V is swimming velocity in m/sec ($r^2 = 0.41$). Between 0.5 and 1.25 m/sec, the cost-of-transport averaged 0.69 ml O_2/kg·m and was independent of V. The independence of cost-of-transport from swimming speed is due to modifications of swimming behavior which increase efficiency with V. As these penguins approach a speed of ca. 0.6 m/sec, they dive.

Furness and Cooper (1982) suggested that swimming underwater and surface swimming could be treated energetically as flap flying and gliding, respectively. A common surface speed seen in captive Peruvian Penguins is ca. 0.5 m/sec (Hui, personal communication). The metabolism at that speed is 1.42 1 O_2/kg·hr or 2572 kj/da (assuming RQ = 0.77) according to equation (6). This is 3.13 × BMR, as determined by Drent and Stonehouse (1971) for this species. This value is indeed similar to that of gliding Herring Gulls (see Table 3), but

TABLE 3

Metabolic Cost of Locomotion

Species	X BMR[a]	Source
Flying (flap)		
Ring-billed Gull	7.5	Berger, Hart, and Roy, 1970
Laughing Gull	9.5-11.1	Tucker, 1972
Sooty Tern	4.8	Flint and Nagy, 1984
Gliding		
Herring Gull	3.1	Baudinette and Schmidt-Nielsen, 1974
Swimming (surface)		
Peruvian Penguin	3.1	Hui, 1983
Walking		
Emperor Penguin	4	Pinshow, Fedak, and Schmidt-Nielsen, 1977
Emperor Penguin	3.7-4.4[b]	Dewasmes, Le Maho, Cornet and Groscolas, 1980
Adelie Penguin	4.5	Pinshow et al., 1977
White-flippered Penguin	3.4[c]	Pinshow et al., 1977

[a] see text for speeds

[b] fasting

[c] probably a multiple of RMR, not BMR

this may be coincidental. Linking underwater swimming to flap flying in terms of energy expenditure is based on the presumption that underwater swimming velocities are limited by metabolic capacity not by hydrodynamic constraints. Nagy et al. (in press) found that during submerged swimming in Jackass Penguins (*Spheniscus demersus*) the metabolic expense was 9.8 × BMR; this value is consistent with computations by Hui (1983) for Peruvian Penguins and is similar to values for flap fliers.

In summary, surface swimming is fairly inexpensive because maximum speeds are low. Future studies of the energetics of underwater swimming in birds should be highly enlightening. It should be noted that although swimming appears metabolically less demanding, flight has a lower cost-of-transport, because at the higher speeds birds can go much further for the same kilocalorie when flying.

Walking

The metabolic costs involved in bipedal locomotion (particularly running in birds) has been analyzed by Fedak et al. (1974). Walking plays an important role in the life histories of many penguins, particularly Emperor Penguins, and studies of walking in seabirds have been limited to these species. Pinshow et al. (1977) and Dewasmes et al. (1980) investigated the cost of walking in several penguin species. Pinshow et al. found walking at speeds of 2 km/hr (0.56 m/sec) to equal 4.0, 4.5, and 3.4 × resting metabolic rates (RMR) in unfasted Emperor Penguins, Adelie Penguins, and White-flippered Penguins (*Eudyptula albosignata*), respectively. Dewasmes et al. found that metabolic rate dropped from about 4.4 to 3.7 × RMR as body mass decreased from 38 to 23 kg in fasted Emperor Penguins walking at 1.4 km/hr (0.39 m/sec). In both studies, the birds were assumed or known to be in thermoneutrality. The RMR values reported by Pinshow et al. for Emperor and Adelie Penguins are close to BMR (see Table 1); however the RMR of White-flippered Penguins exceeds the BMR predicted by equation (1) by 55%. If BMR in the White-flippered Penguin is less then RMR, then 3.4 × RMR would increase to a BMR multiple comparable to that of the other penguin species (Table 3). Walking therefore, is a metabolically undemanding form of locomotion (ca. 4 × BMR) probably due to mechanical limitations on maximum speed. The chapter by LeMaho should shed more light on this subject.

ENERGY BUDGETS

The energetics of free-ranging birds has been approached most often with the use of energy budgets. These budgets have been estimated either indirectly through the collection of time-activity data (for example, Pearson, 1954; Schartz and Zimmerman, 1971; King, 1974; Burger, 1981) or directly using isotopically labeled water (LeFebvre,

1964; Utter and LeFebvre, 1973; Weathers and Nagy, 1980). Because
of their relative ease and economy, time budgets have been the primary means to estimate energy budgets although they pose special
problems with seabirds. The development of detailed energy budgets
for seabirds requires an analysis of how these birds spend their
time. This information has been very hard to acquire (although see
Dunn, 1979). Many seabirds are able to travel long distances while
foraging. Even when breeding, especially where incubation shifts
are long, seabirds may fly very far. Red-tailed Tropicbirds, as
well as many procellariiforme species (among others), may fly long
distances to feed, and Sooty Terns can fly to feeding areas 700 miles
away (Ashmole and Ashmole, 1967). It is nearly impossible to know in
what activities seabirds are engaged when they are out at sea. Consequently, energy budgets usually are based on birds during the
breeding season when at least some activities can be monitored
directly. Nevertheless, seabirds are not as amenable to the development of energy budgets as other birds and little has been done until
recently.

Models Using Basal Metabolism

One recent study of energy budgets (Burger, 1981) utilized time-
activity budgets for Lesser Sheathbills (*Chionis minor*) and trans-
lated the activities into energy estimates utilizing multiples of
BMR. Here BMR was predicted by an allometric equation provided by
Kendeigh et al. (1977: eq. 5.5). This study recognized various
multiples of BMR for flight (10-12×), swimming (4×), hopping (5×),
running (3.5-14×), resting and brooding (1.5×), comfort behavior
(2×), foraging, nest-building and displaying (4×), and active territorial defense (12×). All these multiples came from different
studies of passerine and nonpasserine birds or were estimates. The
cost of thermoregulation was estimated using other predictive
equations (Kendeigh et al., 1977). Burger found peak energy demands
in three pairs of sheathbills to range from 1050 to 1505 kJ/da
(5.4-7.7 × BMR). From this information Burger was able to predict
certain ecological consequences regarding reproduction in these
birds. If this study with its many estimates and assumptions can
be verified by more direct means (doubly labeled water), the use of
time-activity budgets converted with multiples of BMR may be viewed
with greater confidence. Of course, BMR itself must be empirically
determined or at least closely predicted for this analysis to work.

Burger (1981) had the advantage of a seabird which fed close to
its nesting site. Because that is not often the case for seabirds,
other energy budgets proposed for seabirds have been much less am-
bitious. The energy budget of Withers and Timko (1977) concentrated
on estimates of foraging, while Ricklefs et al. (1980) and Ricklefs
and White (1981) estimated the cost of a family unit with special
regard to bringing a chick to fledging. The subjects of these three
studies spend appreciable periods of time away from the nest and
out of sight.

Models Using Existence Metabolism

Another way to convert time-activity budgets to energy budgets involves the predictive equations of Kendeigh et al. (1977) instead of using various multiples of BMR. Kendeigh (1970) provided an allometric equation predicting existence metabolism (EM) on the basis of mass. This equation was revised (Kendeigh et al., 1977) as more information became available and now involves several related equations for passerines and nonpasserines in summer (long photoperiod) and winter (short photoperiod), and at 0°C and 30°C. The nonpasserine equations for summer (breeding season) are as follows:

at 30°C, ca. 15 hr photoperiod (N = 70)

$$EM = 4.55 \ m^{0.664} \pm 5.53 \qquad\qquad (7a)$$

at 0°C, ca. 15 hr photoperiod (N = 70)

$$EM = 17.34 \ m^{0.544} \pm 4.72 \qquad\qquad (7b)$$

where EM is an kJ/da and m is mass in g. The comparable equations for short photoperiods (ca. 10 hr) are very similar. Existence metabolism, as defined by Kendeigh et al. (1977) encompasses basal metabolism, temperature regulation, specific dynamic action, and the energy expended in locomotion while a bird is in a cage. If this cage locomotory energy is equivalent to that near a nest, EM may be a useful measure of the metabolic costs of seabirds near the nest. Kendeigh et al. (1977) also gave temperature corrections for EM which, unlike BMR, shows no thermoneutral zone. For nonpasserine birds in the summer, the temperature coefficient is

$$b = 1.16 \ m^{0.282} \pm 6.27 \qquad\qquad (8)$$

where b is in kJ/da·°C and m is mass in g.

Most seabirds spend appreciable amounts of time feeding away from their nest. With the exception of penguins, much of that time is spent flying. If the amount of time spent flying can be estimated, the energy utilized can be computed using an equation by Kendeigh et al. (1977):

$$P_i = 1.32 \ m^{0.698} \pm 4.85 \qquad\qquad (9)$$

where P_i is power input during flight in kJ/hr and m is mass in g. The estimate of P_i is about 12 × BMR, which may be high for seabirds (Table 3).

Daily energy budgets (DEB) would estimate total metabolism for
free-living birds and would combine EM and major locomotry costs.
Kendeigh et al. (1977) found DEB to exceed EM by 7.1% in the winter
and by 26% in the summer (breeding/molting season). For the summer,
they predicted

at 30°C: DEB = 4.52 $m^{0.67}$ (10a)

at 0°C: DEB = 33.73 $m^{0.50}$ (10b)

where DEB is in kJ/da and m is mass in g. Unfortunately, equations
(10a) and (10b) are based on data from a single species of gallina-
ceous bird (West, 1968). Because Kendeigh et al. recognize the
effects of climate, they suggest that equations (10a) and (10b) may
overestimate DEB for low latitude birds and underestimate it for
high latitude birds (but see below).

Energy budgets based on Kendeigh's equations have been proposed
for seabirds, especially in building models for estimating the energetics of populations. Wiens and Scott (1975) and Furness (1978)
utilized predictions of EM based on Kendeigh (1970) to estimate the
energetics of northern seabirds in two different communities. More
recently, Furness and Cooper (1982) have used the equations of
Kendeigh et al. (1977) to build a model explaining the interactions
of seabird and fish populations off the west coast of southern
Africa. One seabird species considered by Furness and Cooper was
the Jackass Penguin (*Spheniscus demersus*) whose DEB was independently
studied by Nagy et al. (in press) using a doubly labeled water technique. This direct method gave virtually the same results as the
equations of Kendeigh et al. (1977) when applied to a time-activity
budget (Nagy, personal communication).

Energy Budgets from Isotopically Labeled Water

Doubly labeled water has been used as a measure of metabolism
in birds since it was first demonstrated for pigeons by LeFebvre
(1964). The method utilizes isotopes of both hydrogen and oxygen
and, through their turnover in animals, allows a precise determination
of CO_2 production. The theory of the method is described by Lifson
and McClintock (1966) and critically assessed by Nagy (1980).

Whereas time-activity studies depend ultimately on one or more
predictive equations or subjective estimates for their conversion
energy budgets, doubly labeled water studies do not. Weathers and
Nagy (1980) showed that time-activity studies often underestimate
daily energy expenditure by as much as 40% when compared with doubly
labeled water measurements. However, it may be possible to bring
the two methods much closer together by applying heat transfer
analysis and other corrections to the time-activity model (Weathers
and Buttemer, personal communication).

Several labeled water studies have been done recently on
seabirds (Table 4). Those studies provide integrated values for
field metabolism and thereby demonstrate one of the advantages of
the direct (isotope) method: even without time-activity measurements - often difficult or impossible to obtain in seabirds - it
is possible to measure free-ranging energetics. To the extent that
time-activity budgets are available, it is possible to establish
real costs of various activities with accuracy. Such is the case
for Sooty Terns (Flint and Nagy, 1984) and Jackass Penguins (Nagy
et al., 1984), as noted in the flight and swimming sections
above.

Because labeled water studies require the recapture of
experimental animals after a relatively short period, these studies
are usually done during the breeding season with nesting birds. The
values given for Sooty Terns and Brown Noddies in Table 4 represent
DEB for birds foraging and incubating eggs. The values for Wedge-
tailed Shearwaters are for birds foraging and incubating and/or
brooding very young chicks. Birds away from their nests are generally foraging, but they may also be resting on an island or on the
water, or flying to a specific foraging site (as opposed to searching
for food while flying). To partition out foraging costs, it is
necessary to know foraging habits and time-activity budgets as sea.
Among some species of northern seabirds, the time spent foraging
during the breeding season is an inverse function of body size
(Pearson, 1968).

Studies on Gentoo Penguins (*Pygoscelis papua*), Macaroni Penguins
(*Eudyptes chrysolophus*), and King Penguins (*Aptenodytes patagonica*)
have been done, not with doubly labeled water, but with singly
labeled (tritiated) water. Although this method is subject to
several sources of error (Nagy and Costa, 1980), these studies report
DEBs which are consistent with others in Table 4. Davis and Kooyman
treat this separately in this volume.

The significance of DEB as a multiple of BMR is unclear. The
BMR multiples for penguins and Sooty Terns are similar; those for
Wedge-tailed Shearwaters and Brown Noddies are considerably higher.
Two of the ten Shearwaters reported in Table 4 incubated without
relief from a mate and may not have left their nest at all, yet
their DEB differed insignificantly from the eight which left their
nests. The "foragers" were gone at most 0.9-2.4 da, indicating
that they did not travel far (Grant, personal communication), so
the high DEB does not appear to be related to foraging in Wedge-
tailed Shearwaters. A single Wedge-tailed Shearwater measured in
Australia (Nagy, personal communication) had a DEB consistent with
these findings. Current studies (Table 4) do not indicate that
seabird DEBs are simple multiples of BMR, although a larger data
base might provide a clearer picture. Furthermore, DEBs in this
group do not appear adequately predicted by existing equations.

TABLE 4
Daily Energy Budgets Measured by Isotopically Labeled Water

Species	N	m^a	DEB^b	$\%DEB^c$	$\%DEB^{d,e}$	$\times$ BMR	Source
Gentoo Penguin	5	6.2	3803.3	148	143[d]	3.1[f]	Davis, Kooyman, and Croxall, 1983
Macaroni Penguin	3	3.6	2830.4	106	140[d]	3.4[f]	Davis et al., 1983
King Penguin	3	13.0	6851.4	170	178[d]	3.3[f]	Kooyman, Davis, Croxall, and Costa, 1982
Jackass Penguin	10	3.13	1945.0	114	103[d]	2.6[f]	Nagy, Siegfried, and Wilson, 1984
Sooty Tern	14	0.188	263.6	85	175[e]	3.0	Flint and Nagy, 1984
Brown Noddy	9	0.195	352.2	111	228[e]	5.2	Ellis, Pettit, and Whittow, unpub. data
Wedge-tailed Shearwater	10	0.384	613.7	128	252[e]	4.8	Ellis, Pettit, and Whittow, unpub. data

[a] mass in kg

[b] Daily Energy Budget in kJ/da

[c] % expected DEB based on equation (11)

[d,e] expected DEB based on equations (10b) or (10a), respectively

[f] BMR predicted from equation (1)

Walsberg (1983) has developed a predictive equation for DEB based on a variety of studies for 42 avian species:

$$DEB = 13.05 \; m^{0.605} \tag{11}$$

where DEB is in kJ/da and m is mass in g. Table 4 shows DEB as a percent of the value predicted by equation (11). This equation varies in its predictive ability for the seabirds so far known. A comparison of % DEB from equation (11) with %BMR from Table 1 also shows little agreement; that is, seabirds whose DEBs vary from those predicted by equation (11) have BMRs that vary by different amounts or even in different directions from those predicted by equation (1). Because, as was mentioned above, equation (10b) of Kendeigh et al. (1977) closely predicts the DEB of Jackass Penguins (see Nagy et al., 1984) the known DEBs for all seabirds are compared to those predicted by Kendeigh et al. (Table 4). Although equation (10b) closely predicts DEB for Jackass Penguins, it underestimates the DEB for the other three penguin species. Furthermore, equation (10a) considerably underestimates the DEBs for warm latitude seabirds, predicting only about half the measured values. By comparison, equation (11) seems to be the better predictor of DEB, but its efficacy for use with seabirds awaits a larger data base.

Whereas an analysis of DEBs for seabirds as a group may be premature at this time, energy budgets for individual species can be of great value. Energy budgets, whether determined directly or (with discretion) indirectly, can be useful in estimating the reproductive capacity, migratory capability, and other long-term energy demanding functions of a species. They are also of importance in analyzing community and population energetics, a topic addressed separately in this volume.

ACKNOWLEDGEMENTS

Most of the unpublished data of H.I. Ellis, T.N. Pettit, and G.C. Whittow along with data from these authors and M. Maskrey were the result of research supported by a University of Hawaii Sea Grant (N1/R-14). This work was possible due to the cooperation of the U.S. Fish and Wildlife Service, the Division of Forestry and Wildlife of the State of Hawaii, and the U.S. Marine Corps Air Station at Kaneohe (especially its Commanding Officer and D. Drigot, its Environmental Protection Specialist). D.A. Ainley, E.N. Flint, C.A. Hui, E. Knudtson, K.A. Nagy, B. Obst, and G.E. Walsberg kindly provided data and other information not yet published. Discussions with D.A. Ainley, R.W. Davis, E.N. Flint, and C.A. Hui were especially helpful in interpreting some of the material herein provided; final responsibility for these interpretations, of course, rests with the author. R.E. Carpenter, E.N. Flint, C.A. Hui, and B.K. McNab commented an earlier drafts of this paper. R.L. Infantino drew the figure. S. Gossom typed repeated revisions of this manuscript.

REFERENCES

Ainley, D.G., 1977, Feeding methods in seabirds: a comparison of
    polar and tropical nesting communities in the eastern Pacific
    Ocean, *in* "Adaptations Within Antarctic Ecosystems", G.A.
    Llano, ed., Gulf Publishing Co., Houston, Texas.
Ainley, D.G., and Boekelheide, R.J., 1984, An ecological comparison
    of oceanic seabird communities of the south Pacific Ocean,
    in "Tropical Seabird Biology", R.W. Schreiber, ed., <u>Studies
    in Avian Biology</u>, no. 8.
Aschoff, J., and Pohl, H., 1970, Rhythmic variations in energy
    metabolism, <u>Fed. Proc.</u>, 29:1541-1552.
Ashkenazie, S., and Safriel, U.N., 1979, Time-energy budget of the
    semipalmated sandpiper, *Calidris pusilla* at Barrow, Alaska,
    <u>Ecology</u>, 60:783-799.
Ashmole, N.P., 1971, Seabird ecology and the marine environment, *in*
    "Avian Biology", D.S. Farner, J.R. King, and K.C. Parkes,
    eds., vol. 1, Academic Press, New York.
Ashmole, N.P., and Ashmole M.J., 1967, Comparative feeding ecology
    of seabirds of a tropical oceanic island, <u>Peabody Mus. Nat.
    Hist.(Yale Univ.) Bull.</u>, 23:1-131.
Baudinette, R.V., and Schmidt-Nielsen, K., 1974, Energy cost of
    gliding flight in herring gulls, <u>Nature</u>, 248:83-84.
Benedict, F.G., and Fox, E.L., 1927, The gaseous metabolism of
    large wild birds under aviary conditions, <u>Proc. Amer. Philos.
    Soc.</u>, 66:511-534.
Berger, M., Hart, J.S., and Roy, O.Z., 1970, Respiration, oxygen
    consumption and heart rate in some birds during rest and
    flight, <u>Z. vergl. Physiol.</u>, 66:201-214.
Burger, A.E., 1981, Time budgets, energy needs and kleptoparasitism
    in breeding Lesser Sheathbills, *Chionis minor*, <u>Condor</u>,
    83:106-112.
Calder, W.A., 1974, Consequences of body size for avian energetics,
    in "Avian Energetics", R.A. Paynter, ed., <u>Publ. Nuttall
    Ornith. Club</u>, no. 15, Cambridge, Massachusetts.
Davis, R.W., Kooyman, G.L., and Croxall, J.P., 1983, Water flux
    and estimated metabolism of free-ranging gentoo and macaroni
    penguins at South Georgia, <u>Polar Biol.</u>, 2:41-46.
Davydov, A.F., 1972, Seasonal variations in the energy metabolism
    and thermoregulation at rest in the black-headed gull, <u>Sov.
    J. Ecol.</u> 2:436-439.
Dewasmes, G., LeMaho, Y., Cornet, A., and Groscolas, R., 1980,
    Resting metabolic rate and cost of locomotion in long-term
    fasting emperor penguins, <u>J. Appl. Physiol.: Respirat.
    Environ. Exercise Physiol.</u>, 49:888-896.
Drent, R.H., and Stonehouse, B., 1971, Thermoregulatory responses
    of the Pervian penguin, *Spheniscus humboldti*, <u>Comp. Biolchem.
    Physiol.</u>, 40A:689-710.
Dunn, E.H., 1979, Time-energy use and life history strategies of
    northern seabirds, *in* "Conservation of Marine Birds in North

America", J.C. Bartonek and D.N. Nettleship, eds., U.S. Fish and Wildl. Serv., Wildlife Res. Rept. 11.

Ellis, H.I., 1980a, Metabolism and evaporative water loss in three seabirds (Laridae), Fed. Proc., 39:1165.

Ellis, H.I., 1980b, Metabolism and solar radiation in dark and white herons in hot climates, Physiol. Zool., 53:358-372.

Ellis, H.I., Maskrey, M., Pettit, T.N., and Whittow, G.C., 1982a, Temperature regulation in Hawaiian Red-footed Boobies, Am. Zool., 22:916.

Ellis, H.I., Maskrey, M., Pettit, T.N., and Whittow, G.C., 1982b, Temperature regulation in Hawaiian Brown Noddies (Anous stolidus pileatus), Physiologist, 25:279.

Enger, P.S., 1957, Heat regulation and metabolism in some tropical mammals and birds, Acta Physiol. Scand., 40:161-166.

Ettinger, A.O., and King, J.R., 1980, Time and energy budgets of the Willow Flycatcher (Empidonax traillii) during the breeding season, Auk, 97:533-546.

Fedak, M.A., Pinshow, B., and Schmidt-Nielsen, K., 1974, Energy cost of bipedal running, Am. J. Physiol., 227:1038-1044.

Flint, E.N., and Nagy, K.A., 1984, Flight energetics of free-living Sooty Terns, Auk, 101:288-294.

Furness, R.W., 1978, Energy requirements of seabird communities: a bioenergetics model, J. Anim. Ecol., 47:39-53.

Furness, R.W., and Cooper, J., 1982, Interactions between breeding seabird and pelagic fish populations in the southern Benguela region, Mar. Ecol. Prog. Ser., 8:243-250.

Gould, P.J., 1974, Sooty Tern (Sterna fuscata), in "Pelagic Studies of Seabirds in the Central and Eastern Pacific Ocean", W.B. King, ed., Smithsonian Contrib. Zool., no. 158.

Grant, G.S., and Whittow, G.C., 1983, Metabolic cost of incubation in the Laysan albatross and Bonin petrel, Comp. Biochem. Physiol., 74A:77-82.

Greenewalt, C.H., 1975, The flight of birds, Trans. Amer. Philos. Soc., 65:3-67.

Hails, C.J., 1983, The metabolic rate of tropical birds, Condor, 85:61-65.

Hartman, F.A., 1961, Locomotor mechanisms of birds, Smithsonian Misc. Coll., 143:1-91.

Hennemann, W.W., 1983, Environmental influences on the energetics and behavior of anhingas and double-crested cormorants, Physiol. Zool., 56:201-216.

Hui, C.A., 1983, Swimming in penguins, Unpublished Ph.D. diss., Univ. Calif. Los Angeles.

Iverson, J.A., and Krog, J., 1972, Body temperatures and nesting metabolic rates in small petrels, Norw. J. Zool., 20:141-144.

Johnson, S.R., and West, G.C., 1975, Growth and development of heat regulation in nestlings and metabolism in adult Common Murre and Thick-billed Muree, Ornis Scand., 6:109-115.

Johnston, D.W., 1979, The uropygial gland of the Sooty Tern, Condor, 81:430-432.

Kendeigh, S.C., 1970, Energy requirements for existance in relation to size of bird, Condor, 72:60-65.

Kendeigh, S.C., Dol'nik, V.R., and Gavrilov, V.M., 1977, Avian energetics, pp. 127-205 and 363-378 *in* "Granivorous Birds in Ecosystems", J. Pinowski and S.C. Kendeigh, eds., Cambridge University Press, London.

King, J.R., 1974, Seasonal allocation of time and energy resources in birds, *in* "Avian Energetics", R.A. Paynter, ed., Publ. Nuttall Ornith. Club, no. 15, Cambridge, Massachusetts.

King, W.B., 1967, Seabirds of the tropical Pacific Ocean, (Prelim. Smithson. Identification Manual), Smithsonian Inst., Washington, D.C.

Kooyman, G.L., Gentry, R.L., Bergman, W.P., and Hammel, H.T., 1976, Heat loss in penguins during immersion and compression, Comp. Biochem. Physiol., 54A:75-80.

Kooyman, G.L., Davis, R.W., Croxall, J.P., and Costa, D.P., 1982, Diving depths and energy requirements of king penguins, Science, 217:726-727.

Krasnow, L., 1979, Feeding energetics of the Sooty Shearwater *Puffinus griseus* in Monterey Bay, Unpublished M.S. thesis, Calif. St. Univ., Sacramento.

Lasiewski, R.C., and Dawson, W.R., 1967, A re-examination of the relation between standard metabolic rate and body weight in birds, Condor, 69:13-23.

Lechner, A.J., 1978, The scaling of maximal oxygen consumption and pulmonary dimensions in small mammals, Resp. Physiol., 34:29-44.

LeFebvre, E.A., 1964, The use of D_2O^{18} for measuring energy metabolism in *Columba livia* at rest and in flight, Auk, 81:403-416.

LeMaho, Y., Delclitte, P., and Chatonnet, J., 1976, Thermoregulation in fasting emperor penguins under natural conditions, Am. J. Physiol., 231:913-922.

Lifson, N., and McClintock, R., 1966, Theory of use of the turnover rates of body water for measuring energy and material balance, J. Theoret. Biol., 12:46-74.

Lissaman, P.B.S., and Shollenberger, C.A., 1970, Formation flight of birds, Science, 168:1003-1005.

Lustick, S., Battersby, B., and Kelty, M., 1978, Behavioral thermo-regulation: orientation toward the sun in herring gulls, Science, 200:81-83.

MacMillen, R.E., and Carpenter, F.L., 1977, Daily energy costs and body weight in nectarivorous birds, Comp. Biochem. Physiol., 56A:439-441.

MacMillen, R.E., Whittow, G.C., Christopher, E.A., and Ebisu, R.J., 1977, Oxygen consumption, evaporative water loss, and body temperature in the Sooty Tern, Auk, 94:72-79.

Nagy, K.A., 1980, CO_2 production in animals: analysis of potential errors in the doubly labeled water method, Am. J. Physiol., 238 (Regulatory Integrative Comp. Physiol., 7):R466-R473.

Nagy, K.A., and Costa, D.P., 1980, Water flux in animals: analysis of potential errors in the tritiated water method, Am. J. Physiol., 238 (Regulatory Integrative Comp. Physiol., 7): R454-465.

Nagy, K.A., Siegfried, W.R., and Wilson, R., 1984, Energy utilization by free-ranging jackass penguins, Ecology (in press).

Pearson, O.P., 1954, The daily energy requirements of a wild Anna Hummingbird, Condor, 56:317-322.

Pearson, T.H., 1968, The feeding biology of sea-bird species breeding on the Farne Islands, Northumberland, J. Anim. Ecol., 37:521-552.

Pennycuick, C.J., 1975, Mechanics of flight, in "Avian Biology", D.S. Farner, J.R. King and K.C. Parkes, eds., vol. 5, Academic Press, New York.

Pennycuick, C.J., 1982, The flight of petrels and albatrosses (Procellariiformes), observed in South Georgia and its vicinity, Phil. Trans. R. Soc. Lond., B300:75-106.

Pennycuick, C.J., 1983, Thermal soaring compared in three dissimilar tropical bird species, Fregata magnificens, Pelecanus: occidentalis and Coragyps atratus, J. Exp. Biol., 102-307-325.

Pennycuick, C.J., and Bartholomew, G.A., 1973, Energy budget of the lesser flamingo (Phoeniconaias minor geoffroy), E. Afr. Wildl. J., 11:199-207.

Pinshow, B., Fedak, M.A., Battles, D.R., and Schmidt-Nielsen, K., 1976, Energy expenditure for thermoregulation and locomotion in emperor penguins, Am. J. Physiol., 231:902-912.

Pinshow, B., Fedak, M.A., and Schmidt-Nielsen, K., 1977, Terrestrial locomotion in penguins: it cost more to waddle. Science, 195:592-594.

Prange, H.D., and Schmidt-Nielsen, K., 1970, The metabolic cost of swimming in ducks, J. Exp. Biol., 53:763-777.

Raveling, D.G., and LeFebvre, E.A., 1967, Energy metabolism and theoretical flight range of birds, Bird Banding, 38:97-113.

Ricklefs, R.E., and Matthew, K.K., 1983, Rates of oxygen consumption in four species of seabird at Palmer Station, Antarctic Peninsula, Comp. Biochem. Physiol., 74A:885-888.

Ricklefs, R.E., White, S.C., and Cullen, J., 1980, Energetics of post-natal growth in Leach's Storm-petrel, Auk, 97:566-575.

Ricklefs, R.E., and White, S.C., 1981, Growth and energetics of chicks of the Sooty Tern (Sterna fuscata) and Common Tern (S. hirundo), Auk, 98:361-378.

Schartz, R.L., and Zimmerman, J.L., 1971, The time and energy budget of the male Dickcissel (Spiza americana), Condor, 73:65-76.

Schmidt-Nielsen, K., 1972, Locomotion: energy cost of swimming, flying and running, Science, 177:222-228.

Schnell, G.D., and Hellack, J.J., 1979, Bird flight speeds in nature: optimized or a compromise, Am. Nat., 113:53-66.

Scholander, P.F., Hock, R., Walters, V., and Irving, L., 1950, Adaptation to cold in Arctic and tropical mammals and birds in relation to body temperature, insulation and basal metabolic rate, Biol. Bull., 99:259-271.

Stahel, C.D., and Nicol, S.C., 1982, Temperature regulation in the little penguin, *Eudyptula minor* in air and water, J. Comp. Physiol., 148:92-100.

Taylor, C.R., Schmidt-Nielsen, K., and Raab, J.L., 1970, Scaling of energetic cost of running to body size in mammals, Am. J. Physiol., 219:1104-1107.

Tucker, V.A., 1968, Respiratory exchange and evaporative water loss in the flying Budgerigar, J. Exp. Biol., 48:67-87.

Tucker, V.A., 1972, Metabolism during flight in the laughing gull (*Larus atricilla*), Am. J. Physiol., 222:237-245.

Tucker, V.A., 1973, Bird metabolism during flight: evaluation of a theory, J. Exp. Biol., 58:689-709.

Utter, J.M., and LeFebvre, E.A., 1970, Energy expenditure for free flight by the purple martin, *Progne subis*, Comp. Biochem. Physiol., 35:713-719.

Utter, J.M., and LeFebvre, E.A., 1973, Daily energy expenditure of purple martins (*Progne subis*) during the breeding season: estimates using D_2O^{18} and time budget methods, Ecology, 54:397-604.

Walsberg, G.E., 1983, Avian ecological energetics, *in* "Avian Biology", D.S. Farner, J.R. King, and K.C. Parks, eds., vol. 7, Academic Press, New York.

Warham, J., 1977, Wing loadings, wing shapes, and flight capabilities of procellariiformes, New Zealand J. Zool., 4:73-83.

Wasser, J.F., 1979, Comparative energetics of some falconiform birds, Unpublished M.S. thesis, Univ. Florida, Gainesville.

Watson, J.B., and Lashley, K.S., 1915, Homing and related activities of birds, Carnegie Inst. Washington, Publ. 211, Papers from Dept. Marine Biol., 7:5-104.

Weathers, W.W., 1979, Climate adaptation in avian standard metabolic rate, Oecologia, 42:81-89.

Weathers, W.W., and Nagy, K.A., 1980, Simultaneous doubly labeled water ($^3HH^{18}O$) and time-budget estimates of daily energy expenditure in *Phainopepla nitens*, Auk, 97:861-867.

West, G.C., 1968, Bioenergetics of captive willow ptarmigan under natural conditions, Ecology, 49:1035-1045.

Wiens, J.A., and Scott, J.M., 1975, Model estimation of energy flow in Oregon coastal seabird populations, Condor, 77:439-452.

Withers, P.C., 1977, Energetic aspects of reproduction by the Cliff Swallow, Auk, 94:718-725.

Withers, P.C., 1979, Aerodynamics and hydrodynamics of the "hovering" flight of Wilson's storm petrel, J. Exp. Biol., 80:83-91.

Withers, P.C., and Timko, P.L., 1977, The significance of ground effect to the aerodynamic cost of flight and energetics of the black skimmer (*Rynchops nigra*), J. Exp. Biol., 70:13-26.

234

ENERGETICS OF WALKING IN PENGUINS

Yvon Le Maho[1] and Gérard Dewasmes[2]

[1]Laboratoire de Physiologie Respiratoire
CNRS
Associé Université Louis Pasteur
Strasbourg, France
[2]Laboratoire de Thermorégulation
CNRS
Université Claude Bernard
Lyon, France

INTRODUCTION

It is of a particular interest to study the energy budget of penguins since they represent 75 to 85% of the bird biomass of the southern ocean (Croxall, 1984). When considering only Antarctica, the population size of penguins represents 66% of the total bird population - about 180 million birds; however, because penguins are in general birds of a large body mass, their biomass reaches as much as 98% of total bird biomass (see Mougin, 1984).

The cost of walking must be considered as an important proportion of the energy budget of penguins because for most species the breeding colonies are not close to the sea. The distances some species have to walk before reaching their colonies may be considerable. For example there are often hundreds of km between Emperor Penguin (<u>Aptenodytes</u> <u>forsteri</u>) colonies - which are established on well-anchored sea-ice along the coasts of the antarctic continent - and the open sea (see Dewasmes <u>et al</u>., 1980). Adélie penguins (<u>Pygoscelis</u> <u>adeliae</u>) may also have to walk more than 100 km from the sea to their colonies (Sladen, 1958). Emperor tracks and what were probably Adélie tracks have been reported at least 300 km from the nearest known open water (Sladen and Ostenso, 1960). Emperor and Adélie penguins are the only two species of penguins which breed along the coasts of the antarctic continent, excluding the antarctic peninsula.

The variations in size are considerable among the sixteen species of penguins since the smallest species, the Little Penguin Eudyptula minor, weighs only 1 kg, while the largest, the Emperor Penguin, weighs as much as 20 to 40 kg.

Like some other birds - geese and ducks - penguins have a peculiar walking gait which is called waddling. Their legs are relatively short and their body undergoes large lateral displacements at each stride.

In addition, penguins feed only at sea, as do many other marine animals. Thus they are fasting while coming ashore for breeding or molting.

Interesting questions therefore arise within this context. First, what are the changes in the cost of walking with variations in speed of locomotion, and what is the cost of walking at a given speed for a large penguin by comparison with a smaller one? Second, does it cost the same energy for walking at the same speed at the beginning and at the end of the periods of fasting? And third, does it cost more to waddle and what are the distances penguins may cover by walking?

COST OF WALKING VERSUS SPEED OF LOCOMOTION

The large range in body mass of penguins is of a particular interest since it is now well known that the energetic cost of locomotion is relatively higher for smaller animals (see Schmidt-Nielsen, 1972).

Pinshow et al. (1977) studied the energetic cost of walking, at various constant speeds, in penguins which included the smallest and the largest species: White-flippered Penguins (Eudyptula albosignata) - these penguins have a body mass of 1.2 kg, which is therefore very close to that of Little Penguins - and Emperor Penguins. They also studied the cost of walking in Adélie Penguins, these birds are middle-size penguins, they have a body mass of about 4 kg.

All three species were studied over the range of speeds that they would maintain for 20 minutes or more. Over this rather limited range their specific metabolic rate increased approximately linearly with walking speed (Fig. 1). This increase was smallest with the largest species. In addition, at the same speed the specific metabolic rate was lowest for the largest species.

The slope of the line relating the specific metabolic rate $\dot{M}/m$ to the speed of walking S is of interest for comparing the energy cost of activity in animals differing in size, body mass,

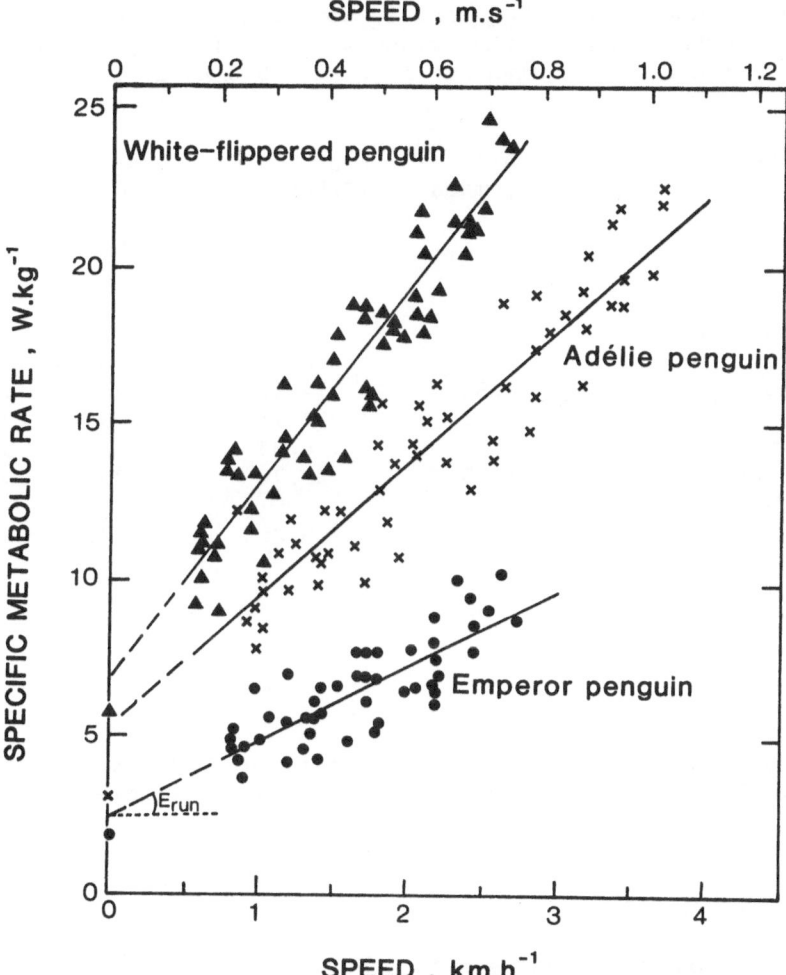

Fig. 1. Specific metabolic rate versus speed of walking in three species of penguins (after Pinshow et al., 1977). The slope of the lines (E_run) is useful for comparing the energetic cost of walking in the various species (see text).

morphology, resting metabolic rate and mechanics of locomotion. This is because in animal species that show a linear increase in metabolic rate with speed of walking (as do penguins and most other animals so far studied), this slope (E_{run}) is constant for each species. In addition, time has conveniently been eliminated in this energy term which is defined as $d\dot{M} / mdS$ and is expressed in $J.m^{-1}.kg^{-1}$ when $\dot{M}$ is in W, m the body mass in kg and S in $m.s.^{-1}$.

Table 1. Comparison of E_{run} and E_t for penguins with those of other birds. After Pinshow et al. (1977), assuming that the energy equivalent of O_2 is 19.6 kJ.l^{-1}.

Species	Mean body mass	E_{run} (J.kg^{-1}.m^{-1})	E_t (J.kg^{-1}.m^{-1})	
			2km.h^{-1}	3km.h^{-1}
Emperor Penguin	20.79	8.43	12.94	11.37
* Rhea	22.00	6.66	10.58	9.21
Adélie Penguin	3.89	14.90	24.11	20.97
Goose	3.81	14.11	19.01	17.44
* Turkey	4.31	8.04	13.92	11.96
White-flippered Penguin	1.15	21.76	33.91	29.79
* Guinea Fowl	1.21	9.21	22.93	18.42

* Birds that do not waddle.

As indicated in Table 1, E_{run} is smaller in larger penguins. The large body mass of the Emperor Penguin is therefore a decisive advantage.

THE COST OF WADDLING

When comparing penguins with other birds of similar body mass which do not waddle (see Table 1), it appears that E_{run} is much higher in penguins. It is of interest that E_{run} is also high in the goose, a bird which waddles when walking, as penguins do. Thus, it costs more energy to waddle.

As was shown by Pinshow et al. (1977), mechanical analysis may explain why E_{run} is higher in birds which are waddling when running. To walk with the body tilting from side to side, as geese or penguins do, may involve large kinetic energy changes with each stride. In addition, to walk at a given speed, short legs require a higher stride frequency than do long legs. The stride frequency of turkeys at a walking speed of 3.9 km.h^{-1} is only 50 percent of that of Adélie penguins (Pinshow et al., 1977). It may well be that the morphology of penguins and geese represents a compromise between swimming and running and that their energy economy of terrestrial transport suffers as a consequence (Fedak et al., 1974).

THE COST OF TRANSPORT

In addition to E_{run}, another energy quantity - the cost of transport, E_t-appears very useful when comparing the energetic cost of locomotion in animals (see Schmidt-Nielsen, 1972; Fedak et al., 1974 and Pinshow et al., 1977). E_t is defined as $\dot{M}/(m.S)$ where $\dot{M}$ is metabolic rate (in watts), m is body mass (in kilograms) and S the speed of walking (in meters per second). This is the amount of fuel (expressed as joules) it takes to transport one unit of body mass (one kilogram) over one unit of distance (one meter) while walking at a particular speed. E_t may change significantly with the speed of locomotion; it has been determined for penguins and other birds of similar body mass (see Table 1).

The interest in E_t is that it expresses a cost effectiveness which might be of interest when considering whole-animal energy budgets, such as the energetic cost of a migration (Fedak et al., 1974). Although it has the same units as E_t, E_{run} cannot be used for this purpose. This is because the calculation of the slope E_{run} involves subtraction of the Y-intercept value of metabolic rate. Thus E_{run} has a lower numerical value than E_t.

The problem, however, when using E_t to determine the energy budgets in "migrating" species (like in Emperor Penguins trekking on sea-ice), is that it must be known if there is a change in E_t when body mass is decreasing during fasting. When Emperor Penguins walk in long files to their colony, at the beginning of their breeding cycle, they weigh about 40 kg. The males, which are entirely responsible for incubation, fast during 4 months in the colony. When they leave it to walk back to the open sea they usually weigh about 23 kg (see Le Maho, 1977; Dewasmes et al., 1980).

During fasting in Emperor Penguins, the part of energy expenditure which is provided by lipids is about 93%, and that from proteins is therefore only 7% (Groscolas et Clément, 1976). Thus protein sparing during fasting in these birds is remarkable. For comparison, proteins provide about 15-20% of the energy in fasting man (Grande, 1964; Cahill, 1970). It is only in some small migratory birds (Odum et al., 1964) and in hibernating bears (Nelson et al., 1973) that protein sparing is better than in Emperor Penguins, since these animals totally save proteins.

The changes in E_t during fasting have been determined under natural conditions in Emperor Penguins, their body mass decreasing between 38 and 18 kg (Dewasmes et al., 1980). To calculate E_t, the metabolic rate of the birds was regularly measured during the course of fasting, while they were walking on a treadmill for at least 20 min at the speed of 1.4 km.h^{-1}.

The data which were obtained in this study indicate that E_t

tends to remain constant between 38 and 23 kg body mass and decrease below 23 kg body mass (Fig. 2). During most of the fast of Emperor Penguins (as indicated above, in the colony it takes them about 4 months to reach a body mass of 23 kg) the cost of transport tends to remain constant per unit of body mass. This means that the metabolic rate for walking at a particular speed decreases approximately in proportion to body mass. The metabolic rate of an Emperor Penguin of 38 kg walking at 1.4 km.h^{-1} is about 330 W, while it is of only 220 W when the bird has reached a mass of 23.3 kg (Dewasmes et al., 1980).

Considering the change in E_t below 23-kg, it is remarkable that, by leaving their colony when they reach this body mass, Emperor Penguins start a long trek on sea-ice when a period of decreased cost of transport begins. The metabolic rate of a 18-kg Emperor Penguin walking at 1.4 km.h^{-1} is only 140 W (Dewasmes et al., 1980). This decreased cost of transport below a 23-kg body mass might be due to mechanical factors: as a consequence of the decrease in the mass of body fat and muscles (particularly in the lower abdomen region). It indeed seems from observations of Emperor Penguins that they waddle much less at the end of their fast. A study of their mechanics of locomotion would therefore be of a great interest.

The cost of transport in the other species of penguins is not known. To compare the distance various species may theoretically cover by walking we therefore have to assume that - as for Emperors before they reach 23 kg - their cost of transport per unit of body mass tends to remain constant.

Data are available in the literature on the amount of fat and protein stores in several penguin species. It may be assumed that all the energy used for the walk is derived from the catabolism of fat stores. The amount of fat stores was found to be about 20-25% of body mass in Macaroni Penguins, Eudyptes chrysolophus, Rockhopper Penguins, Eudyptes chrysocome (Williams et al., 1977), Adélie Penguins (Johnson et West, 1973) and Emperor Penguins (Groscolas and Clément, 1976). It may be assumed that the proportion of fat is also similar in White-flippered Penguins and that fat tissue in penguins has an energy density of 3.77 . 10^7 J.kg^{-1}, similar to that of migrating birds (Johnson et al., 1970).

Considering that their fat stores represent about 25% of body mass, let us now estimate how far penguins would be able to cover if they were walking on non-stop treks until their energy stores were totally depleted. At the same speed of 1.4 km.h^{-1}, a 1 kg White-flippered Penguin might walk 230 km, which would take about 7 days; a 4 kg Adélie Penguin might walk 400 km in 12 days and a 40 kg Emperor Penguin 630 km in 19 days. Of course, adult penguins usually do not walk until their energy stores are depleted. The

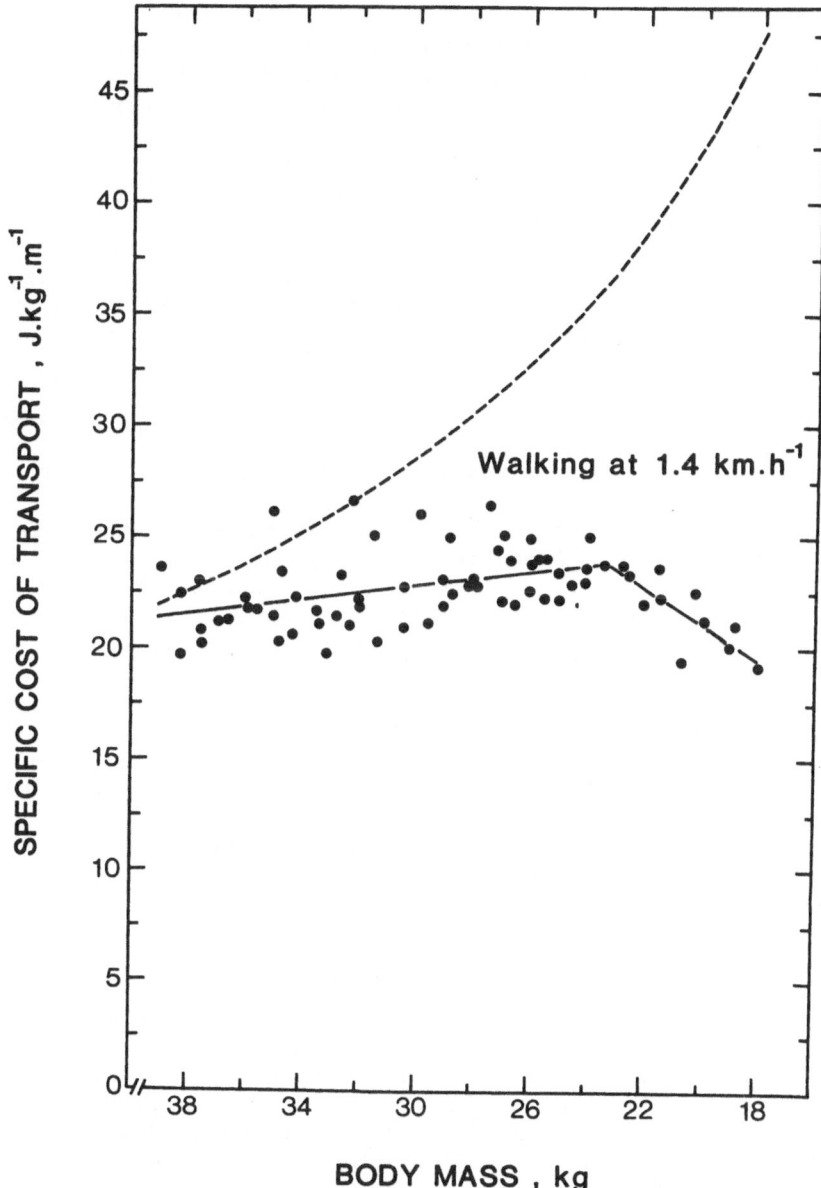

Fig. 2. Specific cost of transport (E_t), in 4 fasting Emperor Penguins when walking for 20 min. at 1.4 km.h^{-1}, plotted against body mass. The dotted line corresponds to the theoretical data that would have been obtained if walking metabolic rate had not decreased during the fast. After Dewasmes et al., 1980.

goal of their walk is generally the breeding colony or open sea. When in their colony the birds must still have enough energy reserves to go back to the sea and feed. We know that Emperor Penguins leaving their colony when they reach 23 kg are still able to walk about 200 km (see Dewasmes et al., 1980). However, as indicated above, Emperor tracks, and what were probably Adélie tracks, have been reported on sea-ice at least 300 km from the nearest known open water (Sladen and Ostenso, 1960). Thus penguins, although their waddling gait gives them a high cost of walking, may engage in remarkably long treks. This may be because the distance between the colony and the sea is unusually large; penguins may also be lost. It may be that immature birds (Emperor Penguins only start to breed at about 4-5 years of age, see Isenmann, 1971) take time for traveling.

REFERENCES

Cahill, G.F., Jr., 1970, Starvation in man, New Engl. J. Med., 282: 668.

Croxall, J.P., 1984, Seabird ecology, in: "Ecology of the antarctic", R.M. Laws, ed., Academic Press, London.

Dewasmes, G., Le Maho, Y., Cornet, A. and Groscolas, R., 1980, Resting metabolic rate and cost of locomotion in long-term fasting emperor penguins, J. Appl. Physiol.:Respirat. Environ. Exercise Physiol., 49: 888.

Fedak, M., Pinshow, B., and Schmidt-Nielsen, K., 1974, Energy cost of bipedal running, Am. J. Physiol., 227: 1038.

Grande, F., 1964, Man under caloric deficiency , in: "Handbook of Physiology. Adaptation to the environment", Am. Physiol. Soc., Washington D.C., sect. 4, p. 911.

Groscolas, R., and Clément, C., 1976, Utilisation des réserves énergétiques au cours du jeûne de la reproduction chez le manchot empereur, Aptenodytes forsteri, C.R. Acad. Sci., 282: 297.

Isenmann, P., 1971, Contribution à l'éthologie et à l'écologie du manchot empereur (Aptenodytes forsteri Gray) à la colonie de Pointe Géologie (Terre Adélie), L'Oiseau et R.F.O., 41: 9.

Johnson, S.R. and West, G.C., 1973, Fat content, fatty acid composition and estimates of energy metabolism of Adélie penguins (Pygoscelis Adéliae) during the early breeding season fast, Comp. Biochem. Physiol., 45 B: 709.

Johnston, D.W., 1970, Caloric density of avian adipose tissue, Comp. Biochem. Physiol., 34: 827.

Le Maho, Y., 1977, The emperor penguin: a strategy to live and breed in the cold, Am. Sci., 65: 680.

Mougin, J.L., 1984, Ecology of the emperor penguin, in: "The emperor penguin", Y. Le Maho, ed., Yale Univ. Press, Newhaven.

Nelson, R.A., Wahner, H.W., Jones, J.D., Ellefson, R.D., and Zollman, P.E., 1973, Metabolism of bears before, during, and after winter sleep, <u>Am. J. Physiol.</u>, 224: 491.

Odum, E.P., Rogers, D.T., and Hicks, D.L., 1964, Homeostasis of
    the nonfat components of migrating birds, <u>Science</u>, 143: 1037.

Pinshow, B., Fedak, M.A., and Schmidt-Nielsen, K., 1977, Terrestrial
    locomotion in penguins: it costs more to waddle, <u>Science</u>,
    195: 592.

Schmidt-Nielsen, K., 1972, Locomotion: energy cost of swimming,
    flying and running, <u>Science</u>, 177: 222.

Sladen, W.J.L., 1958, The Pygoscelids penguins. I. Methods of study;
    II. The Adélie penguin, <u>Pygoscelis Adeliae</u> (Hombron and
    Jacquinot), <u>Falkland Isl. Depend. Surv. Sci. Rep.</u>, 17: 1.

Sladen, W.J.L., and Ostenso, N.A., 1960, Penguin tracks far inland
    in the Antarctic, <u>Auk</u>, 77: 466.

Williams, A.J., Siegfried, W.R., Burger, A.E., and Berruti, A.,
    1977, Body composition and energy metabolism of moulting
    Eudyptid penguins, <u>Comp. Biochem. Physiol.</u>, 56 A: 27.

FREE-RANGING ENERGETICS OF PENGUINS

Randall W. Davis and Gerald L. Kooyman

Physiological Research Laboratory
Scripps Institution of Oceanography
La Jolla, CA   92093

Penguins exhibit the greatest degree of adaptation to the aquatic environment of all seabirds. They are powerful swimmers that pursue their prey at depth allowing them to exploit the water column more than any other seabird. Dives 240 m deep have been recorded for King Penguins (Kooyman et al., 1982) and as deep as 265 m for Emperor Penguins (Kooyman et al., 1971). Intermediate sized Gentoo and Chinstrap Penguins dive from 70-100 m (Conroy and Twelves, 1972; Lishman and Croxall, in prep) and even the smallest species, the Fairy Penguin, can dive to 60 m (Montague, 1982). How does swimming as a primary mode of locomotion at sea influence the average daily metabolic rate (ADMR) and foraging energetics of penguins relative to seabirds that fly to and from feeding areas?

Information on the energy metabolism of penguins has come primarily from studies of birds in captivity or free-ranging birds while ashore. Usually the energetics of a particular behavior or physiological state such as resting, incubating, walking, fasting, or moulting has been measured (Mougin, 1974; Barre, 1975; Le Maho, Delclitte, and Chatonnet, 1976; Pinshow et al., 1976; Pinshow, Fedak and Schmidt-Nielsen, 1977; Dewasmes et al, 1980; Croxall, 1982; Stahel and Nicol, 1982; Copestake, Croxall, and Prince, 1983; Brown, in press). In most cases, metabolic rate has been measured by indirect calorimetry or calculated from measurements of weight loss during fasting. These conventional procedures are not suitable for measuring energy budgets at sea, and consequently no measurements were made until recently. Techniques for measuring water turnover rate and carbon dioxide production using isotopes of hydrogen and oxygen have been refined to the degree that it is now economically possible to determine the free-ranging energy expenditure of birds ranging in size to over 30 kg (Nagy, 1980; Schoeller and van Santen, 1982). As a result, two preliminary reports have been published in which only tritiated water was used to estimate free-ranging energetics (Kooyman et al., 1982; Davis, Kooyman, and Croxall, 1983) and a more detailed analysis using the doubly labeled water method will be reported in the near future (Nagy, pers. com.). The purpose of this report is to review past studies, compare them to other studies of free-ranging bird energetics, and comment on the most likely direction of future studies of penguin energetics.

Published reports on the free-ranging energetics of penguins are based on three species that breed sympatrically on South Georgia Island. The species are Macaroni, Eudyptes chrysolophus, Gentoo, Pygoscelis papua, and King Penguins, Aptenodytes patagonica, with average body weights of 3.6, 6.2, and 13 kg, respectively. Although the breeding season for these species overlaps, they are ecologically separated in their nesting sites and foraging behavior (Croxall and Prince, 1980). Macaroni Penguins breed in a few vast colonies, lay two eggs of unequal size, and never hatch more than one egg. Gentoo Penguins breed in numerous small colonies, lay two eggs, and at South Georgia, frequently raise two chicks. King Penguins breed in large and small colonies. They lay one egg, raise a single chick, and continue to feed the chick throughout the winter. The longer fledging time for King Penguins apparently limits breeding to two out of every three years.

Field work for these studies was conducted from January to March 1980 on Bird Island and Schlieper Bay, South Georgia (54°S, 36°W), South Atlantic. The climate is cold, wet, and cloudy with strong winds and little seasonal variation (Richards and Tickell, 1968). The mean temperature in January and February is +4°C with a maximum of +9°C and a minimum of -2°C. Average rainfall at this time of year is 8-15 cm. The ocean temperature is about +2°C reflecting South Georgia's position 250 km south of the Antarctic Convergence.

At the start of these studies, Macaroni Penguin

chicks were about 18 days old and just beginning to form small creches. For about the first 10 days after creche formation, the male Macaroni, which has just completed a 35 day fast attending the nest, takes only a small part in feeding the chick. Thereafter both parents share in feeding the chick. Gentoo Penguin chicks were about 41 days old, had been in creches for 14 days, and both parents were making regular foraging trips. The King Penguin chicks were 8-12 weeks old. The male and female of each pair took turns tending the chick while the other member foraged at sea.

Detailed time budgets of penguins foraging at sea and returning to feed their chicks have not been determined. Macaronis spend about 30 hours at sea on a single trip and take both large (5.3 cm) and small (2.0 cm) krill, Euphausia superba. Croxall and Prince (1980) suggested that Macaronis may forage up to 50 km offshore where small krill have been seen to concentrate in waters at the edge of the South Georgia continental shelf. Gentoos appear to forage closer to shore on large krill (5.4 cm) and immature nototheniid fish, particularly Notohenia rossi (15-25 cm). An average trip at sea lasts about 10 hours. King Penguins spend 2-5 days on a single foraging trip and feed primarily on squid rather than the immensely abundant krill. The deep diving ability of King Penguins (Kooyman et al, 1982) sets them apart from other seabirds at South Georgia and may enable them to exploit a resource in which: 1) there is less interspecific competition, 2) the prey is larger, and 3) the prey may be regularly available throughout the year. The latter factor may be important because this species remains near the breeding areas throughout the year.

Energy metabolism of birds fasting in the rookery and the metabolic rate and prey consumption of foraging birds were estimated by following the decline in the specific activity of tritiated water (HTO) in the total body water after initial labeling. Adult Macaroni and Gentoo Penguins were captured after feeding their chicks and King Penguins just before they departed for sea. Water turnover rates and the calculated prey consumption and energy metabolism based on the equations of Shoemaker et al (1976) have been published (Kooyman et al, 1982; Davis, Kooyman, and Croxall, 1983).

The metabolic rate of fasting, male Macaroni Penguins brooding their chicks was 5.6 W/kg or 1.8 x standard metabolic rate (SMR) based on calculations of

247

Table 1. Fasting metabolism and ADMR of free-ranging Macaroni, Gentoo, and King Penguins. Mean values ± S.D.

	Macaroni	Gentoo	King
n	3	5	3
Mass (kg)	3.6±0.3	6.2±1.3	13.0+0.8
Fasting on rookery (W/kg)	5.6	4.9	4.0
ADMR (W/kg)	9.1±2.9	7.1±1.0	5.0±1.0
$(W/kg^{0.729})$	12.8±3.8	11.6±1.4	10.1±2.0
SMR (W/kg) [1]	3.1±0.1	2.7±0.1	2.2±0.4
ADMR/SMR	2.9±0.8	2.6±0.3	2.3±0.5
Existence Metabolism [2] (W/kg)	4.7	3.7	2.6
Predicted ADMR [3] (W/kg)	10.6	9.0	7.2

1) $SMR\ (W/kg) = (4.4097 \cdot Mb^{0.729})/Mb$ where Mb = body mass in kg (Ashoff and Pohl, 1970).

2) $EMR\ (W/kg) = 0.048458\ (a-bT)/Mb$, where EMR equals the existence metabolism for non-passerine birds, at ambient temperature, $a = 4.142 \cdot Mb^{0.5444}$, Mb is body mass in g, b is the temperature coefficient = $0.2761 \cdot Mb^{0.2818}$, and T = ambient temperature (4°C) (Kendeigh et al., 1977).

3) $(\log ADMR = \log 317.7 + 0.7052 \log Mb) \times 0.048458/Mb$, where Mb is body mass in kg and 0.048458 is the conversion factor of kcal/day to watts (King, 1974).

energy metabolism from water flux (Table 1). Assuming the same ratio of 1.8 x SMR for Gentoo and King Penguins brooding their chicks, the metabolic rate for these two species is 4.9 and 4.0 W/kg, respectively. Similar estimates based on weight loss over a 6-39 day period averaged 4.8 W/kg (range 4.0-6.6) for Macaroni Penguins and 4.3 W/kg (range 3.2-5.3) for King Penguins depending on the assumed composition of the mass lost (e.g. 56% fat and 9% protein or 90% fat)(from Croxall, 1982, from Stonehouse, 1960). The metabolism of resting and incubating Macaroni Penguins measured by indirect calorimetry at 13.5°C was 3.5 W/kg and 2.5 W/kg, respectively (Brown, in press). These values are less than the fasting level of a Macaroni on the rookery brooding the chick and may result from less activity and a warmer temperature in the metabolic chamber. The predicted existence metabolism at 4°C is 4.7, 3.7, and 2.6 W/kg for Macaroni, Gentoo, and King Penguins, respectively (Kendeigh et al., 1977; Table 1). The metabolism of birds brooding their chicks on the rookery is therefore 1.2-1.5x greater than the predicted existence metabolism and probably results from the bird's activity on the rookery (e.g. walking and intraspecific aggression at nest sites) and climatic factors.

The ADMR for Macaroni and Gentoo Penguins over a 5-25 day period while the birds were brooding their chicks and making regular foraging trips to sea was 7.1 W/kg (2.9 x SMR) and 9.1 W/kg (2.6 x SMR), respectively (Table 1). The metabolic rate of King Penguins during 2-5 day absences from the rookery (most of this time was probably spent foraging at sea) was 6.1 W/kg or 2.8 x SMR. The estimated ADMR of King Penguins assuming that the male and female of each pair divide their time equally between the rookery and foraging at sea (one member tending the chick while the other is a sea) was 5.1 W/kg or 2.3 x SMR. This calculation is based on the mean of the estimated metabolic rate on the rookery (4.0 W/kg; Table 1) and the measured metabolic rate during foraging trips (6.1 W/kg). Such a calculation is necessary for comparison with the estimated ADMR of Macaroni and Gentoo Penguins that covered the entire period of time spent on the rookery and foraging at sea. These estimates of ADMR represent the consumption of 210 g krill/kg·day for Macaronis and 100 g krill/kg·day and 51 g fish/kg·day for Gentoos (Davis, Kooyman, and Croxall, 1983). For King Penguins, the ADMR is equivalent to the consumption of 161 g squid/kg day (Kooyman et al., 1982).

ADMR for the three species is comparable when
expressed as the ratio of SMR or when scaled with the
allometric coefficient for the mass dependent metabolism
(0.729; Aschoff and Pohl, 1970; Table 1). This suggests
that despite pronounced differences in the duration of
individual foraging trips to sea, the energy expended
per day by the adults relative to the mass dependent
metabolic rate is approximately equivalent.

Previous estimates of ADMR in birds weighing 5-406 g
range from 2.2-5.9 x SMR based on 1) pellet analysis, 2)
crop content, 3) time-activity budgets combined with
laboratory data, 4) observations of feeding rate and
excretion rate and 5) doubly labeled water (King, 1974,
Hails and Bryant, 1979). The predicted ADMR for the
three species of penguins based on the regression
equation of King (1974) for passerines and
non-passerines (Table 1) is within one standard
deviation of the measured value for Macaroni Penguins
but greater than the values for Gentoo and King
Penguins. However, there are too few measurements to
say whether the ADMR of aquatic birds such as penguins
is less than flying birds.

We have demonstrated the feasibility of a
capture-recapture technique using isotope turnover
methods to measure metabolism in free-ranging penguins.
Future studies should use the doubly labeled water
method (Nagy, 1980) which, although more expensive in
terms of the isotopes and analysis, does not require
detailed knowledge of the diet or drinking habits of the
animal to estimate energy metabolism. Further studies
must also acquire more detailed information on time and
energy budgets throughout the reproductive season.
Specifically, how much time and energy is spent in
courtship, incubation, brooding the chicks, and foraging
at sea? It would also be interesting to examine the
division of time and energy between the two sexes during
incubation and chick rearing. Information should also
be sought on time budgets at sea and diving behavior.
The development of new and smaller depth recorders and
swim velocity recorders that monitor the time spent at a
particular depth or swimming velocity will enable us to
partition time budgets at sea and examine foraging
strategies and indirectly monitor prey behavior (Wilson
and Bain, in press). A swimming velocity recorder will
also provide an estimate of routine swimming speed and
complement studies of swimming metabolism measured in a
water flume under laboratory conditions. We will then

be able to develop a more detailed picture of time and energy partitioning in penguins, examine the differences in breeding and foraging strategies amongst the different species, and better assess the role of penguins in the antarctic marine ecosystem. This assessment is essential to understanding the broad interrelationships among top predators within the antarctic ecosystem because penguins comprise 75-85% of the bird biomass in the Southern Ocean (Croxall, 1982). This means that, as a group, they are one of the major consumers of antarctic marine living resources. Finally, these estimates of free-ranging energetics will provide valuable information for future assessment of the impact of krill fisheries on krill stocks and those species dependent on krill.

REFERENCES

Aschoff, J., and Pohl, H., 1970, Rhythmic variations in energy metabolism, Fed. Proc., 29:1541.

Barre, H., 1975, Le jeune du manchot royal (Aptenodytes patagonica J. F. Miller) a l'ile de la Possession, Comptes rendus, Academie des sciences, Paris, Serie D, 280:2885.

Brown, C. R., In press, Resting metabolic rate and energetic cost of incubation in macaroni penguins (Eudyptes chrysolophus) and rockhopper penguins (E. chrysocome), Comp. Biochem. Physiol.

Conroy, J. W. H., and Twelves, E. L., 1972, Diving depths of the gentoo penguin Pygoscelis papua and blue-eyed shag Phalacrocorax atriceps from the South Orkney Islands, Br. Antarc. Surv. Bull., 30:106.

Copestake, P. G., Croxall, J. P., and Prince, P. A., 1983, Food digestion and energy consumption experiments on a king penguin Aptenodytes patagonicus , Br. Antarct. Surv. Bull., 58:83.

Croxall, J.P., 1982, Energy costs of incubation and moult in petrels and penguins, J. Anim. Ecol., 51:177.

Croxall, J. P., and Prince, P. A., 1980, The food of gentoo penguins, Pygoscelis papua , and macaroni penguins, Eudyptes chrysolophus , at South Georgia, Ibis, 122:245.

Davis, R. W., Kooyman, G.L., and Croxall, J.P., 1983, Water flux and estimated metabolism of free-ranging gentoo and macaroni penguins at South Georgia, Polar Biol., 2:41.

Dewasmes, G., Le Maho, Y., Cornet, A., and Groscolas, R., 1980, Resting metabolic rate and cost of locomotion in long-term fasting emperor penguins, J. Appl. Physiol., 49:888.

Hails, C. J., and Bryant, D. M., 1979, Reproductive energetics of a free-living bird, J. Anim. Ecol., 48:471.

Kendeigh, S. C., Dol'nik, V. R., and Gavrilov, V. M., 1977, Avian energetics, In : "Granivorous Birds in Ecosystems," J. Pinowski and S. C. Kendeigh, ed., Cambridge University Press, Cambridge, New York, pp. 127-204.

King, J. R., 1974, Seasonal allocation of time and energy resources in birds. In : "Avian Energetics," R. A. Paynter, Jr., ed., Nuttal Ornithological Club, Cambridge, Mass.

Kooyman, G. L., Drabek, C. M., Elsner, R., and Campbell, W. B., 1971, Diving behavior of the emperor penguin, Aptenodytes forsteri, The Auk, 88:775.

Kooyman, G. L., Davis, R. W., Croxall, J. P., and Costa, D. P., 1982, Diving depths and energy requirements of king penguins, Science, 217:726.

Le Maho, Y., Philippe, D., and Chatonnet, J., 1976, Thermoregulation in fasting emperor penguins under natural conditions, Am. J. Physiol., 231:913.

Lishman, G. S., and Croxall, J. P., In Prep., Diving depths of the chinstrap penguin Pygoscelis antarctica.

Montague, T. L., 1982, The food and feeding ecology of the little penguin (Eudyptula minor) at Phillip Island, Victoria, Australia, Doctoral Thesis, Monash University.

Mougin, J. L., 1974, Enregistrements continus de temperatures internes chez quelques Spheniscidae. 2. Le manchot royal Aptenodytes patagonica de l'ile de la Possession (archipel Crozet), Com. Nat. Franc. Recherch. Antarct., 33:29.

Nagy, K. H., 1980, CO_2 production in animals: analysis of potential errors in the doubly labeled water method, Am. J. Physiol., 238:R466.

Pinshow, B., Fedak, M. A., Battles, D. R., and Schmidt-Nielsen, K., 1976, Energy expenditure for thermoregulation and locomotion in emperor penguins, Am. J. Physiol., 231:903.

Pinshow, B., Fedak, M. A., and Schmidt-Nielsen, K., 1977, Terrestrial locomotion in penguins: It costs more to waddle, Science, 195:592.

Richards, P. A., and Tickell, W. L. N., 1968, Comparison between the weather at bird island and King Edward Point, South Georgia, Br. Antarc. Surv. Bull., 15:63.

Schoeller, D. A., and van Santen, E., 1982, Measurement of energy expenditure in humans by doubly labeled water method, J. Appl. Physiol.:Respirat. Environ. Exercise Physiol., 53:955.

Shoemaker, V. H., Nagy, K. A., and Costa, W. R., 1976, Energy utilization and temperature regulation by jackrabbits Lepus californicus in the Mojave desert, Physiol. Zool., 49:364.

Stahel, C. D., and Nicol, S. C., 1982, Temperature regulation in the little penguin, Eudyptula minor in air and water, J. Comp. Physiol., 148:93.

Stonehouse, B., 1960, The king penguin Aptenodytes patagonica of South Georgia I. Breeding behaviour and development, F.I.D.S. Scientific Reports, 23:1.

Wilson, R. P., and Bain, C. A. R., In press, An inexpensive depth gauge for penguins, J. Wildl. Manage.

MODELLING THE ENERGY REQUIREMENTS OF SEABIRD POPULATIONS

John A. Wiens

Department of Biology
University of New Mexico
Albuquerque, New Mexico 87131

INTRODUCTION

The previous chapters in this volume have considered aspects
of the physiology of marine birds: the energetics of egg formation,
embryonic metabolism, activity costs, incubation and developmental
patterns, and the like. In this chapter I approach seabird energetics from an ecological rather than a physiological perspective,
asking "How can information on the metabolism and energetics of
seabirds be used to address questions about their behavior, population dynamics, or role in the trophic pathways of marine ecosystems?".

Consideration of such questions requires a focus on how seabirds relate to their environment. More specifically, we must
determine the demands that individuals and populations place upon
their resources, how the environment may affect those demands, and
the likelihood that those demands will be satisfied. These relationships are perhaps best visualized in the context of "resource
systems" (Wiens, in press) (Fig. 1). Resources exist in nature at
various abundance levels. Not all of a resource that is present,
however, is likely to be available to an individual: features of
the abiotic environment, of the resources themselves, or of the
consumer organisms may affect the translation of resource abundance
into resource availability. Organisms use resources from this
available pool in a fashion determined by their morphological
limitations, behavioral preferences, foraging patterns, and metabolic requirements.

There is a fundamental difference in the way ecologists and
physiologists view the metabolism-resource use coupling of Fig. 1.

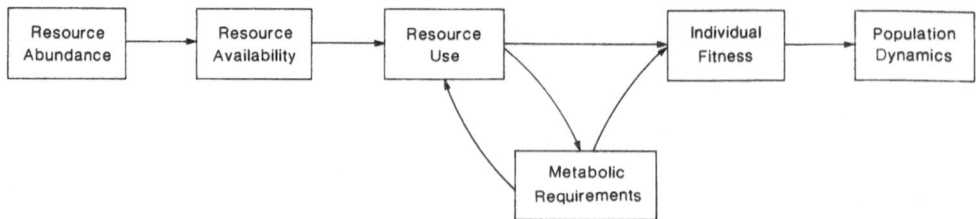

Fig. 1. Components of a "resource system," illustrating how organisms are tied to environmental resources and how that linkage affects population dynamics. From Wiens (in press).

The usual physiological perspective takes resource use (intake) as the starting point and then inquires how the energy or nutrients are allocated to various physiological functions, such as growth, storage, reproduction, or thermoregulation. An ecologist is more likely to structure these flows in the opposite direction. The physiological functions that an organism must perform under normal (or extreme) conditions incur energetic costs, which collectively determine overall metabolic requirements. These create a demand for energy or nutrients, which must be satisfied by intake of resources from the environment. Resource use then feeds back on metabolism, through intake of adequate or inadequate resources to support the metabolic functions. The balance or imbalance of intake and demand in turn determines the fitness of individuals, through effects on reproduction and/or survival (Fig. 1). Variations in these parameters among individuals produce the patterns of recruitment and mortality that give populations their dynamics.

The metabolism-resource use coupling thus represents an important interface of physiological and ecological investigations. In the remainder of this chapter I will explore how the information that comes from physiological studies may be used to project the demands that individuals and populations place upon their resources, which drives resource use. Ideally, we should then assess the likelihood that these demands might or might not be met under natural conditions, and what the consequences might be. This requires information on the other components of resource systems shown in Fig. 1, but such information is generally lacking, especially for marine systems.

MODELLING SEABIRD ENERGETICS

Because of the environment that seabirds occupy and their frequently wide-ranging behavior, it is difficult or impossible to obtain direct measures of the energetic demands that they place upon their resources. As a consequence, various indirect approaches have been developed. Generally these have involved the use of models to integrate estimations of individual metabolism with estimations of population sizes or dynamics. Hunt et al. (1981), for example, projected food consumption of Bering Sea seabird communities as a simple function of body mass. Puttick (1980), McLachlan and Wooldridge (1981), and Schneider and Hunt (1982) used equations relating basal metabolism to body mass, adjusted in a simple fashion for the energetic costs of activity and for assimilation efficiency, to estimate energy budgets of Curlew Sandpipers (Calidris ferruginea), South African shorebirds, and Bering Sea seabirds, respectively. Other studies (Furness, 1978; Grant 1981; Croxall and Prince, 1982; Furness and Cooper, 1982; Pienkowski, 1982; see also Chapter 1 of this volume) have based energetic estimations on existence metabolism, adjusted for a complex of other functions affecting the energy budgets of individuals. Because this is the approach that we have followed (Wiens and Innis, 1974; Wiens and Scott, 1975; Ford et al., 1982), I will use it as a framework to relate physiological information to the estimation of energy demands. The basic structure of such models is shown in Fig. 2.

Existence metabolism

Existence metabolism (EM) is the rate at which energy is used by caged birds provided with food and water ad libitum and maintaining a constant mass over several days when not undergoing reproduction, molt, migratory unrest, growth, or fat deposition (Kendeigh et al., 1977). It is a measure that integrates basal metabolism (BM), temperature regulation, the heat increment of feeding (SDA), and the energy expended in activity while in the cage. Kendeigh et al. (1977) calculated that EM of House Sparrows (Passer domesticus) was approximately 2.78 X BM at 0°C and 1.49 X BM at 37°C.

Kendeigh (1970) presented equations that estimated EM as a function of the body mass of a bird and ambient (air) temperature. These equations were based upon a limited number of studies, and their use in energetics models led to confidence intervals on estimates of population energy demands of as much as ±50% (Furness, 1978). Pimm (1976) reanalyzed Kendeigh's data and derived an equation for EM by stepwise multiple regression analysis that explained more of the variance than did the equations that Kendeigh produced. The relationships that Kendeigh et al. published in 1977, however, were based upon a substantially larger sample size; this had the effect of reducing the confidence interval of energy demand estimates to ±30% (Furness, 1981). In these equations, EM is scaled separately for passerines and nonpasserines as a function of body mass, ambient temperature, and photoperiod.

The Kendeigh et al. (1977) equations thus permit calculation of estimated EM for individuals or species of different mass under

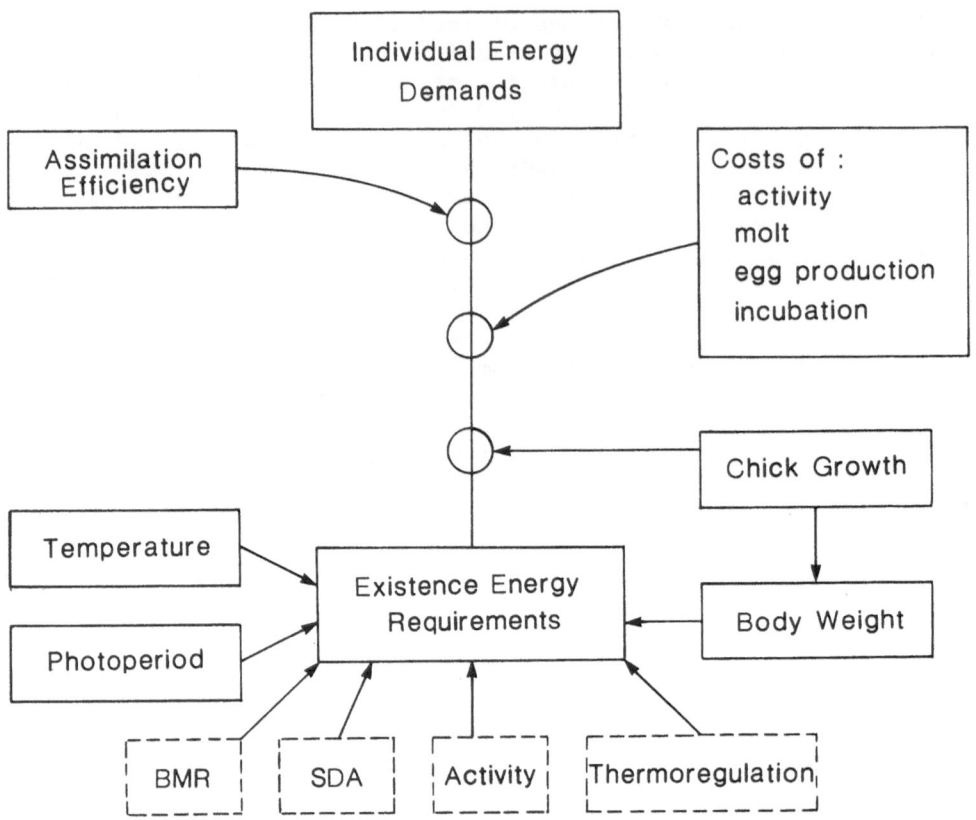

Fig. 2. A general framework for models relating components of individual metabolism to the energy demands placed on environmental resources.

a variety of ambient conditions. Despite this versatility, however, the approach does suffer drawbacks. First, it derives EM as a function of air temperature, and thus ignores the complicating effects of other avenues of heat exchange that are not necessarily closely correlated with air temperature, such as radiation or convection (King, 1974). Lustick (Chapter 9, this volume) has documented the substantial effects that behavioral posture and microhabitat selection may have on heat exchange and individual energetics, and Weathers et al. (in press) have suggested that models based on heat-transfer theory may provide more accurate estimates of daily energy demands than those based on general existence metabolism regressions. Second, of the 91 nonpasserines used by Kendeigh and his colleagues to derive their equations, only 8 were seabirds (2 penguins, 1 shorebird, 3 gulls, and 2 alcids). There is some question regarding how closely seabirds may fit general nonpasserine metabolic equations. Several of the contributions to this

volume, for example, have documented how procellarids may depart
from the metabolic relationships that characterize nonpasserines in
general. Adams and Brown (in press), on the other hand, calculated
BM for 15 procellarid species, and found that they were not metabolically atypical; ratios of measured BM to that predicted from
the Kendeigh et al. (1977) equations averaged 1.11 for the sub-
Antarctic species, 1.26 for the other species. The measured values,
however, varied substantially between studies, reflecting at least
in part differences in the experimental procedures used.

Seabird metabolic patterns may also vary geographically.
Weathers (1979) suggested that BM may be less in tropical climates
and greater in northern latitudes than expected on the basis of
general mass-metabolism relationships. His analysis included only
penguins among seabirds, and his consideration of penguin metabolism was overly simplistic (see Croxall, 1982). More recently,
Pettit, Ellis, and Whittow (pers. comm.) have obtained BM measurements for eight tropical seabird species (see also Chapters 1 and
10, this volume). Six of these were considerably lower than predicted by general mass-metabolism models; they suggested that in hot
tropical areas a reduced BM (and a consequent reduction in endoge-
nous heat production) may be adaptive in reducing heat loads and
the amount of water that must be used in evaporative cooling. To
the extent that seabirds do depart from the relationships contained
in the Kendeigh et al. (1977) EM equations, estimations of energy
demands obtained using this base will be in error.

Additional energetic costs

Calculations of EM include only the energy costs incurred by
birds exposed to ambient conditions in a small cage or aviary.
Free-living individuals, however, engage in a variety of activities
and functions that bear additional energetic costs, and attempts to
determine demands placed upon resources must include consideration
of these costs. For adults, these costs are associated primarily
with locomotor activity, molt, egg production, and incubation (Fig.
2). Walsberg (1983) has reviewed much of the available information
on these metabolic costs.

For birds, the most obvious form of locomotion is flight.
Intuitively, one would expect flight to require substantial amounts
of energy, despite the many flight-related adaptations of birds.
Garland (1983) has documented that the "ecological cost of transport" (the percentage of daily energy intake devoted to transport)
scales as a function of body size in mammals, with large animals
bearing a proportionately greater cost than smaller ones; for small
organisms, he suggests that these costs may be negligible. This is
not likely to be the case for most seabirds. Unfortunately, direct
measurements of flight costs are extremely difficult to obtain,
especially for wide-ranging individuals, and theoretically derived

relationships (e.g. Pennycuick, 1969; Tucker, 1973) do not appear to be substantiated by empirical values (King, 1974; Flint and Nagy, in press). King estimates that flight costs approximately 9.3 X BM for passerines and 15.2 X BM for nonpasserines, regardless of body weight. Such general approximations, however, mask important components of variation in flight costs. Sustained flapping flight, for example, undoubtedly incurs greater costs than gliding or soaring. Furness and Cooper (1982) used equations from Kendeigh et al. (1977) for flapping and gliding flight together with approximations of the proportions of these activities in the total flight time of species in their model of seabird bioenergetics. I am unable to discover an equation for gliding flight costs in Kendeigh et al. (1977), however. Croxall and Prince (1982) considered the equations of Kendeigh et al. (1977) (and of Tucker, 1973) to be inappropriate for procellariform seabirds, as they were based primarily upon flapping flight rather than gliding flight and thus overestimated actual flight costs substantially. They used Baudinette and Schmidt-Nielsen's (1974) estimate of gliding flight costs in Herring Gulls (Larus argentatus) to derive a flight cost of 1.85 X EM. They noted, however, that by their calculations the costs of gliding among gliding species are relatively greater for large than for small species. This is consistent with Garland's (1983) theoretical suggestion; Croxall and Prince (1982), however, felt that theory would predict exactly the opposite, as small species do more flapping than large species; they therefore applied the gull value uniformly to all species. Flint and Nagy (in press), however, used $D_2^{18}O$ techniques to determine that the metabolic rate of Sooty Terns (Sterna fuscata) during flight was 4.8 X BM. This was much lower than predicted by theory, despite the predominance of flapping flight in this species. Flint and Nagy attributed the deviation to the high aspect ratio and low wing loading of the terns, and cautioned that species-specific aerodynamic adaptations might influence flight costs substantially, reducing the usefulness of general models of flight costs.

Other variables may affect flight costs as well. In species that fly close to a surface, such as skimmers, the "ground effect" may substantially reduce the power requirements of flapping flight (Withers and Timko, 1977); how this might influence the flight costs for near-surface soaring species, such as shearwaters or albatrosses, is uncertain. Short-term increases in body mass, such as by ingestion of large quantities of food during a foraging episode (Sibly and McCleery, 1983) or formation of a relatively large egg prior to laying (as in many procellarids), may alter flight costs as well. The effects of such changes in mass on flight energetics is not known, but as mass is changed while wing area remains constant, wing-loading will increase, increasing the power requirements of flight.

For many seabirds, swimming and diving are as important modes of transport as is flight, but even less is known about the energetic costs of such activities. In the absence of empirical values, Furness and Cooper (1982) used equations for the costs of flapping and gliding flight to estimate the costs of underwater and surface swimming, respectively. Croxall and Prince (1982) instead used a preliminary estimate of 2.4-2.5 X EM as the energy cost of swimming and/or diving for South Georgia seabirds. This value was derived from the studies of Kooyman et al. (1982) and Davis et al. (1983), in which labeled water procedures were used to obtain direct estimates of energy expenditures by free-living birds. Kooyman et al. (1982) reported a metabolic rate of 2.8 X BM for King Penguins (Aptenodytes patagonica) while the birds were at sea. Davis et al. (1983) estimated that feeding Gentoo Penguins (Pygoscelis papua) required 2.6 X BM. The metabolic rate of feeding Macaroni Penguins (Eudyptes chrysolophus) averaged 2.9 X BM, while for fasting individuals the estimated rate was 1.8 X BM.

To use equations or coefficients relating the costs of specific activities such as flapping flight or surface swimming to EM, one must also have time-activity budget information, to indicate the proportion of a time period spent in each of the activities (e.g. Wiens and Innis, 1973). Furness (1978) followed such an approach in his seabird energetics model, although he had actual data only for inshore feeders and was compelled to estimate activity budgets for offshore-feeding species. We have used a similar procedure in our model projections of seabird sensitivity to oil spills (Ford et al., 1982). Furness (1978) found that overall activity costs varied substantially between species, reflecting basic differences in their time-activity budgets. For Great Skuas (Stercorarius skua), activity costs were 0.36 X EM, while costs for Arctic Terns (Sterna paradisaea) were 1.13 X EM. Furness felt that these might represent extremes of activity levels for seabird species, although one would expect activity in many alcids to be even more expensive, given their inefficient flight and extensive diving (J. Croxall, pers. comm.). It is obvious, in any case, that use of a single value to adjust EM for activity costs (e.g. Wiens and Scott, 1975; Ford et al., 1982), although necessary in the absence of species-specific information, may obscure important detail. In one of our models (Innis and Wiens, 1977), we employed general activity-cost coefficients, but varied the magnitude of this adjustment over the year to reflect seasonal variations in activity levels associated with reproduction.

Although locomotion is perhaps the most conspicuous of the energy-demanding functions not included in EM, it is not the only one (Fig. 2). Molt, for example, involves the production of new tissues, and thus must incur a cost. Whether or not this cost is significant may depend on how abrupt or gradual the molt is, and on whether other energy demands may be reduced during the molt period through lowered activity levels. Ricklefs (1974) and King (1980) suggest that additional energy costs during molt are generally minor, on the order of 0.2-0.4 X BM. Croxall (1982) calculated molt costs for several species of penguins from weight losses during associated fasts, obtaining a value of 2.0 X BM. This contrasts with the value of 1.3 X BM for fasting, non-molting birds; molt thus elevated energy demands by 0.7 X BM. Furness and Cooper (1982) adjusted EM by 10-50% (depending on the rate of molt) to reflect molt costs; in our earlier model (Wiens and Scott, 1975) we used an adjustment factor of 0.12 X EM.

The process of reproduction involves additional energy costs, particularly for females. The production of a clutch of eggs, for example, creates both energy and nutrient demands that are super-imposed upon normal EM and activity costs. Nutrient demands may be especially critical, as the condition of a female immediately prior to egg formation (e.g. protein reserve levels) or the availability of food of specific nutrient characteristics during egg formation may affect the size of the clutch that is produced, the quality of the eggs, or the survival probabilities of the female (Jones and Ward, 1976; Ankney and MacInnes, 1978; Raveling, 1979; Grau, Chapter 2 in this volume). Model-based approaches have not considered such nutrient constraints, however, but have focused on the energetics of egg production. Generally, the costs of production of a clutch may be modelled by considering the number of eggs produced, the weight of an average egg, the caloric value of egg material, the efficiency of the egg-production process, the rate of ovarian growth, and the laying interval (King, 1973; Ricklefs, 1974; Wiens and Innis, 1974; Walsberg, 1983). Given the wide range of egg sizes and yolk contents as proportions of adult body mass that are displayed among seabirds (see Chapters 1, 2, 5, and 6 of this volume), it is unlikely that a single general regression equation (e.g. Kendeigh et al., 1977) will provide very accurate species-specific estimations of egg production costs. In addition, such equations do not consider the phenology of egg development, which can have important effects on the rate of energy demand. Rahn et al. (Chapter 5, this volume), for example, calculate that formation of a single egg of a petrel may require 229 kcal (959 kJ), while that of a cormorant requires only 40 kcal (167 kJ). Because a cormorant may have a clutch of as many as five eggs, however, its clutch may cost approximately as much as the one-egg clutch of the petrel. The times required for egg formation and laying in the two seabirds may differ markedly, and the degree to which egg-production costs elevate daily energy demands may differ accordingly.

Based on model analyses, it seems doubtful that egg production costs represent an important component of individual or population energetics, at least over the duration of a breeding season. Croxall and Prince (1982) simply presumed egg costs to be "trivial," while Furness and Cooper (1982), using an equation that

262

incorporated clutch size, egg weight, and caloric value, found that egg production costs accounted for no more than 0.1% of the total population energy demands of the three seabird species they considered. The model of Wiens and Scott (1975) additionally considered the rate of egg formation and laying; they estimated that egg production accounted for less than 1% of the breeding-season energy flows through several seabird populations. Although these estimated values are quite low when considered over a season, they may represent a significant increase in individual energy demands during the period of egg formation. Ricklefs (1974), for example, reported estimated peak daily energy costs of egg production of approximately 1.4 X BM in shorebirds and 1.7 X BM in gulls. These peak demands may be especially critical if resource availability is limited. Energy deprivation generally does not influence the size of eggs, but rather affects the number of eggs laid (Ricklefs, 1974). The costs of egg production may thus be large enough to preclude egg development and laying of a clutch when resource conditions are extremely adverse, especially for seabirds in which eggs are large relative to body mass (e.g. some alcids, procellarids).

Once eggs are formed and laid they must be incubated, and this involves energetic demands by both the developing embryos and the incubating adults. Several of the previous chapters in this volume (1, 5, and 6) have documented embryonic energetics in detail. These energetics are supported entirely by the energy stores sequestered in the yolk during egg formation by the female. Once an egg is formed, therefore, its subsequent development to the time of hatching imposes no additional direct energy demands upon environmental resources, and consideration of embryonic energetics is irrelevant to models that attempt to estimate such energy demands. Energy costs of incubation to adults, however, are another matter. For passerines, in which there is a premium on rapid development of embryos and a short nesting period, incubation is continuous and may elevate egg temperature significantly above ambient levels. Kendeigh et al. (1977) have considered the factors that may influence incubation energy costs in such species in detail. In many seabirds, on the other hand, embryonic development is prolonged, especially in pelagic-feeding or tropical species (Whittow, 1980). In some of these species, the pattern of incubation is frequently interrupted, incubation temperatures are considerably lower than those characterizing more continuously incubating species (Boersma, 1982), and the embryos are unusually tolerant of chilling (Wheelwright and Boersma, 1979; Vleck and Kenagy, 1980; Roby and Ricklefs, in press). Many seabirds also incubate in prolonged bouts, during which they are inactive and fasting; their overall energy expenditures thus might be expected to fall somewhere between the requirements for basal and for existence metabolism (Croxall and Ricketts, 1983) or even below basal (see Chapter 3, this volume). For fasting seabirds, the energy costs of incubation may be derived from rates of mass loss. Based on this procedure, incubation costs have been estimated at 1.2-2.0 X BM (probably closer to the lower value) for albatrosses (Croxall and Ricketts, 1983) and 1.3-1.4 X BM for petrels and penguins (Croxall, 1982). Direct measurements made by Grant and Whittow (1983), however, revealed that the metabolic rate of the Laysan Albatross (Diomedea immutabilis) and Bonin Petrel (Pterodroma hypoleuca) during incubation was less than the resting metabolic rate of birds that were not incubating, and less than or equal to the predicted BM. Brown (in press), calculating metabolic rate on the basis of oxygen consumption, found that energy costs for fasting, incubating individuals of two species of sub-Antarctic penguins were significantly lower than resting metabolic rates, and were close to the calculated values of BM for both species. On the other hand, King (1974) suggested that incubation in passerines elevated BM by 1.65. In our previous model analysis of seabird energetics (Wiens and Scott, 1975), incubation costs were not calculated separately but were simply presumed to be included in the general coefficient used to adjust EM for overall activity. Furness (1978) and Furness and Cooper (1982) did not consider incubation costs in their models and suggested that the costs were not easily calculated, but were probably small. This is an acceptable procedure for most seabirds if estimations are desired over an entire breeding period, but if attention is focused on the nesting period and/or the birds exhibit prolonged or interrupted incubation or fast, more detailed consideration of incubation costs may be required. As Walsberg (1983) has noted, it is difficult to generalize about avian incubation costs.

Chick energetics

The factors contributing to the energy demands of nestlings or chicks differ in important ways from those affecting adult energetics, and chicks must thus be considered separately in bioenergetic models. In general, two approaches have been followed to estimate chick energetics. One involves using information on food (energy) intake rates of growing chicks under laboratory or field conditions to calculate an overall daily energy demand. Kendeigh et al. (1977) generated a regression relationship of chick daily energy budgets to body mass from studies of one passerine and one anatid, for example, which Furness and Cooper (1982) adopted to calculate chick energetics in their analysis of seabirds. Croxall and Prince (1982) and Williams (1982), on the other hand, based their energetics projections on direct measures of food intake rates by chicks of the species they studied. The advantage of estimating chick energetics from intake rates is that the information can often be obtained directly from field observations on the species of interest under known environmental conditions; the disadvantage is that only overall chick energy demand is estimated, and the contributions of various components of metabolic functioning to this demand cannot be separated. This may not represent an important drawback, however, unless one wishes to explore the partitioning of chick energetics. A more substantial problem may be the need to obtain data on meal size and feeding frequency throughout the rearing period (J. Croxall, pers. comm.).

The other approach to estimating chick energetics considers the components of energy demand separately, following the general approach applied to adult energetics (Fig. 2). The calculations may be based on mass-dependent estimations of either EM or BM, although Kendeigh et al. (1977) have indicated that their equations for EM may not apply to chicks. To some extent, this is because EM as calculated includes the costs of thermoregulation outside of the thermoneutral zone. Chicks, however, are generally unable to thermoregulate until they are at least several days old. Murres (Uria spp.) do not become homeothermic until they are 9-10 days old (Johnson and West, 1975), kittiwakes (Rissa) until they are 6-8 days of age (Barrett, 1978), storm-petrels (Oceanodroma) until they are 5-6 days old (Wheelwright and Boersma, 1979). Thus, prior to this time chicks may allocate little of their metabolic demand to temperature maintenance, allowing their body temperature to vary with ambient conditions or being brooded by adults. Young of at least some seabirds have a capacity to withstand considerable cooling (Wheelwright and Boersma, 1979). Once thermoregulation does begin, its costs are not extraordinarily large: from laboratory studies, Dunn (1980) calculated that thermoregulation by chicks of altricial Double-crested Cormorants (Phalacrocorax auritus) cost 80% of daily BM, while for semi-precocial Herring Gull chicks thermoregulation elevated BM by 20%.

If a suitable metabolic base for chick energetic calculations can be established, it must then be adjusted for various additional energy costs. Prior to fledging the young of many seabirds are relatively inactive, and existence metabolism (which includes a limited amount of activity) may be elevated relatively little in this regard. Dunn (1980) suggested that activity in nestling cormorants and gulls increased energy costs above BM by 100-105%. Based on studies of nestling passerines, Wiens and Innis (1974) chose not to include an activity-cost component in their calculations of nestling energetics.

The other costs that adults may incur, such as for molt, egg formation, or incubation (Fig. 2), are obviously inapplicable to chicks, but another cost, that of growth, is of critical importance. Considerable information exists on the patterns of gain in mass of growing seabird chicks (e.g. Ricklefs, 1973), but our knowledge of the energetic costs accompanying this growth is less certain. A growing chick must not only support the metabolic requirements of its existing mass, but must also take in sufficient food to meet the requirements of the addition of new tissues during growth. This bears an obvious energy cost, which is related to the energy content of the new tissues and the efficiency with which food is converted into new tissue. Ricklefs (1974) has considered aspects of the energetics of chick growth and development in detail, noting that the energy value of tissues and the proportion of total energy devoted to growth both change during the growth period. He used a production efficiency (a measure of the metabolic energy cost of synthesizing new tissue; production/metabolizable energy) of 75% in his calculations. From field studies of murre chicks, Birkhead (1977) estimated that 31% of the actual food intake was converted into new tissues. The latter value includes energy ingested but not assimilated (i.e. production/gross energy intake), and is thus not directly comparable with Ricklefs' efficiency measure. Montevecchi et al. (in press) calculated an overall net growth efficiency (after assimilation) of 33% for Northern Gannet (Sula bassanus) chicks. In our model analyses of seabird energetics, we used an overall adjustment of 0.20 X EM to account for growth costs in nestlings, although this value was a coarse estimate based on studies of a small number of passerine species (Wiens and Innis, 1974).

Assimilation efficiency

Not all of the food that is eaten by individuals is available for metabolism, due to inefficiencies in the digestive process. A measure of assimilation efficiency [(metabolizable energy/gross energy intake) X 100] must therefore be used to convert calculations of overall metabolic requirements into estimates of the energy intake that is needed to meet those requirements (Fig. 2). Ricklefs (1974) and Kendeigh et al. (1977) summarized the limited information available on avian assimilation efficiencies (mostly derived from laboratory studies of passerines). It is apparent that assimilation efficiency varies rather widely, as a function of food type, bird species, and ambient temperature. Ricklefs (1974) cited information suggesting that assimilation efficiencies on a fish diet were on the order of 79-82%. Modelling analyses have used various values: Wiens and Scott (1975) used 70% on the basis of unpublished information of Kendeigh, but in later exercises (Wiens et al. unpublished) a value of 75% was used; Schneider and Hunt (1982) used 70% and Croxall and Prince (1982) 75%, both citing Kendeigh et al. (1977) as their source [I am unable to find either value suggested as appropriate in Kendeigh et al. (1977)]; and Puttick (1980), Furness (1978), and Furness and Cooper (1982) assumed a value of 80%. Kooyman et al. (1982) based their calculations of King Penguin energetics on an assimilation efficiency of 80%, derived in part from D.P. Costa's unpublished measurements on Adelie Penguin (Pygoscelis adeliae) chicks. Cooper's (1977, 1978) field measures on young gannets and penguins suggest that assimilation efficiency varies with age, but is generally on the order of 74-79% on diets of fish or krill.

Overall energy demands

The approach to modelling seabird energy demands reviewed here involves the partitioning of metabolism into separate components, calculation of costs for each component, aggregation of these cost estimates, and adjustment of the overall estimate for assimilation efficiency (Fig. 2). The resulting estimate of individual energy demand can then be combined with information on or model projections of population structure to produce estimates of overall population energy demand. To the degree that the available data or the objectives of an analysis warrant, such estimations may be produced for different segments of the population (sexes, age classes) or may be projected over time periods of various lengths, with varying levels of resolution (Wiens and Innis, 1974).

Such models may thus provide estimates of individual or population energetics, but how can such estimates be used to address questions of interest in ecology or evolution? The following section provides some examples.

MODEL APPLICATIONS

Population and community energy flows

Models of seabird energetics have been used most frequently to estimate the energy flowing into or through seabird populations occupying some area over a specified period of time (Wiens and Scott, 1975; Furness, 1978; Puttick, 1980; Croxall and Prince, 1982; Furness and Cooper, 1982). Here I will develop an example of this approach, using our own studies of Alaskan seabirds (Wiens, Ford, and Heinemann, unpublished). We have analyzed the energy demands of these communities for two areas: the Gulf of Alaska about Kodiak Island and the vicinity of the Pribilof Islands, in the Bering Sea.

We used two sources of data for our Kodiak analyses. First, we calculated pelagic densities of seabirds recorded in 59 shipboard transect surveys covering 631 km. These densities were then combined with model projections of individual energy demands for the different species to obtain an estimate of daily energy demands per km^2 for each species. There were substantial seasonal shifts in energy demands, characterized by a peak in May, associated with movement of Black-legged Kittiwakes (*Rissa tridactyla*) and Sooty Shearwaters (*Puffinus griseus*) into the area. Another peak in August-September corresponded with the fall movement of large numbers of shearwaters. Peak daily energy demand, 10.2×10^4 kJ km^{-2} day^{-1}, occurred at this time; shearwaters contributed 92% of this total. In comparison, Wiens and Scott (1975) estimated peak daily demand for a 4-species seabird community on the Oregon coast of

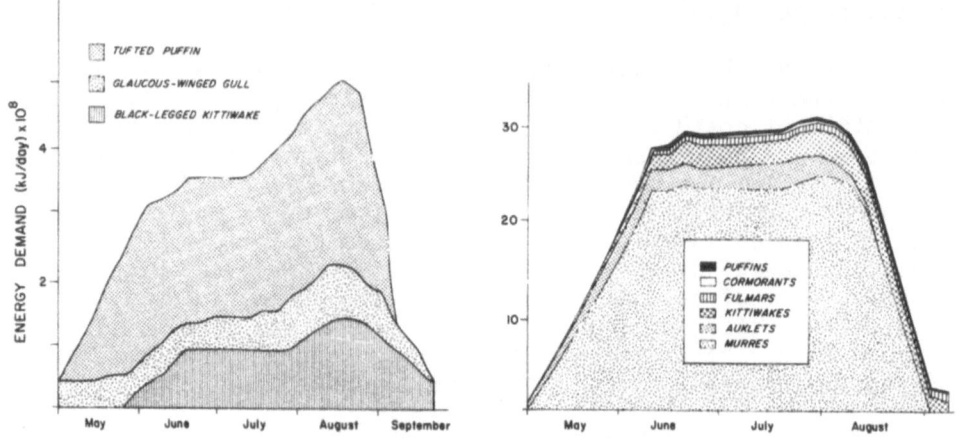

Fig. 3. Patterns of estimated daily energy demands by the dominant
species of seabirds breeding on Kodiak Island (left) and
the Pribilof Islands (right). Note that the two graphs have
different scales.

22.6×10^4 kJ km^{-2} day^{-1}, again dominated by the fall passage of
shearwaters.

The other Kodiak analysis was based on density estimates of
the three dominant seabird species breeding at colonies on the
island. Because more detailed population information was available
for this analysis, we could model the changes in population and
community energy demands over the duration of the breeding season
(Fig. 3). Peak daily demand for the three-species community ($5.2 \times
10^8$ kJ/day) occurred in late August, with the fledging of young.
Over the entire breeding season, we estimated that the populations
of these three species breeding on Kodiak required 4.1×10^{10} kJ,
60% of which was accounted for by Tufted Puffins (<u>Fratercula</u>
<u>cirrhata</u>).

These latter values take on some meaning when they are compared with values obtained from a similar analysis on another
system. The breeding-season pattern of energy demands for the 11-
species community on the Pribilof Islands (Fig. 3) shows both a
different form and different magnitudes. Common and Thick-billed
murres (<u>Uria</u> <u>aalge</u> and <u>U.</u> <u>lomvia</u>) dominate the community dynamics,
accounting for 80% of the overall energy flow. The absence of a
late-season peak in energy flow is a reflection of the low recruitment rates of these species. Peak daily energy demands were estimated at 2.9×10^9 kJ/day; total breeding-season demand was $2.7 \times
10^{11}$ kJ, nearly an order of magnitude greater than that of the
Kodiak breeding seabirds.

Several other studies have used similar procedures to arrive
at estimates of energy demands by seabird populations. Furness and
Cooper (1982), for example, documented the bioenergetics of three
species breeding in the southern Benguela region, along the west
coast of South Africa. The three species, which comprised over 95%
of the breeding seabird biomass in the area, consumed on the order
of 1.3×10^8 kJ/year. Using an earlier version of the same model,
Furness (1978) estimated an annual demand of 5.0×10^{10} kJ for a
nine-species community of seabirds on the Shetland Islands. Croxall
and Prince (1982) employed slightly different procedures to estimate that a 12-species community on South Georgia demanded $5.1 \times
10^{12}$ kJ over their breeding season. These estimates vary widely,
but it seems more likely that the differences are consequences of
the number or types of species or the time period included in the
analyses, rather than reflections of gross inaccuracies in the
modelling procedures.

Because we had detailed information on the distributional
patterns of the Pribilof seabirds in the area about the islands (G.
Hunt, pers. comm.), we were able to explore how the energy demands
of the birds were distributed in space as well as time. There was
substantial variation in the spatial distributions of the birds at
different census times, but the overall pattern (Fig. 4) indicated
a peak in energy flow in a zone 11-40 km from the islands, trailing
off with greater distances out to 200 km. Murres, the energetically
dominant species in this system, concentrated their feeding activities in the relatively close intervals; 77% of their energy demand
occurred within 40 km of the nearest island (Fig. 4). Northern
Fulmars (_Fulmaris glacialis_), which accounted for only a moderate
proportion of the total community energy demand, assumed greatest
importance at greater distances, beyond 120 km. Energy flow through
shearwater populations was concentrated at intermediate distances.

Trophic dynamics

Abstract notions of energy flow through populations or communities become real when they are expressed in terms of quantities
of food consumed, for it is through this trophic linkage that
seabirds may assume an important position in the functioning of
marine ecosystems. Obviously, to project these patterns in any
detail requires information on the composition of the diets of
seabird populations in an area and the energy content of the
various prey types. Wiens and Scott (1975) used such information to
estimate the magnitudes and seasonal patterns of prey consumption
for Oregon murre populations. The analysis revealed distinct seasonal shifts, with the birds consuming large quantities of herring
in fall and winter, smelt in winter and early spring, and predomi-
nately anchovies in late spring and summer. Over the year, we
estimated that the murres consumed 5,700 t of herring, 3,500 t of
anchovies, and 3,000 t of smelt.

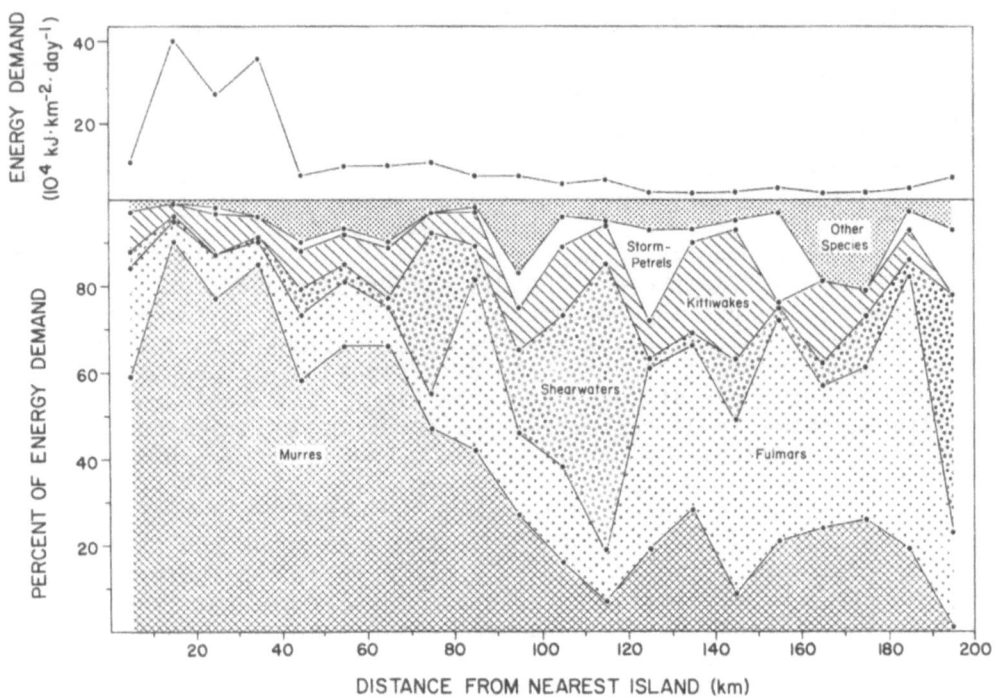

Fig. 4. Distribution of total community energy demand (above) and of the percentage contribution of each species group to that total energy flow (below), as functions of distance from the nearest breeding island in the Pribilof Islands, Bering Sea.

Usually the diets of the birds are not known in adequate detail, and calculations of trophic flows must be more general. In our analysis of the Oregon seabirds, for example, we used crude information on the diets of the other species to calculate that the four-species assemblage consumed 62,500 t of prey per year. Of this total, 43% was anchovies, and 86% of the consumption of anchovies was by shearwaters during their relatively brief spring and fall passages along the coast (Wiens and Scott, 1975). For the three dominant breeding species on Kodiak Island discussed above, we derived an estimate of 8,100 t consumed during the 4-month breeding season, while a similar analysis of the Pribilof Islands seabird community indicated a consumption of 53,600 t of prey over the same period of time. By way of comparison, Furness and Cooper (1982) estimated that the Benguela seabirds they studied consumed 16,500 t of prey per year, 72% of them anchovies. Most of this food was obtained within 50 km of the breeding colonies. Annual food consumption by the South Georgia seabird community analyzed by Croxall and Prince (1982) was estimated to be 1,700,000 t, mostly

(88%) krill; Macaroni Penguins accounted for 80% of the annual consumption. Inclusion of nonbreeding individuals and failed breeders and consideration of the extra energy costs of accumulating fat reserves increased the estimated consumption to 2,700,000 t/yr. Croxall et al. (in press) extended this analysis to include the breeding seabirds of the Scotia Sea. There, seabirds were estimated to consume 13,167,800 t of food, 82% of it krill. Penguins accounted for 76% of the total breeding seabird food consumption. Using simpler estimation procedures, Hunt et al. (1981) calculated that the seabirds occupying the shelf waters of the eastern Bering Sea consumed 580,000-1,150,000 t/yr, while Threlfall (1983) projected an annual consumption of fish (mostly capelin) on the order of 313,000 t by the dominant seabirds (shearwaters, murres, puffins) off Newfoundland.

The magnitudes of these estimated prey consumptions are impressive at face value, but how do they relate to the resource systems that must supply such demands? The most obvious way to determine this is to compare the estimated prey consumption by the seabirds with a measure of the standing crop or (better) annual production of the prey populations. Unfortunately, such measures are available for few of the areas in which seabirds forage. Wiens and Scott (1975) used information from the Oregon coast to suggest that the four seabird species they considered might consume 22% of the net annual production of small pelagic fish within 185 km of the coast, while Furness (1978), following similar calculations, determined that the Shetland seabird community consumed approximately 29% of the annual fish production within 45 km of the colony. Furness and Cooper (1982) calculated that the seabirds in the South African colony they studied might consume about 20% of the anchovy and herring biomass each year, and they analyzed data from Peru to suggest that seabirds there were responsible for roughly 20% of the fish mortality in that system. In a single-species study, Puttick (1980) used energetic estimations to conclude that Curlew Sandpipers might consume as much as 34-51% of the standing crop of benthic invertebrates in their estuarine habitats, approximately 12% of the annual production. For a South African shorebird community, McLachlan and Wooldridge (1981) estimated that the birds consumed 12-36% of the production of the major intertidal prey groups. Overall, these estimates would seem to suggest a rather significant role for birds in the trophic and production dynamics of most nearshore marine ecosystems.

One additional functional relationship of seabirds to marine ecosystems may also be explored using model analyses. Seabirds often breed in vast colonies, foraging widely about the colony to return with food. They and their offspring defecate there, producing large quantities of guano that may enrich the nutrient levels of the terrestrial system and/or the adjacent ocean areas (Tuck, 1961; Burger et al., 1978; Smith, 1979). If the food consumption rate can be estimated and the assimilation efficiency is known, one may calculate the quantity of material that will not be assimilated by the birds and that thus will be returned to the system. Wiens and Scott (1975) used this approach to estimate that the Oregon-coast seabirds returned 6.2×10^{10} kJ to the system in excrement during the breeding season. If one knew the chemical composition of the prey, nutrient returns could be estimated as well. Threlfall (1983), for example, cited J. Piatt's unpublished analysis that estimated that Newfoundland seabirds return approximately 7,700 t of excrement to the marine system annually. Using information on the nutrient composition of Peruvian seabird guano, Threlfall then calculated that 1,227 t of free phosphate and 775 t of free ammonia are voided into the Newfoundland ocean areas annually. This is equivalent to an input of 25 mg PO_4/m^3 and 15 mg NH_4/m^3 in a layer of water 50 m deep over 1,000 km^2.

Effects of resource depression

Energetic models may also be used to assess the potential impacts of changes in food resource levels on seabirds. Most sea-birds breed in colonies from which they radiate out to forage in adjacent ocean areas, returning with food for their chicks. The situation has many attributes of a central-place foraging system, and may be modelled accordingly (Ford et al., 1982). Central-place foraging theory assumes that individuals forage in an optimal manner, minimizing the time spent in flying to and from foraging areas and/or maximizing the rate of food delivery to the young (Orians and Pearson, 1979; Norberg, 1981), and there is indeed some observational support for the notion that some seabirds do feed their young at close to the maximum rate possible (Gaston and Nettleship, 1981). This means that disruptions in the abundance or availability of food resources are likely to have direct effects on adult foraging behavior and, perhaps, delivery rates to young. Such disruptions may take a variety of forms: cold, windy weather or rough sea conditions (Pienkowski, 1982; Birkhead, 1976), seasonal changes in the size or composition (or nutrient/energy quality) of the prey base (Hedgren and Linnman, 1979; Gaston and Nettleship, 1981), oil pollution (Ford et al., 1982), or large-scale oceano-graphic changes such as El Niño (Duffy, 1983; see also Chapter 1, this volume). If we accept the premise that feeding patterns are optimized under normal conditions, such changes have the conse-quence of forcing the birds to forage farther from the colony and spend more time searching for and capturing prey (Fig. 5). This will inevitably increase the energy expenditures of the adults, and with an appropriately structured model such energy costs can be calculated.

Seabird species differ in their normal foraging distances from their breeding sites, and thus operate on different spatial scales. A localized resource depression may therefore have a much greater

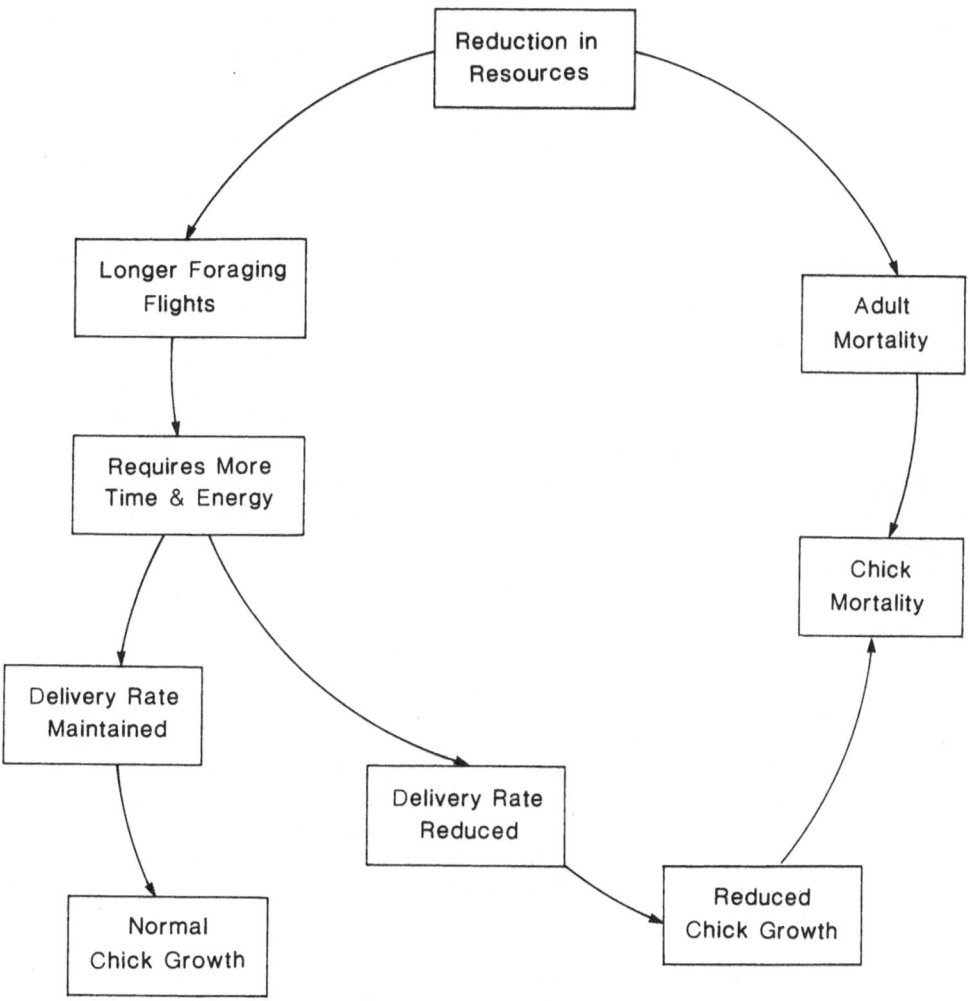

Fig. 5. The pathways of effects on chick growth and survivorship associated with a reduction in food resources to the adults. See text.

effect on a short-distance forager (e.g. cormorant) than on a species that covers a much larger area (e.g. fulmar or storm-petrel). The capacity for adults to adjust their foraging behavior so as to maintain an adequate rate of food delivery to the young is thus likely to vary considerably between species (Crawford and Shelton, 1978). Lloyd's (1977) experiments on Razorbills (Alca torda) provide some indirect evidence of the ability of adults to compensate in their feeding patterns. An additional chick was added to experimental nests, increasing the energy demands placed upon the adults (the counterpart of a reduction in resource availability); the adults responded by nearly doubling their feeding rates and the amount of food delivered to the young. Even so, the compensation was apparently inadequate, as chicks in twinned nests grew significantly slower than in the control (single-chick) nests and generally were less successful in fledging.

The most likely consequence of reductions in resource levels is a reduction in the rate of food delivery to the young. This, in turn, is likely to cause a reduction in the growth rate of the young (Fig. 5). Birkhead (1976), for example, documented a 30% reduction in the rate of food delivery to young murres in rough versus calm sea conditions, and Stempniewicz (1980) noted how variations in food delivery rates affected growth rates of Dovekie (Alle alle) chicks. Pettit et al. (in press) noted substantial annual variation in growth rates of Wedge-tailed Shearwater (Puffinus pacificus) chicks on Hawaii, which they suggested might be related to variation in prey abundance. If the duration or magnitude of food deprivation to the chick is relatively short, the consequences on eventual fledging mass and chick survival are likely to be slight. Chicks of some species, such as gannets, may accumulate fat reserves during growth, and be able to withstand periods of starvation of as long as 2 weeks without deleterious effects, at least toward the end of the nestling period (Nelson, 1978). As the degree of deprivation increases, however, growth may no longer be adequate to bring the chick to the range of acceptable fledging masses at the proper time. Fledging may then be delayed too long, sharply decreasing survival probabilities. Under condtions of severe food deprivation, chicks may respond by shutting down growth processes and reducing body temperature, although such responses represent a "last-ditch attempt to extend survival time" (Ricklefs, 1974).

Food deprivation of growing chicks represents an imbalance between energy intake and energy demand. The latter can be estimated from energetic models, while the former may be obtained from direct observations of food delivery rates or from models of foraging behavior (e.g. Ford et al., 1982). Determining the effects of such imbalance on growth rates requires consideration of the efficiency of the growth process as well as of the nature of the growth response to changes in the intake:demand ratio (which is likely to vary in a nonlinear manner). Such information may then be converted into a function expressing how survival probability is affected by energy-related reductions in growth rate (Ford et al., 1982). Unfortunately, although such relationships are amenable to modelling, the information needed to specify the form of the relationships in a quantitative manner is almost entirely lacking, and we must therefore proceed by educated intuition (or not at all).
There is one other avenue relating resource depression to

chick mortality (Fig. 5). Most seabird chicks are attended by both adults during the nestling phase. If one of the adults disappears during this period, the young are likely to die, as a single parent is unable to maintain adequate delivery rates while protecting the young from predators (Stempniewicz, 1980; Gaston and Nettleship, 1981).

Seabird energetics models structured in the context of chick demands and adult delivery rates shown in Fig. 5 may be used to explore evolutionary as well as ecological questions, by simulation of alternative adaptive strategies. Consider, for example, the timing of fledging. Seabird species vary widely in the extent to which their chicks are precocial or altricial. Murre chicks are intermediate, fledging at roughly 21 days of age (Birkhead, 1977), at 18-21% of adult body mass (Johnson and West, 1975). The chick then goes to sea, accompanied by a single adult (Hedgren, 1981; Gaston and Nettleship, 1981). It has been suggested that as the chick grows the adults may no longer be able to keep up with its increasing energy demands, as foraging trips may take over an hour (Birkhead, 1977) and adults carry but a single prey item per trip. The evolutionary "solution" is for the chick to go to the food supply with the adult, minimizing adult travel time and energy costs of transport (Birkhead, 1977; Gaston and Nettleship, 1981).

How plausible is this evolutionary argument? How much would the energy budget of the adults be altered if, say, the chick remained at the nest site until it attained adult body weight? Scott and I (unpublished) explored this question by calculating the daily energy demands of growing chicks in two scenarios: in one, the chick fledged normally, at a mass of 250 g, while in the other the chick delayed fledging until it grew to 1,000 g. Knowing the chick energy demands, we then used features of the prey delivery and foraging patterns of adults to determine the daily activity budgets for adults as a function of chick age in the two scenarios. By assigning the adult activities appropriate energy costs, we then calculated how adult energy demands might differ in the two situations (largely a consequence of the differences in time spent in flight and diving). Preliminary results of this exercise indicated that adult energy demands did indeed rise sharply soon after the chick passed the normal fledging weight. It was apparent that if the adults had to forage very far from the colony (more than 10 km), the likelihood that they would be able to meet the energy demands of a large chick at the nest site was remote.

DISCUSSION

The preceeding sections of this chapter have shown how information on seabird energetics may be integrated into models, and how the models may be used to generate estimates or predictions of energy flow or food consumption of populations or communities or may function as tools to explore the energetic consequences of environmental perturbations or of alternative adaptive strategies. Models are of course no better than the empirical values or assumptions on which they are based, and no responsible modeller would claim that models represent the best way to study seabird ecology, or that they obviate the need for careful physiological or behavioral investigations. The models that have been developed and used on seabirds in the past are all relatively unrefined, reflecting the state of our knowledge. This has led some to doubt the validity or usefulness of modelling approaches. In his discussion of model-derived estimates of the proportions of fish production consumed by seabirds, for example, Bourne (1983) concludes that "it seems likely that these calculations are exaggerated by many wrong assumptions about both the food which the birds eat and the distance which they travel to catch it, and that the true total is only about one-tenth as high, so that fisheries present a more serious threat to birds than birds do to fisheries." Although Bourne's final statement is undoubtedly true, his conclusion that the model estimates of consumption are too high by 1-2 orders of magnitude (Bourne, 1981) is based on disputes regarding the distance over which seabirds are distributed from their colonies. Model estimations of seabird impacts have clearly been spatially restricted, and their projections have been applied only to the areas they considered. Nonetheless, Bourne's criticisms clearly emphasize the need for detailed distributional information as a foundation for model analyses.

In any event, the fact remains that much information about seabirds, such as their overall food consumption or the energetic costs of adopting different breeding strategies, is difficult or impossible to obtain directly. Well-structured models based on sound logic, reliable information, and realistic assumptions may provide one way to obtain at least rough estimates of such values, which may be better than none at all. Moreover, such models may also play an important role in revealing areas in which our present knowledge is especially poor (Wiens et al., 1983). Future research may then be focused on such areas.

Models differ in their complexity or detail. To some degree, this reflects basic differences in modelling philosophy. Weathers (1983) has argued that, until our knowledge of ecological energetics of free-living birds is more fully developed, it seems prudent to use simple models involving the fewest assumptions. Certainly complex models that contain a great many variables and assumptions carry more potential sources of error, and errors in some variables may be compounded by their interactions with other variables in the model. Further, by their very complexity such models may foster a false sense of security in the reality of the estimates they produce. On the other hand, model simplicity by no means assures that the estimates are any more accurate or less prone to errors. Complex models are more powerful as simulation tools, as one may explore the effects of altering various parameters while holding others constant or test different model scenarios in greater detail.

These considerations bear on the contributions that physiological information such as that filling this volume can make to modelling efforts. If one adopts a simple-model philosophy, rather little detailed physiological information is required. Weathers (1983), for example, based his admittedly coarse approximations of community energy demands on general equations of daily energy expenditure (Walsberg, 1980), multiplied by estimated population densities. A modeling approach such as that outlined here, on the other hand, makes heavy use of detailed information on a variety of physiological functions of seabirds. Indeed, virtually all of the components of the energetics model could profit from increased precision of physiological information. Thus, for example, estimations of fulmar energy demands using the models of Wiens and Scott (1975) or of Furness (1978) and Furness and Cooper (1982) would be improved by substitution of a set of seabird-derived EM equations for the more general nonpasserine equations, or even more by use of a set of procellarid-specific EM equations.

One way to assess the need for detail and precision in model parameters is to structure a complex model and then use its simulation powers to conduct sensitivity tests on the parameters or model functions. Sensitivity tests of the model used by Wiens and Scott (1975) indicated that population energy demand estimates were most strongly affected by variations in values of assimilation efficiency, population density, breeding phenology, and ambient temperature (Wiens and Innis, 1974). Furness (1978) conducted more appropriate Monte Carlo sensitivity tests on his version of this model, finding that values of the EM equations, population size, flight and swimming activity levels and energy costs, and assimilation efficiency were of critical importance. Croxall and Prince (1982) and Croxall et al. (in press) similarly emphasized the importance of information on population sizes (especially of nonbreeders), activity budgeting, and the energetic costs of flying, swimming, and diving, although they did not conduct formal sensitivity tests.

These tests indicate that both physiological and population parameters may play key roles in model performance. Among the physiological variables, accuracy in determination of the costs of locomotion (and in the amount of time spent flying, diving, etc.) seems especially critical. Accurate specification of several population parameters, however, may be even more important to estimating population energy demands. Population size figures importantly in the tests mentioned above, and the demographic model analysis of Ollason and Dunnett (1983) suggests that recruitment (of both young and emigrants) into a population may have a greater effect upon population size than survival of breeding adults. The model analyses of Ford et al. (1982), on the other hand, drew attention to the importance of adult survivorship rather than fecundity in determining long-term population dynamics. Drawing from their modelling experiences, Wiens et al. (1983) defined 15 parameters for which gathering detailed information was of high or intermediate priority to the development of accurate models of seabird population energetics, especially as related to oil-spill effects on populations. Of these, 11 were parameters related to features of the demography or behavior of populations; only two (flight energy costs and the energetics of chick growth and survival) were physiological. While this does not mean that much of the physiological information contained in this volume is irrelevant or that future physiological studies should concentrate only on locomotion costs or chick energetics, it does draw attention to the critical need for detailed information on both physiology and demography in future modelling efforts aimed at documenting the energetics of seabird populations and their role in the functioning of marine ecosystems.

No matter how detailed the information used to construct a model or how complex the model is, however, the utility of modelling approaches to seabird energetics must ultimately be gauged by the accuracy of the predictions they generate. It may be heuristically satisfying that different models produce generally similar estimations of energy demands or food consumption rates, but this is to be expected when the models employ much of the same information and follow the same logical structure. The major problem is to obtain independent estimates of energy demands or food consumption that can be used to verify the model projections. Recent studies using $D_2^{18}O$ techniques in tandem with model or time-budget estimations of energy expenditures (e.g. Weathers and Nagy, 1980; Bryant and Westerterp, 1983) offer encouragement. Williams and Nagy (in press) compared doubly labelled water estimates of daily energy expenditure in Savannah Sparrows (Passerculus sandwichensis) with activity-budget-based estimates obtained from several models, and found differences of only ±5% for several of the model estimates. In a similar study, Weathers et al. (in press) found that estimates of Loggerhead Shrike (Lanius ludovicianus) daily energetics obtained from the models of Kendeigh et al. (1977) were 10-20% greater than the $D_2^{18}O$ determinations. This technique is most readily applied to birds leaving and returning to a breeding site or colony fairly frequently, however, and is not generally amenable to the more difficult task of determining the energetics of free-ranging birds at sea. Until such independent verification of model predictions is possible, we must continue to hope that models based upon accurate and detailed information and structured following clear logic will provide estimates that are reasonably close approximations of reality, but we must exercise caution in interpreting or uncritically accepting these estimates.

ACKNOWLEDGMENTS

The approach to modelling seabird energetics was developed in collaboration with Glenn Ford and Dennis Heinemann, under support of the U.S. Bureau of Land Management through interagency agreement with the National Oceanic and Atmospheric Administration, as part of the Alaskan Outer Continental Shelf Environmental Assessment Program (OCSEAP). George Hunt also contributed to the modelling of the Pribilof Islands system. He and David Schneider, John Croxall, John Cooper, Peter Prince, and Ted Pettit generously provided unpublished manuscripts and information.

REFERENCES

Adams, N.J., and Brown, C.R., in press, Metabolic rates of sub-Antarctic procellariiformes: a comparative study, Comp. Biochem. Physiol.

Ankney, C.D., and MacInnes, C.D., 1978, Nutrient reserves and reproductive performance of female Lesser Snow Geese, Auk, 95:459.

Barrett, R.T., 1978, Adult body temperatures and the development of endothermy in the Kittiwake (Rissa tridactyla), Astarte, 11:113.

Baudinette, R.V., and Schmidt-Nielsen, K., 1974, Energy cost of gliding in herring gulls, Nature, 248:83.

Birkhead, T.R., 1976, Effects of sea conditions on rates at which Guillemots feed chicks, Brit. Birds, 69:490.

Birkhead, T.R., 1977, Adaptive significance of the nestling period of Guillemots Uria aalge, Ibis, 119:544.

Boersma, P.D., 1982, Why some birds take so long to hatch, Amer. Natur., 120:733.

Bourne, W.R.P., 1981, Some factors underlying the distribution of seabirds, in: "Proceedings of the Symposium on Birds of the Sea and Shore, 1979," J. Cooper, ed., African Seabird Group, Cape Town.

Bourne, W.R.P., 1983, Reappraisal of threats to seabirds, Mar. Pollution Bull., 14:1.

Brown, C.R., in press, Resting metabolic rate and energetic cost of incubation in Macaroni Penguins (Eudyptes chrysolophus) and Rockhopper Penguins (E. chrysocome), Comp. Biochem. Physiol.

Burger, A.E., Lindeboom, H.J., and Williams, A.J., 1978, The mineral and energy contributions of guano of selected species of birds to the Marion Island terrestrial ecosystem, S. Afr. J. Antarct. Res., 8:59.

Bryant,D.M., and Westerterp, K.R., 1983, Short-term variability in energy turnover by breeding House Martins <u>Delichon urbica</u>: a study using doubly-labelled water ($D_2^{18}O$), <u>J. Anim. Ecol.</u>, 52:525.

Cooper, J., 1977, Energetic requirements for growth of the Jackass Penguin, <u>Zool. Africana</u>, 12:201.

Cooper, J., 1978, Energetic requirements for growth and maintenance of the Cape Gannet (Aves; Sulidae), <u>Zool. Africana</u>, 13:305.

Crawford, R.J.M., and Shelton, P.A., 1978, Pelagic fish and seabird interrelationships off the coasts of South West and South Africa, <u>Biol. Conserv.</u>, 14:85.

Croxall, J.P., 1982, Energy costs of incubation and moult in petrels and penguins, <u>J. Anim. Ecol.</u>, 51:177.

Croxall, J.P., and Prince, P.A., 1982, A preliminary assessment of the impact of seabirds on marine resources at South Georgia, <u>Com. Nat. Franc. Recherch Antarct.</u>, 51:501.

Croxall, J.P., and Ricketts, C., 1983, Energy costs of incubation in the Wandering Albatross <u>Diomedea exulans</u>, <u>Ibis</u>, 125:33.

Croxall, J.P., Prince, P.A., and Ricketts, C., in press, Relationships between prey life-cycles and the extent, nature and timing of seal and seabird predation in the Scotia Sea, <u>in</u>: "Nutrient Cycles and Food Chains in the Antarctic," W.R. Siegfried and R.M. Laws, eds., Proceedings of the Fourth Symposium on Antarctic Biology.

Davis, R.W., Kooyman, G.L., and Croxall, J.P., 1983, Water flux and estimated metabolism of free-ranging Gentoo and Macaroni Penguins at South Georgia, <u>Polar Biol.</u>, 2:41.

Duffy, D.C., 1983, Environmental uncertainty and commercial fishing: effects on Peruvian guano birds, <u>Biol. Conserv.</u>, 26:227.

Dunn, E.H., 1980, On the variability in energy allocation of nestling birds, <u>Auk</u>, 97:19.

Flint, E.N., and Nagy, K.A., in press, Flight energetics of free-living Sooty Terns, <u>Auk</u>, 101.

Ford, R.G., Wiens, J.A., Heinemann, D., and Hunt, G.L., 1982, Modelling the sensitivity of colonially breeding marine birds to oil spills: guillemot and kittiwake populations on the Pribilof Islands, Bering Sea, <u>J. Appl. Ecol.</u>, 19:1.

Furness, R.W., 1978, Energy requirements of seabird communities: a bioenergetics model, <u>J. Anim. Ecol.</u>, 47:39.

Furness, R.W., 1981, Estimating the food requirements of seabird and seal populations and their interactions with commercial fisheries and fish stocks, <u>in</u>: Proceedings of the Symposium on Birds of the Sea and Shore, 1979," J. Cooper, ed., African Seabird Group, Cape Town.

Furness, R.W., and Cooper, J., 1982, Interactions between breeding seabird and pelagic fish populations in the southern Benguela region, <u>Mar. Ecol. Prog. Ser.</u>, 8:243.

Gaston, A.J., and Nettleship, D.N., 1981, The Thick-billed Murres of Prince Leopold Island, <u>Canad. Wildl. Serv. Monogr. Ser.</u>,

No. 6.

Garland, T., Jr., 1983, Scaling the ecological cost of transport to body mass in terrestrial mammals, Amer. Natur., 121:571.

Grant, G.S., and Whittow, G.C., 1983, Metabolic cost of incubation in the Laysan Albatross and Bonin Petrel, Comp. Biochem. Physiol., 74A:77.

Grant, J., 1981, A bioenergetic model of shorebird predation on infaunal amphipods, Oikos, 37:53.

Hedgren, S., 1981, Effects of fledging weight and time of fledging on survival of Guillemot Uria aalge chicks, Ornis Scand., 12:51.

Hedgren, S., and Linnman, A., 1979, Growth of Guillemot Uria aalge chicks in relation to time of hatching, Ornis Scand., 10:29.

Hunt, G.L., Jr., Burgeson, B., and Sanger, G.A., 1981, Feeding ecology of seabirds of the Eastern Bering Sea, in: "The Eastern Bering Shelf: Oceanography and Resources," D.W. Hood and J.A. Calder, eds., U.S. Department of Commerce, National Oceanic and Atmospheric Administration, Office of Marine Pollution Assessment, Rockville, Maryland.

Innis, G.S., and Wiens, J.A., 1977, BIRD model - version II: description and documentation, Wildlife Sci. Dept. Report Ser., No. 3. Utah State Univ., Logan, Utah.

Johnson, S.R., and West, G.C., 1975, Growth and development of heat regulation in nestlings, and metabolism of adult Common and Thick-billed Murres, Ornis Scand., 6:109.

Jones, P.J., and Ward, P., 1976, The level of reserve protein as the proximate factor controlling the timing of breeding and clutch-size in the Red-billed Quelea Quelea quelea, Ibis, 118:547.

Kendeigh, S.C., 1970, Energy requirements for existence in relation to size of bird, Condor, 72:60.

Kendeigh, S.C., Dol'nik, V.R., and Gavrilov, V.M., 1977, Avian energetics, in: "Granivorous Birds in Ecosystems," J. Pinowski and S.C. Kendeigh, eds. Cambridge Univ. Press, Cambridge.

King, J.R., 1973, Energetics of reproduction in birds, in: "Breeding biology of birds," D.S. Farner, ed., National Academy of Sciences, Washington, D.C.

King, J.R., 1974, Seasonal allocation of time and energy resources in birds, in: "Avian energetics," R.A. Paynter, Jr., ed., Publ. Nuttall Ornithol. Club No. 15, Cambridge, Massachusetts.

King, J.R., 1980, Energetics of avian moult, Acta XVII Congr. Intern. Ornithol., Berlin,:312.

Kooyman, G.L., Davis, R.W., Croxall, J.P., and Costa, D.P., 1982, Diving depths and energy requirements of King Penguins, Science, 217:726.

Lloyd, C.S., 1977, The ability of the Razorbill Alca torda to raise an additional chick to fledging, Ornis Scand., 8:155.

McLachlan, A., and Wooldridge, T., 1981, The role of birds in the ecology of eastern Cape sandy beaches, South Africa, in: "Proceedings of the Symposium on Birds of the Sea and Shore, 1979," J. Cooper, ed., African Seabird Group, Cape Town.

Montevecchi, W.A., Ricklefs, R.E., Kirkham, I.R., and Gabaldon, D., in press, Growth energetics of nestling Northern Gannets, Sula bassanus, Auk, 101.

Nelson, J.B., 1978, "The Sulidae: gannets and boobies," Oxford Univ. Press, Oxford.

Norberg, R.Å., 1981, Optimal flight speed in birds when feeding young, J. Anim. Ecol., 50:473.

Ollason, J.C., and Dunnett, G.M., 1983, Modelling annual changes in numbers of breeding Fulmars, Fulmarus glacialis, at a colony in Orkney, J. Anim. Ecol., 52:185.

Orians, G.H., and Pearson, N.E., 1979, On the theory of central place foraging, in: "Analysis of Ecological Systems," D.J. Horn, G.R. Stairs, and R.D. Mitchell, eds., Ohio State Univ. Press, Columbus, Ohio.

Pennycuick, C.J., 1969, The mechanics of bird migration, Ibis, 111:525.

Pettit, T.N., Byrd, G.V., Whittow, G.C., and Seki, M.P., in press, Growth of the Wedge-tailed Shearwater in the Hawaiian Islands, Auk, 101:

Pienkowski, M.W., 1982, Diet and energy intake of Grey and Ringed plovers, Pluvialis squatarola and Charadrius hiaticula, in the non-breeding season, J. Zool., Lond., 197:511.

Pimm, S.L., 1976, Existence metabolism, Condor, 78:121.

Puttick, G.M., 1980, Energy budgets of Curlew Sandpipers at Langebaan Lagoon, South Africa, Estuar. Coast. Mar. Sci., 11:207.

Raveling, D.G., 1979, The annual cycle of body composition of Canada Geese with special reference to control of reproduction, Auk, 96:234.

Ricklefs, R.E., 1973, Patterns of growth in birds. II. Growth rate and mode of development, Ibis, 115:177.

Ricklefs, R.E., 1974, Energetics of reproduction in birds, in: "Avian energetics," R.A. Paynter, Jr., ed., Publ. Nuttall Ornithol. Club No. 15, Cambridge, Massachusetts.

Roby, D.D., and Ricklefs, R.E., in press, Observations on cooling tolerance of the embryos of the Diving Petrel Pelecanoides georgicus, Auk, 101:

Schneider, D., and Hunt, G.L., Jr., 1982, Carbon flux to seabirds in waters with different mixing regimes in the southeastern Bering Sea, Mar. Biol., 67:337.

Sibly, R.M., and McCleery, R.H., 1983, Increase in weight of Herring Gulls while feeding, J. Anim. Ecol., 52:35.

Smith, V.R., 1979, The influence of seabird manuring on the phosphorus status of Marion Island (Subantarctic) soils, Oecologia, 41:123.

Stempniewicz, L., 1980, Factors influencing the growth of the Little Auk, _Plautus alle_ (L.), nestlings on Spitsbergen, _Ekol. Polska._, 28:557.

Threlfall, W., 1983, Seabirds, _in_: "Biogeography and Ecology of the Island of Newfoundland," G.R. Smith, ed., Dr W. Junk Publishers, The Hague.

Tuck, L.M., 1961, The murres, _Canad. Wildl. Serv. Monogr._, No. 1.

Tucker, V.A., 1973, Bird metabolism during flight: evaluation of a theory, _J. Exp. Biol._, 58:689.

Vleck, C.M., and Kenagy, G.J., 1980, Embryonic metabolism of the Fork-tailed Storm Petrel: physiological patterns during prolonged and interrupted incubation, _Physiol. Zool._, 53:32.

Walsberg, G.E., 1980, Energy expenditure in free-living birds: patterns and diversity, _Acta XVII Congr. Intern. Ornithol., Berlin_,:300.

Walsberg, G.E., 1983, Avian ecological energetics, _in_: "Avian Biology, Vol. 7," D.S. Farner, J.R. King, and K.C. Parkes, eds., Academic Press, New York.

Weathers, W.W., 1979, Climatic adaptation in avian standard metabolic rate, _Oecologia_, 42:81.

Weathers, W.W., 1983, "Birds of Southern California's Deep Canyon," Univ. California Press, Berkeley, California.

Weathers, W.W., and Nagy, K.A., 1980, Simultaneous doubly labelled water (^{3}HH^{18}O) and time-budget estimates of daily energy expenditure in _Phainopepla nitens_, _Auk_, 97:861.

Weathers, W.W., Buttemer, W.A., Hayworth, A.M., and Nagy, K.A., in press, An evaluation of time-budget estimates of daily energy expenditure in birds, _Auk_, 101:

Wheelwright, N.T., and Boersma, P.D., 1979, Egg chilling and the thermal environment of the Fork-tailed Storm Petrel (_Oceanodroma furcata_) nest, _Physiol. Zool._, 52:231.

Whittow, G.C., 1980, Physiological and ecological correlates of prolonged incubation in sea birds, _Amer. Zool._, 20:427.

Wiens, J.A., in press, Resource systems, populations, and communities, _In_: "A New Ecology: Novel Approaches to Interactive Systems," P.W. Price, C.N. Slobodchikoff, and W.S. Gaud, eds., John Wiley & Sons, New York.

Wiens, J.A., and Innis, G.S., 1973, Estimation of energy flow in bird communities. II. A simulation model of activity budgets and population bioenergetics, _Proc. 1973 Summer Computer Simulation Conf., Montreal, Canada_,:739.

Wiens, J.A., and Innis, G.S., 1974, Estimation of energy flow in bird communities: a population bioenergetics model, _Ecology_, 55:730.

Wiens, J.A., and Scott, J.M., 1975, Model estimation of energy flow in Oregon coastal seabird populations, _Condor_, 77:439.

Wiens, J.A., Ford, R.G., and Heinemann, D., 1983, Information needs and priorities for assessing the sensitivity of marine birds to oil spills, _Biol. Conserv._, 27:

Williams, A.J., 1982, Chick-feeding rates of Macaroni and Rock-hopper Penguins at Marion Island, Ostrich, 53:129.

Williams, J.B., and Nagy, K.A., in press, Daily energy expenditure of Savannah Sparrows: Comparison of time-energy budget and doubly labeled water estimates. Auk, 101.

Withers, P.C., and Timko, P.L., 1977, The significance of ground effect to the aerodynamic cost of flight and energetics of the Black Skimmer (Rhynchops nigra), J. Exp. Biol., 70:13.

IMPACT OF SEABIRDS ON MARINE RESOURCES, ESPECIALLY KRILL,

OF SOUTH GEORGIA WATERS

John P Croxall, Christopher Ricketts and Peter A Prince

British Antarctic Survey
Natural Environment Research Council
Madingley Road, Cambridge CB3 OET, UK

INTRODUCTION

Quantitative assessments of the energy and food requirements of
seabird communities are few and mainly recent (Wiens and Scott,
1975; Furness, 1978; Croxall and Prince, 1982a; Ford et al., 1982;
Schneider and Hunt, 1982; Sanger, 1972, 1983). Most have concerned
northern hemisphere sites, particularly those of northwest Alaska.
Such communities, and also that of the South African Benguela
system (Furness and Cooper, 1982), are mainly dominated by species
that feed inshore (usually within 50-100 km of, and often much
closer to, their breeding colony), such as auks Alcidae, gulls
Larus, kittiwakes Rissa and shags Phalacrocorax. This situation
confers some useful advantages. First, inshore feeding birds can
easily be observed and their distribution and density at sea often
realistically assessed. Second, most species feed mainly diurnally
and feeding ranges may be determined by direct observation from
land or sea, and it may even be possible to estimate general
activity budgets. Third, there is extensive information available
on the biology, ecology and sometimes breeding numbers and
demography of many species. Disadvantages stem principally from,
first, inaccessibility because of cliff nesting habits, making
handling (for collecting food samples, growth data, bioenergetic
research) difficult; second, offspring of some species are
precocial, departing to sea in the early stages of growth; and
third, many species have broods of more than one chick, complicating studies of chick energy budgets, meal size and feeding
frequency.

Studies of seabird communities containing many seabirds that are essentially pelagic even during the breeding season are rare. In this chapter we consider the relationships of the seabirds breeding at South Georgia with the surrounding marine environment. The species comprising the seabird community there (see Table 1) are four penguins, four albatrosses, seven petrels, two storm petrels, two diving petrels, one shag and one tern (the skua feeds exclusively on land) and is thus dominated by Sphenisciformes and Procellariiformes, the two most marine of all orders of birds.

There are, inevitably, numerous problems in studying such species. In particular, it is very difficult to get realistic assessments of the distribution and density of breeding birds at sea because (a) there is a vast area of ocean to be covered, (b) most species feed at night and observations are therefore of birds traveling to and from feeding grounds, (c) the main consumers are penguins which are largely invisible at sea. Also, many species are little studied and it has been necessary to commence research with investigations of basic biology and ecology, including breeding distribution and abundance. Nevertheless, there are some important advantages. Most species are approachable and access-ible, especially the 12 surface-nesting species. Chicks stay in their nests or burrows until independence and can be studied throughout their rearing period. Only the shag, tern and two of the penguins have broods of more than one. All this greatly assists studies of energy budgets and provisioning rates, most of which are still unknown for northern hemisphere species. Also, the only published data on swimming bioenergetics, flight speeds and patterns and activity budgets of pelagic seabirds at sea come from the studies at South Georgia. Furthermore, the Southern Ocean system in which krill, the crustacean Euphausia superba, has such a central role, is characterised by short food chains and a small number of possible prey species for seabirds. This removes some of the potential for dietary variability and prey switching that seems significant in Arctic and north temperate systems (Ainley and Sanger, 1979).

In estimating the impact of the seabird community on South Georgia's marine resources and producing a preliminary model of its seasonal changes, our Chapter is complementary to that of Wiens, whose examples are taken mainly from other types of seabird communities and who has reviewed fully the fundamental requirements of the theory and practice of modeling seabird population ener-getics. We shall also examine briefly the dynamics of the inter-actions between seabirds and their prey, especially krill, at South Georgia. The general principles of this are of wider interest as there have been few attempts to compare the characteristics of the prey taken with those presumed to be available or exploited commercially.

DATA BASE AND METHODS

To estimate the food consumption of a seabird community
throughout the year requires knowledge of the population size,
dietary composition and energy requirement on each day of the year
for each species in the community. Because the sizes, activity
patterns and hence energy requirements of the two sexes may differ,
the food consumption of each sex needs to be assessed separately.
Moreover, although most species complete one breeding cycle in one
year, Wandering, Grey-headed and Light-mantled Sooty Albatrosses
breed biennially (when successful in rearing a chick) and have
cycles lasting two years. The King Penguin typically has two
breeding attempts in three years at South Georgia. Thus our
estimation procedure is based on an activity cycle which can be
one, two or three years long and within which there can be two
successful breeding attempts. Here we illustrate the estimation
process for a species which breeds annually; the extension to more
complicated activity cycles is straightforward.

We discuss here only breeding populations: our demographic
information, especially for penguins, is inadequate to assess the
size of the pre-breeding and non-breeding populations, and we have
data neither on the proportion of them present around South Georgia
nor on their patterns of activity and distribution. For most speci
all data come from studies at Bird Island, at the north-western tip
of South Georgia (chiefly Croxall, 1979; Payne and Prince, 1979;
Croxall and Prince, 1980a; Prince, 1980a; Croxall and Hunter, 1982;
Thomas, 1982; Hunter, 1983; Roby and Ricklefs, 1983; Thomas et al.,
1983; Croxall, 1984; Hunter, 1984; Prince, in press, and additional
unpublished data). Most information for Chinstrap Penguins
(Lishman, in press a, b), Cape Pigeons (Pinder, 1966; Beck, 1969),
Snow Petrels (Beck, 1970), Black-bellied Storm Petrels (Beck and
Brown, 1971) and some for Wilson's Storm Petrels (Beck and Brown,
1972) and Blue-eyed Shags (Shaw, 1984) come from Signy Island,
South Orkney Islands. Essentially, however, all the main empirical
data come from the study site and we have reduced extrapolation
from elsewhere to a minimum. For most parameters the (mean) values
we have used in the model are given in Appendices 1 and 2.

Population Size

Population estimates are taken from Croxall et al. (in press a)
based chiefly on extensive research at Bird Island (see Croxall and
Prince, 1980a; Hunter et al., 1982) and general surveys elsewhere
at South Georgia (Prince and Payne, 1979; Croxall and Prince, 1979;
Prince and Croxall, 1983). Estimates for most surface-breeding
species (all penguins, albatrosses, giant petrels) are reasonably
accurate, except that we have poor data on the very large Macaroni
Penguin population at the Willis Islands. Burrowing species were
surveyed in three seasons at Bird Island using 600 randomly located

$36m^2$ quadrats, where the contents of over 5000 burrows were examined (Hunter et al., 1982). Figures for the whole of South Georgia are extrapolations based on survey records, the extent of suitable habitat and corrections for the presence of Brown Rats Rattus norvegicus, an important predator of small petrels whose breeding densities are significantly reduced where rats occur. The accuracy of these estimates cannot be assessed at present.

Typical losses of eggs and chicks are recorded in the papers cited above; 5-8 years data are available for surface-nesters, usually only 1-2 years for burrowing species. On each day throughout the year the breeding population is divided into active breeders and failed breeders. The total breeding population P is divided in $P_1(i)$ active breeders and $P_2(i)$ failed breeders on day i. The number of active breeders is calculated from the rate of egg loss between the laying date (day 1) and the hatching date (day h) and the rate of chick loss between hatching and fledging date (day f). Both egg loss and chick loss are assumed to be exponential at rates k_1 and k_2 respectively. Thus the populations on day 1 are given by.

before laying	: $P_1(i) = P$	$1 < i < 1$
laying to hatching	: $P_1(i) = P_1(1) \exp(-k_1 (i-1))$	$1 < i < h$
hatching to fledging	: $P_1(i) = P_1(h) \exp(-k_2 (i-h))$	$h < i < f$
after fledging	: $P_1(i) = P$	$f < i < 365$
always	: $P_2(i) = P - P_1(i)$	$1 < i < 365$

Dietary Composition and Energy Content

Diets were studied by quantitative analysis (proportions by weight, by frequency of occurrence and by number) of samples, taken throughout the chick rearing period, mainly from adults intercepted prior to feeding their chicks. Detailed analyses are given in Payne and Prince (1979), Croxall and Prince (1980a, b), Prince (1980a, b), Thomas (1982); Hunter (1983); these and other studies are summarised in Croxall and Prince (1980a, 1982a). Considerable additional data are now available confirming that diet does not vary greatly during the rearing period or generally between seasons. We know very little of diets during the early part of the breeding season and less still of seabird food in winter. Here we specify the dietary composition of each species as the proportion by weight of each seven different food types. These are krill, copepod, amphipod, other crustaceans, fish, squid and "other". Only free-living marine prey are considered; the carrion element, which is only important in the diets of the giant petrels, is ignored. We denote the proportion of the diet of food type k by $d(k)$.

Energy content data are rather few, mainly summarised in Clarke and Prince (1980) and Croxall and Prince (1982b). Values used (kJ g^{-1}) are: krill and other crustaceans 4.35, fish 3.97 and squid 3.47. The energy content of each food type is assumed to be the same for each species and to be constant throughout the year. We denote the energy content of food type k by c(k). The mean energy content of the diet is then

$$\overline{c} = \sum_{k=1}^{7} c(k)d(k) \text{ kJ g}^{-1}$$

Activity Patterns

For simplicity, we assume that all members of a population are performing the same activity on any one day and that they do it for the same (mean) length of time, while allowing for differences in activity between active and failed breeders. Although Antarctic seabirds are highly synchronised (in most species c. 90% of eggs are laid within 2-3 weeks) this is a considerable over-simplification and results in graphs with sudden "steps", rather than smooth curves. Adequate data to produce stochastic models are available for some species but the extra precision, given the much more complex model, seems inappropriate at present. We have arbitrarily divided a bird's activity pattern into seven categories: absent from the population, attending at the nest site, incubating, brooding, feeding chick, foraging for self and molting (penguins only). Failed breeders are assumed to forage for themselves only.

Energetic Costs

The daily energy costs of various activities are estimated or calculated as follows: attending at nest site and brooding as equivalent to existence energy requirements (EER), as defined and derived for non-passerines by Kendeigh et al. (1977); incubation from data on weight loss during fasting (Prince et al., 1981; Croxall and Ricketts, 1983) and the generalised equations for Procellariiformes and Sphenisciformes derived by Croxall (1982), the appropriateness of which have recently been supported by determinations of oxygen consumption under field conditions for a variety of Procellariiformes (Adams and Brown, 1984); swimming (in penguins only) from empirical data for King, Gentoo and Macaroni Penguins at South Georgia (Kooyman et al., 1982; Davis et al., 1983); molt (penguins only) from fasting weight loss data (Croxall, 1982). There are no data on the molt costs of Procellariiformes and because their molt is chiefly a slow, gradual process we make no specific provision here. The allocation of an energy cost to flight is much more difficult (and important); the background was well reviewed by Wiens (this volume). We have argued (Croxall and Prince, 1982a), for the species with which we are dealing (dependent largely on gliding flight), that the only realistic empirical value is that for gliding flight in Herring Gulls <u>Larus argentatus</u> of 1.9-2.4 x resting metabolic rate or 3.2-4.1 x basal metabolic rate (Baudinette and Schmidt-Nielsen, 1974). We believe also that large species, such as albatrosses, are more efficient gliders, both spending a greater proportion of flight time gliding and using proportionately less energy when doing so (see Pennycuick (1982)), than smaller ones and apply a coefficient of 1.85 x EER equally to all species. This is a conservative value, in line with some of the recent work (though none on gliding species) summarised by Wiens (this volume). It will presumably underestimate the costs for species which rely largely or exclusively on flapping flight (e.g. diving petrels, terns, shags). With shags, however, because the time spent flying (and swimming/diving) is such a small proportion of their activity budget, our overall daily values are close to those of Bernstein (1982) who allocated different energy costs to each element of a very detailed activity budget.

The energetic cost of foraging for a chick is calculated as the sum of the cost of foraging for self plus the energy content of the food delivered to the chick. Thus if $g(c)$ is the daily energetic cost of feeding a chick, $g(s)$ is the cost of foraging for self and W (grams) of food is delivered to the chick at frequency ϕ (meals per adult per day) then

$$g(c) = g(s) + \overline{Wc}\phi \text{ kJ d}^{-1}$$

Previously chick energy budgets have been estimated using very limited generalised equations (Kendeigh et al., 1977) and other assessments of the metabolizable energy requirements for growth (e.g. Ford et al., 1982). We have compared our direct estimates with those based on chick growth data (for all the 12 species for which we have daily weights) using the best-fitting of the Kendeigh et al. (1977) equations (that for the Black-bellied Tree Duck Dendrocygna autumnalis). This gives reasonable agreement for penguins and diving petrels but serious overestimates of the energy requirements of petrels, reinforcing the need to use empirical data, appropriate to the group under study, if realistic results are to be obtained.

Food Requirements

If the energy cost of the j th activity is denoted by $g(j)$, the total food required to fulfil activity j is $F(j) = g(j)/\overline{c}$ g d^{-1}. The total food requirements of the population performing activity j on day i is then $C(i) = P_1(i) F(j) + P_2(i) F(s)$ grams where $F(s)$ is food required when foraging for self. The amount of food of type k required on day i is $C(i) d(k)$. The food required between days t_1 and t_2 is simply $\sum_{i=t_1}^{t_2} C(i)$, or, for food type k $\sum_{i=t_1}^{t_2} C(i) d(k)$.

Foraging Range

The potential mean maximum (i.e. greatest distance using mean values of parameters) foraging range of a species during the chick rearing period is calculated from the travel speed (corrected for indirect (zigzag) flight pattern), the length of the foraging trip and proportion of the foraging trip spent on the water (feeding or resting). If the travel speed is v m s^{-1} and proportion of the trip spent feeding is τ then the maximum range is 86.4 $v(1-\tau)/2\phi$ where again ϕ is the feeding frequency: the factor 86.4 converts travel speed from m s^{-1} to km d^{-1} and the factor 2 corrects for outward and return journeys. If z is the correction factor for indirect flight, that is the distance flown to achieve a unit distance forward, then the corrected foraging range is $R = 86.4$ v $(1-\tau)/2z\phi$ km.

This is essentially just one expression of the complex relationship that exists between variables such as wind speed (and direction), flight speed, body mass and lifting capacity, which all affect the rate at which seabirds can deliver energy to their offspring and which are investigated in much greater detail in Pennycuick et al. (1984). The data on flight speeds and patterns of Procellariiformes are from direct field observations at South Georgia by Pennycuick (1982, mainly from fig. 18). Shag data come from Bernstein (1982). Lengths of foraging trips are calculated from the interval between feeds in a variety of ways (discussed with full references in Pennycuick et al. (1984)) and also from direct observations of individually marked birds (unpublished data). Knowledge of simple activity budgets at sea, including time spent flying and on water on foraging trips, is confined to three albatrosses (Prince and Francis, in press and unpublished) – the only data for any pelagic seabird. For other species we have guessed, influenced by extensive data on their diets and habits, but within the range of values for the albatrosses. Swimming speeds of penguins are from Kooyman (1975) and Clark and Bemis (1979). Their mode of travel (and that of diving petrels, shags and terns) does not automatically require them to deviate from a straight line progression and therefore we do not incorporate a correction factor for indirect travel.

Distribution of Foraging Effort and of Food Taken

Our data on observations of South Georgia seabirds at sea are unsuitable for quantitative estimation of the distribution of their foraging effort. To get some idea of possible distributions we proceed as follows. For birds rearing chicks, the data on mean flight speeds, patterns, durations of foraging trips (and time spent in flight on these) define a mean maximum foraging range. We then classify each species as feeding primarily inshore, primarily offshore, or neither and potentially intermediate, perhaps feeding uniformly over its range (see Appendix 2). Further discussion of the distinction between inshore and offshore species can be found in Croxall and Prince (1980a) and Croxall (1984). There are six intermediate species, in four groups: Chinstrap Penguin (in all characters intermediate between Gentoo and Macaroni); two diving petrels (closer to inshore species but with rapid direct flight and probably foraging over an appreciably greater range); two storm petrels (relatively long trips but a very different foraging pattern and progression from other types of petrel); Dove Prion (less offshore than e.g. Blue Petrel (Prince, 1980a) but observed feeding both inshore and at least as far as shelf edge waters (c. 100 km distant)). We assume that these intermediate species forage uniformly out to maximum range (R), based partly on the above information and partly because we have no data to suggest the contrary. For such species the proportion of foraging effort expended between r_1 and r_2 kms from the shore is

$$f(r_1,r_2) = (r_2-r_1)/R \qquad\qquad 0 < r_1 < r_2 < R$$
$$f(r_1,r_2) = 0 \qquad\qquad\qquad R < r_1 < r_2$$

For species that forage primarily inshore we assume that the foraging effort at the inshore end of the range is 10 times that for uniform foragers and declines exponentially over the range so that the total foraging effort remains unity. The foraging effort between distances r_1 and r_2 offshore is then

$$f(r_1,r_2) = \int_{r_1}^{r_2} 10/R \exp(-10r/R)\, dr$$
$$f(r_1,r_2) = \exp(-10r_1/R) \quad \exp(-10r_1 < /R) - \exp(-10r_2/R) \qquad \begin{array}{l} 0 < r_1 < r_2 < R \\ 0 < r_1 < r_2 = R \end{array}$$
$$f(r_1,r_2) = 0 \qquad\qquad\qquad\qquad\qquad\qquad R < r_1 < r_2$$

This is based on the premise that inshore feeders attempt to satisfy their requirements as close to the breeding site as possible, only foraging further afield when unable to do so. Field data for Blue-eyed Shags (Bernstein, 1982; Shaw, 1984; pers. obs.), Gentoo Penguins (including data from individuals radio-tracked from Bird Island (Bengtson, pers. comm.)) and giant petrels (Hunter, 1983 and unpublished) support this and suggest that effective foraging ranges are very significantly less than the attainable mean maximum.

Species which forage primarily offshore are treated as the mirror image, over their range, of inshore feeders. We assume that offshore species usually travel to feeding grounds at or near their mean maximum range and only if they then fail to find food in this area do they forage closer to home. We have a little anecdotal information in support of this assertion. First, retraps (Prince and Francis, in press) and resightings (unpublished data) of albatrosses known still to be rearing chicks at Bird Island are

mainly of birds near their mean maximum range. Second, there are
aerodynamic considerations why continuous flight to a feeding
ground followed by intensive feeding might be a preferable strategy
(Pennycuick et al., 1984). Third, flocks of Macaroni Penguins seen
at sea departing from their breeding colonies in the vicinity of
Bird Island did not deviate from continuous swimming at a more or
less constant heading. Also, data from Antarctic Fur Seals
Arctocephalus gazella with time/depth recorders (Kooyman et al., in
press) and from radio-tracked individuals (Bengtson et al., in
prep.) indicate a lengthy period of swimming, more or less in a
straight line to the limit of shore-station tracking (c. 25 km),
well before their first feeding dive. Macaroni Penguins have
similar diet, basic foraging direction and, at least by analogy
with Chinstrap Penguins (Lishman and Croxall, 1983), feeding and
diving habits probably similar to Antarctic Fur Seals. Fourth, a
disproportionate number of sightings of Light-mantled Sooty
Albatrosses are near their mean maximum range (Thomas, 1982).

While we believe it is correct not to treat these species as
uniform foragers, we have no information on the intensity of
feeding at, or at various distances short of mean maximum range and
the relationships we choose here may exaggerate the true situation.
However, in this very preliminary attempt we do not wish to do more
than indicate the general results of developing this approach and
make no claim that the precise configurations of the patterns are
correct. To simplify the calculation of foraging range and of
distribution of effort we have considered only the population of
seabirds breeding at Bird Island and on the adjacent part of
north-west South Georgia (collectively NWSG in Table 1; treating it
as a "point" source) during their chick rearing periods. It is
likely that the patterns at other times (e.g. during incubation)
and of the non-breeding or failed breeding population would be
rather different but we have no data whatsoever for modeling these
distributions.

Within the chick rearing period, that is from day h to day f,
the amount of food taken on day i in the range r_1 to r_2 km offshore
is

$$D(i,r_1,r_2) = C(i) \ f(r_1,r_2)$$

Community Food Requirements and Distribution

The food requirements of the multi-species seabird community on
day i are calculated by summing the values of $C(i)$ over all species.
Similarly, the amount of food of type k taken on day i is the sum of
the values of $C(i) \ d(k)$ over all species. The distribution of the
food taken by birds feeding chicks on day i, between r_1 and r_2 km
offshore is the sum of the $D(i,r_1,r_2)$ summed over all species
rearing chicks on that day.

A copy of the FORTRAN program used to perform the estimation is available from the authors.

Limitations of the Data

Like most models of seabird food consumption this one must be regarded as preliminary and incomplete. We are particularly handicapped by insufficient data for certain important elements of the model, notably the energy costs of flight, activity budgets at sea and realistic population figures for burrowing seabirds away from Bird Island and for the vast population of Macaroni Penguins at the Willis Islands.

There are other potentially important sources of inaccuracy. We have neglected some of the seasonal changes in the weight of birds. Weights of breeding birds are usually highest on first arrival at the colony and decrease during the pre-laying attendance period, except for an increase during any pre-laying exodus. Most Procellariiformes and penguins lose weight steadily during long incubation fasts, regaining it in the intervening foraging periods at sea and then gradually lose weight during the chick-rearing period (Brooke, 1978; Croxall and Ricketts, 1983). The energy costs of incubation and molt fasts are incorporated in the model but the greater costs of gaining weight and acquiring body reserves are not. These costs must be met by increased consumption during the appropriate periods at sea, that is prior to first arrival ashore (when they may be greater for females because of the costs associated with egg formation) and between incubation shifts, thus enhancing the disparity between the costs of incubation and foraging for self at sea. Also, seasonal changes in the energy content of prey have not been considered: for example, female krill increase energy content by up to 60% during ovarian lipogenesis (Clarke, in press), though increases for males and immatures are much less. Similar, though probably smaller, changes presumably occur in other zooplankton. Therefore, we may have overestimated consumption of krill in February and probably underestimated it in spring and autumn. Changes in the kind of prey taken also create problems but at South Georgia in summer these appear to be rare.

The two most obvious deficiencies are, first, that the level of winter impact is largely hypothetical. We have little dietary information for any species at this time and few indications of distributions at sea. We assume that all species, other than Wilson's Storm Petrel and Black-browed Albatrosses which leave the Southern Ocean altogether in winter, continue to forage in the same vicinity of South Georgia. Although South Georgia is not ice bound in winter, the amount of open water to the south is greatly reduced and this must introduce substantial changes in the foraging areas of resident species. Obviously a proportion of the population of species that remain in high latitudes move outside South Georgia

Table 1. Annual food consumption (tonnes x 10^3)[a] of seabirds breeding at South Georgia.

Species	Krill	Squid	Fish	Copepod	Amphipod	Other	Total
King Penguin Aptenodytes patagonicus	0.0	44.8	19.2	0	0	0	64
Chinstrap Penguin Pygoscelis antarctica	2.5	0	0	0	0	0	2.5
Gentoo Penguin P. papua	61.2	0	28.8	0	0	0	90.0
Macaroni Penguin Eudyptes chrysolophus	3872.7	0	79.0	0	0	0	3951.7
Wandering Albatross Diomedea exulans	0.3	2.4	0.3	0	0	0	3.0
Black-browed Albatross D. melanophrys	8.3	4.6	8.7	0.2	0.2	0	22.0
Grey-headed Albatross D. chrysostoma	6.3	20.6	14.7	0	0.4	0	42.0
Light-mantled Sooty Albatross Phoebetria palpebrata	1.8	2.3	0.6	0	0	0.2b	4.9
Southern Giant Petrel Macronectes giganteus	2.7	0.4	0.2	0	0	0	3.3
Northern Giant Petrel M. halli	1.5	0.6	0.2	0	0	0	2.3
Cape pigeon Daption capensis	3.1	0	0.5	0	0	0	3.6
Snow Petrel Pagodroma nivea	0.4	0.05	0.05	0	0	0	0.5
Dove Prion Pachyptila desolata	1345.4	23.2	46.4	719.1	185.6	0	2319.7
Blue Petrel Halobaena caerulea	6.6	0.08	0.6	0.3	0.4	0	8.0
White-chinned Petrel Procellaria aequinoctialis	210.1	365.8	186.8		7.8	0	770.5
Wilson's Storm Petrel Oceanites oceanicus	4.3	0	0.5	3.8	0.9	0	9.5
Black-bellied Storm Petrel Fregetta tropica	0.2	0	0.03	0.2	0.05	0	0.5
Common Diving Petrel Pelecanoides urinatrix exsul	52.8	0	0	239.2	59.8	0	351.8
South Georgia Diving Petrel P. georgicus	125.6	0	0	33.0	6.6	0	165.2
Blue-eyed Shag Phalacrocorax atriceps	0	0.9	3.3	0	0	0.5c	4.7
Antarctic Tern Sterna vittata	0.04	0	0.1	0.04	0.05	0	0.2
Total	5706	466	390	996	262	1	7820

a Does not include carrion and other non-marine derived prey.   b Decapods.   c Isopods.

waters, which may further reduce local impact. However, there will
be some replacement by individuals and species breeding at higher
latitudes that move north to winter around South Georgia or at the
edge of the pack ice, usually some 100-200 km further south. If
all individuals of all species left South Georgia for five months
in winter this would reduce overall consumption by nearly 40%.
Second, we do not consider the post-fledging juveniles, immatures
or non-breeding adults. Although there is evidence that juveniles
disperse rapidly after fledging, even their temporary presence
would produce an increase in population food requirements and,
usually, an extension of peak consumption beyond fledging date.
Inclusion of immatures and adult non-breeders would substantially
increase population food consumption, particularly because seabirds
have considerably deferred sexual maturity and immatures tend to
frequent the breeding site for several seasons prior to breeding.
In penguins, all immatures come ashore annually to molt (mainly
during the chick-rearing period) and probably spend an appreciable
time at sea in the vicinity. It is therefore likely that the total
summer food consumption of seabirds is underestimated by at least 30%

RESULTS AND DISCUSSION

Overall Food Consumption

We estimate that 7.8 million tonnes of prey are consumed each
year by breeding seabirds of which 73% is krill, 13% copepods, 6%
squid, 5% fish and 3% amphipods (Table 1). Of the 21 species of
seabird only six take 1% or more of this total: namely, Macaroni
Penguin (51%), Dove Prion (30%), White-chinned Petrel (10%), Common
Diving Petrel (4.5%), South Georgia Diving Petrel (2%), and Gentoo
Penguin (1%). Thus penguins, forming 13% of the breeding numbers
but 76% of the biomass eat 53% of the food.The Macaroni Penguin
takes 68% of the krill eaten by seabirds, followed by Dove Prion
(24%), White-chinned Petrel (4%), South Georgia Diving Petrel (2%)
and Gentoo Penguin and Common Diving Petrel (both 1%). Krill is
also, however, of primary importance in the diet of Chinstrap
Penguins and all petrels except the giant petrels (mainly
scavengers of seal and penguin carcasses (Hunter, 1983)) and the
Common Diving Petrel.

Copepods are mainly taken by Dove Prions (72%), by forcing
water through the comb-like palatal lamellae, which act as filters
along the sides of the bill, and Common Diving Petrels (23%), which
have exactly the stature of the small northern hemisphere alcids
which also feed on small zooplankton. Copepods are also important
in the diet of the storm petrels. Amphipods, although occurring as
a trace in the diet of almost every seabird, are only eaten in
quantity by Dove Prions (71%) and Common Diving Petrels (23%).
Squid and fish are of much less importance than krill to South

296

Georgia seabirds. The main squid predators are White-chinned
Petrel (79%), King Penguin (10%), Dove Prion (5%) and Grey-headed
Albatross (4%); squid also predominate in the diet of Wandering
Albatross and Light-mantled Sooty Albatross. Fish are very widely
taken but only Blue-eyed Shag and Antarctic Tern depend on them for
their livelihood. The main consumers are White-chinned Petrel
(48%), Macaroni Penguin (20%), Dove Prion (12%), Gentoo Penguin
(7%), King Penguin (5%), Grey-headed Albatross (4%), Black-browed
Albatross (2%) and Blue-eyed Shag (1%).

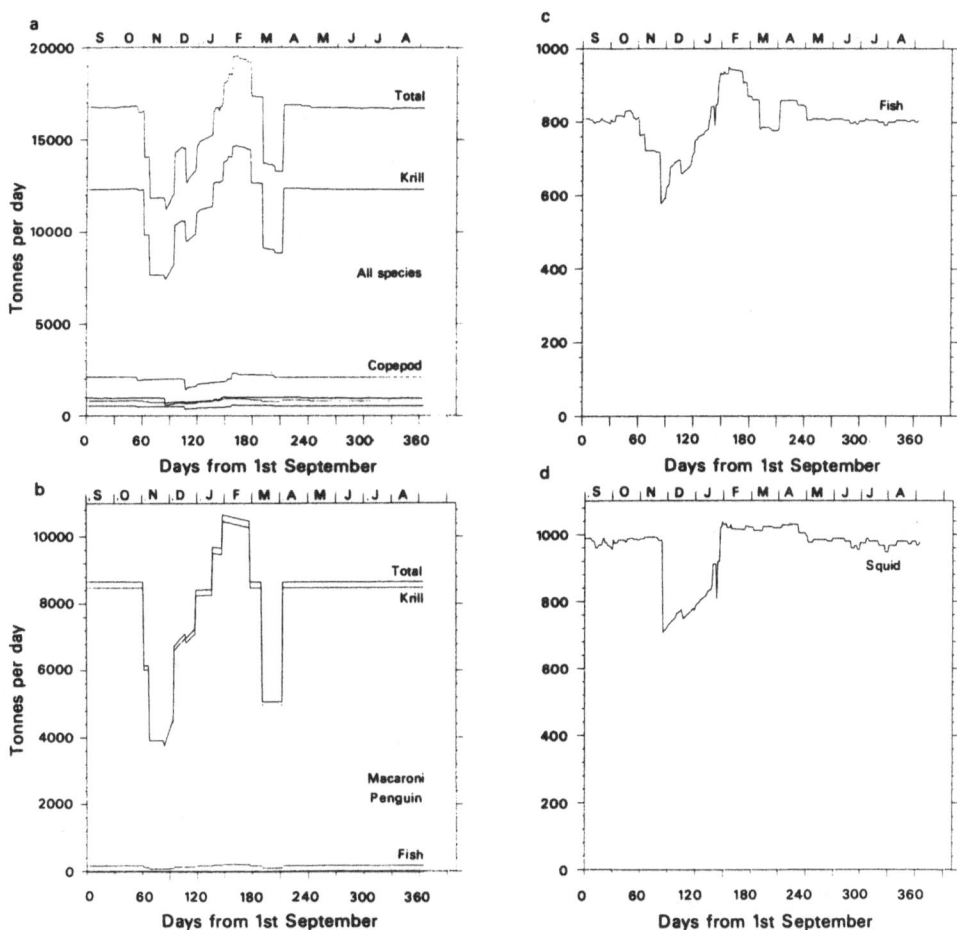

Fig. 1.   Seasonal changes in food consumption by breeding
          populations of South Georgia seabirds. a) all species,
          b) Macaroni Penguin, c) all species, fish only, d) all
          species, squid only.

Seasonal Changes in Food Consumption

The broad pattern of changes in the food requirements of the breeding seabird community is shown in Fig. 1a. This emphasises the importance of krill and shows a general increase in its consumption from spring to late summer with some reduction thereafter. The very marked drop in March is due to the highly synchronised on-shore molt fast of the Macaroni Penguins at this time (Fig. 1b). Their absence then as consumers will be partly compensated for by the newly fledged juveniles (c. 20% of the adult population biomass), which will still be in the vicinity. However, the consumption in late February - early March is, in reality, much greater than shown because Macaroni Penguins double their body weight in a 14 day period at sea in preparation for molt (see Croxall, 1984, fig. 24) and the model does not allow for the substantial extra food consumption that this entails. The seasonal pattern of avian krill consumption thus largely reflects that of its principal consumer, the Macaroni Penguin.

Consumption of fish (Fig. 1c) and squid (Fig. 1d) shows no clear seasonal pattern, other than a reduction during spring and early summer when the main consumers are incubating. The rise in fish consumption in Jan-Feb is solely a consequence of the increase in Macaroni Penguin requirements, even though fish only form 2% by weight of their diet. This illustrates the disproportionate importance of minor elements in the diet of abundant consumers - and the need to have good quantitative data on these.

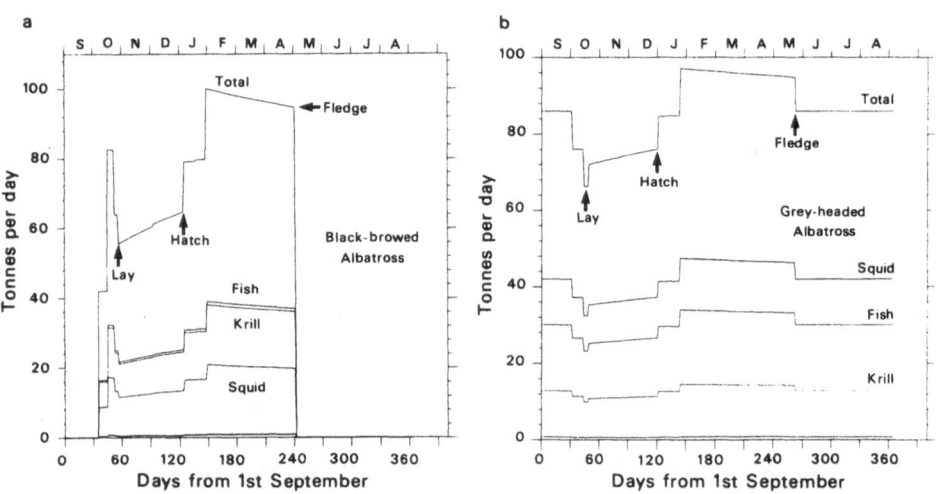

Fig. 2. Seasonal changes in food consumption by a) Black-browed Albatross, b) Grey-headed Albatross.

298

Comparisons Between Related or Similar Species

South Georgia seabirds all breed during the summer months, apart from Wandering Albatrosses which rear chicks throughout winter, and King Penguins, whose peculiar breeding cycle involves mature chicks overwintering. The range of patterns of food consumption can be portrayed well by considering certain pairs of species. There are three true species-pairs of seabirds; the giant petrels, the small Diomedea albatrosses and the diving petrels. The giant petrels are largely omnivorous scavengers, the key to whose coexistence seems to be their dietary flexibility and the six week difference in their breeding timetables (Hunter, 1983, 1984).

The two albatrosses (Fig. 2a, b) commence breeding at a similar time (although they have different patterns and durations of pre-laying attendance) but the proportions of the main prey types in the diet are very significantly different and the Grey-headed Albatross breeding period lasts a month longer. Furthermore, Grey-headed Albatrosses remain in the Southern Ocean, whereas South Georgia Black-browed Albatrosses winter in South African waters (Tickell, 1967). Though superficially very similar, these two mollymawks have many important biological and ecological differences, of which the different duration and pattern of their roles as predators is but one facet. The dietary differences, however, play an important part in determining chick growth rates (Prince and Ricketts, 1981) which themselves contribute to differences in basic breeding frequency (annual in Black-browed, biennial in Grey-headed Albatross) and population demography (Prince, in press).

The diving petrels (Fig. 3a, b) are sibling species, indistinguishable except in the hand. They show distinct differences in diet, but not necessarily just because the Common Diving Petrel breeds seven weeks earlier than its congener. The difference in timing of breeding means that the time of greatest food demand (during chick-rearing) by Common Diving Petrels coincides with the period of lowest demand (during incubation) by South Georgia Diving Petrels. A very similar pattern is shown by two other species, the Blue Petrel and Dove Prion (Fig. 3c, d) which, although not congeneric, are otherwise similar in size and general biology. They also have a seven week difference in breeding schedule and show exactly offset periods of peak food consumption. Dietary differences also exist but it is the Blue Petrel, the earlier breeding species, that takes more krill, in contrast to the diving petrels.

Distribution of Feeding Effort

The estimated mean maximum foraging ranges during the chick-rearing period of the main seabird species breeding at Bird Island are shown in Fig. 4. Species 10-17 are the offshore foragers (with the addition of species 3 from the (non-flying) penguins), species 1, 2, 4 and 5 the inshore foragers and species 6-9 are treated as uniform foragers.

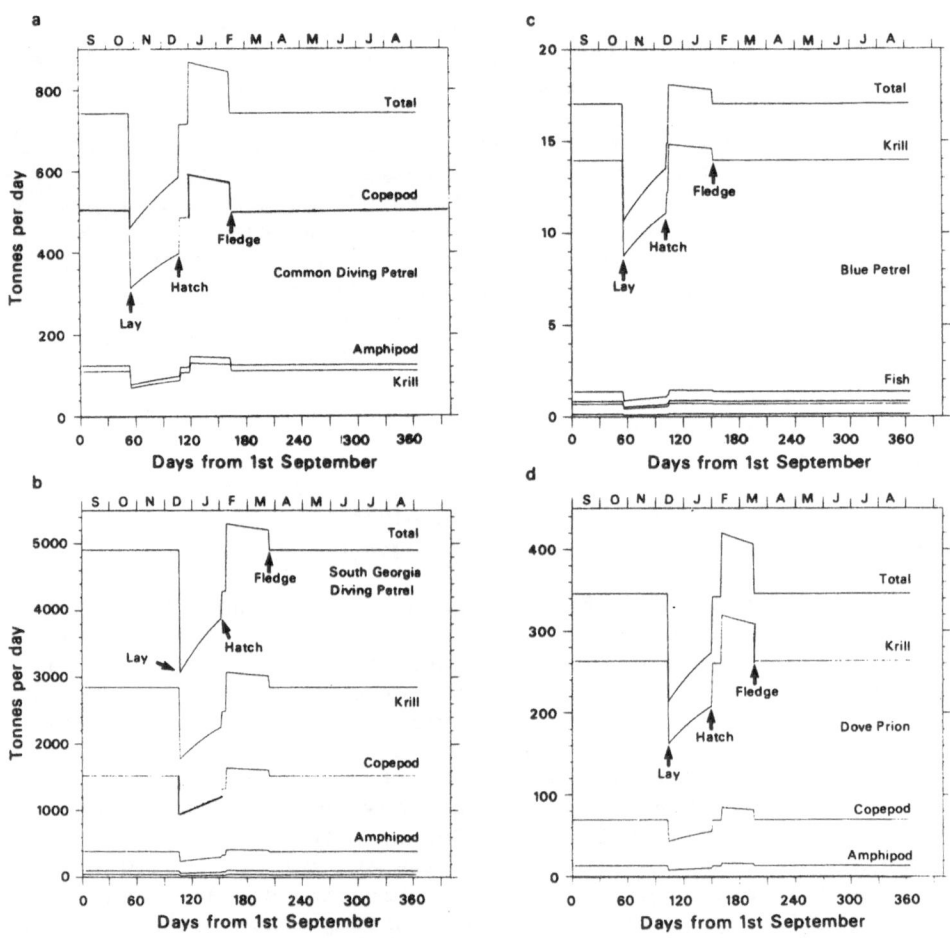

Fig. 3.   Seasonal changes in food consumption by a) Common Diving
          Petrel, b) South Georgia Diving Petrel, c) Blue Petrel,
          d) Dove Prion.

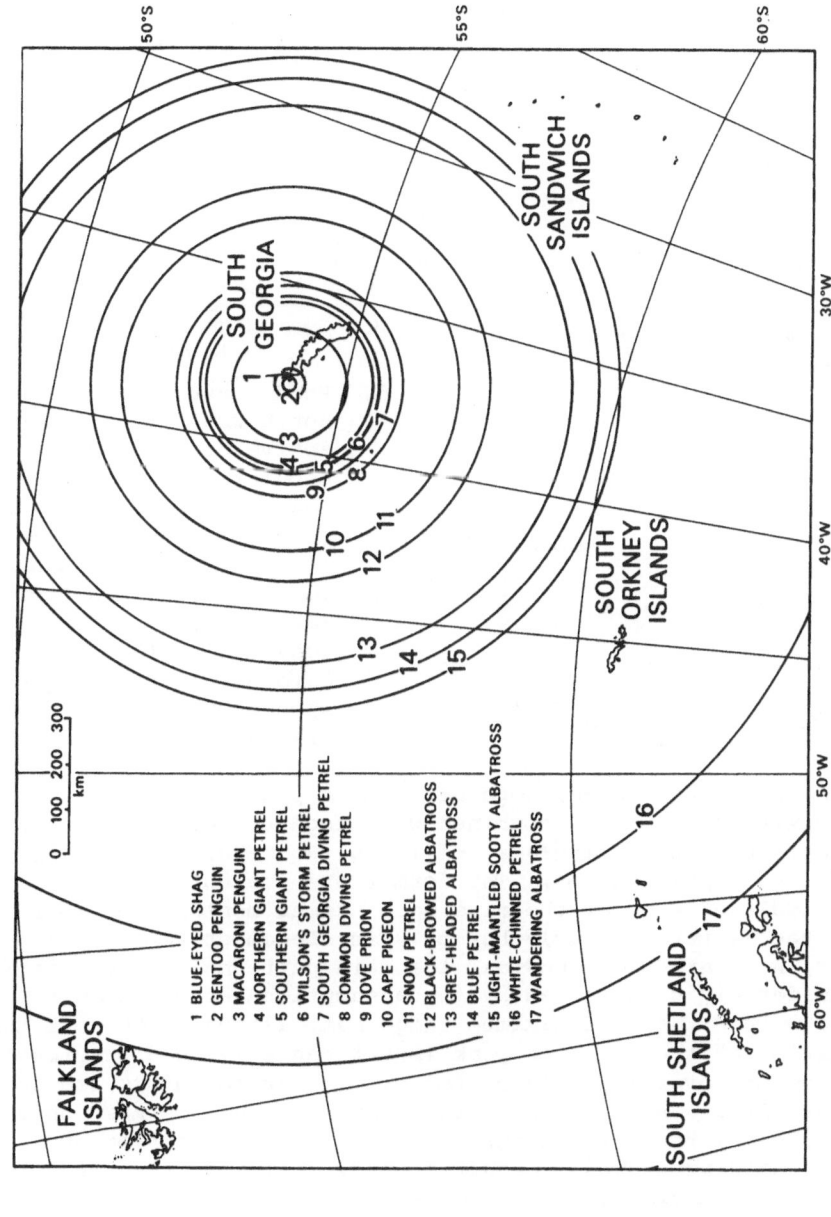

Fig. 4 Mean maximum foraging ranges (during chick-rearing period) of seabirds breeding at Bird Island. (Map is Polar Stereographic Projection.)

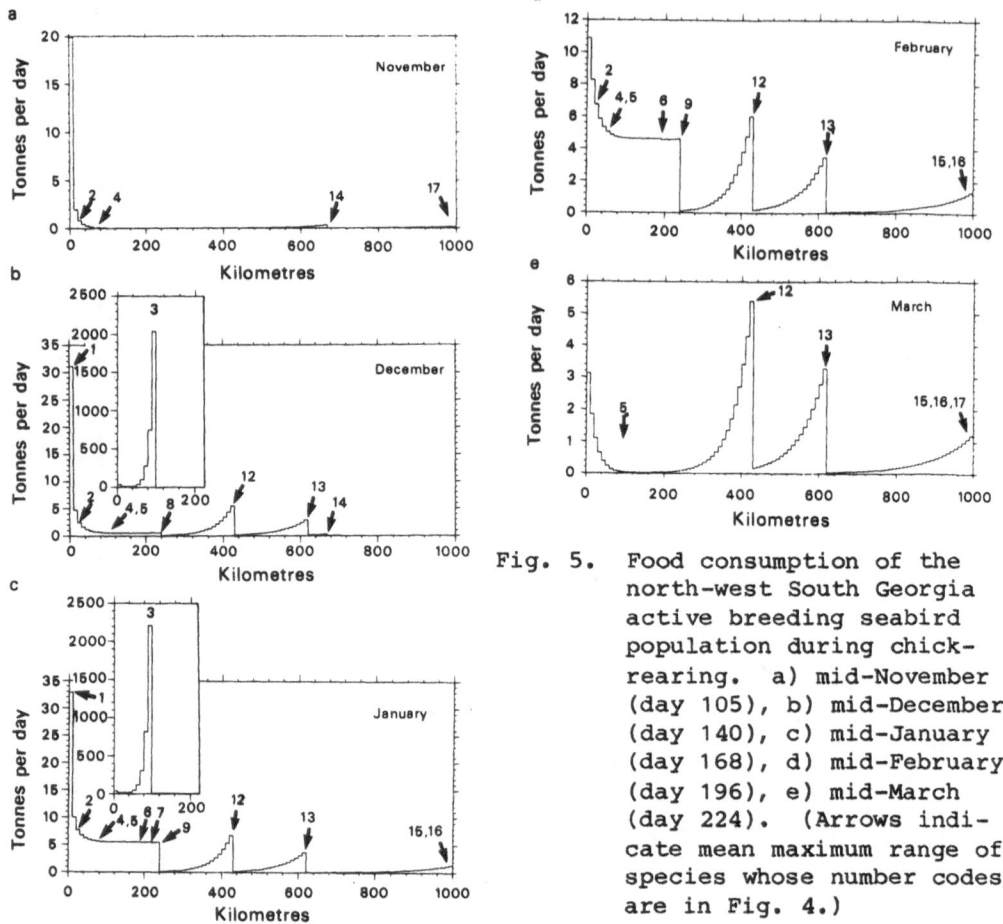

Fig. 5. Food consumption of the north-west South Georgia active breeding seabird population during chick-rearing. a) mid-November (day 105), b) mid-December (day 140), c) mid-January (day 168), d) mid-February (day 196), e) mid-March (day 224). (Arrows indicate mean maximum range of species whose number codes are in Fig. 4.)

These foraging ranges, together with a possible distribution of foraging within them, are combined with food consumption data and the results presented diagramatically in two ways; as profiles (Fig. 5) and maps (Fig. 6). The former are presented for one day in the middle of each of the five main summer months (November-March); the latter similarly but only for December-February. Together, however, they show all the main patterns. These diagrams are considerably constrained by the assumptions about foraging patterns within the mean maximum range but, although the amplitude and location of the areas of peak impact can only be approximate, we believe the overall patterns for birds rearing chicks are not too unrealistic. However, foraging distributions are most unlikely to be uniformly radial, but our evidence for the location of preferred feeding grounds is sufficiently scanty that we prefer to produce diagrammatic maps based on the simplest assumption.

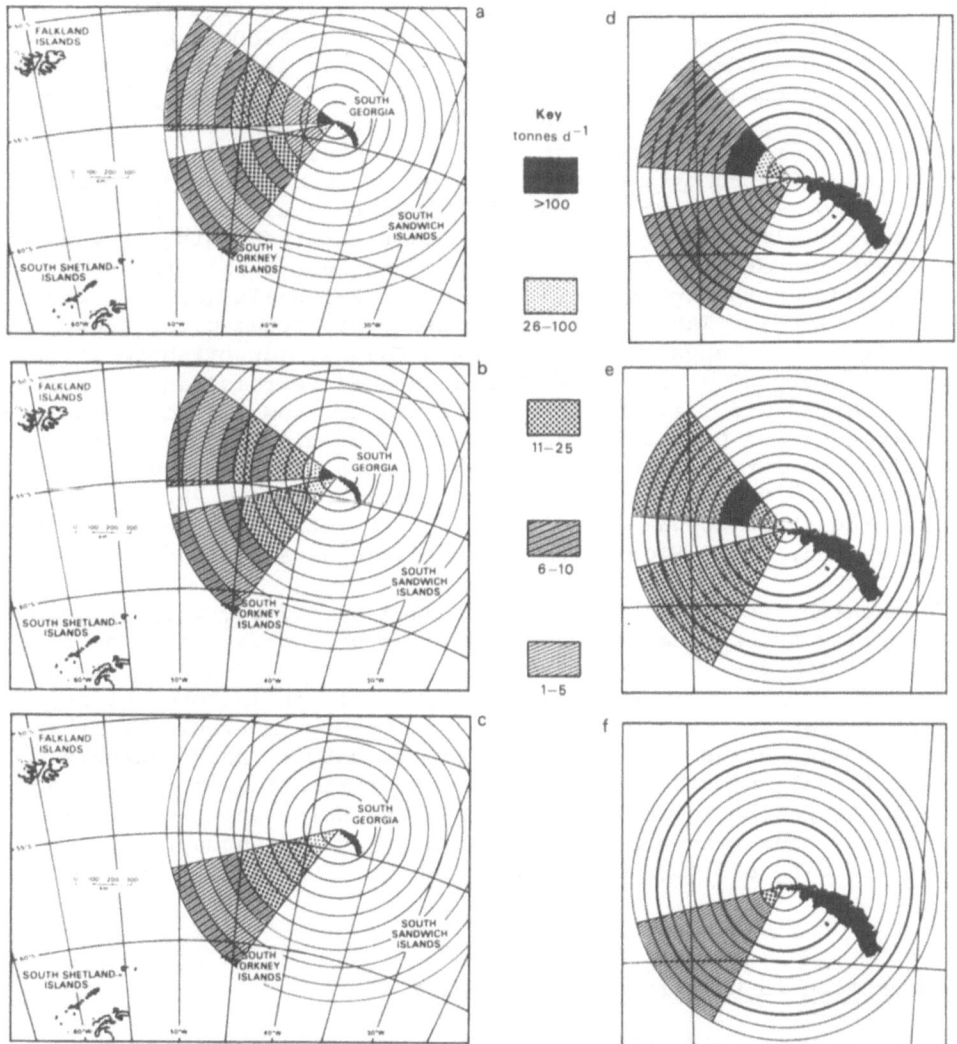

Fig. 6.   Food consumption of the north-west South Georgia active
breeding seabird population during chick-rearing. a,d)
mid-December; upper sector with, lower sector without,
Macaroni Penguins, b,e) mid-January; upper sector with,
lower without, Macaroni Penguins, c,f) mid-February.
a,d,e) total range 1000 km, bands of 100 km; d,e,f) total
range 240 km, bands of 20 km.

In mid-November only four species are rearing chicks. Gentoo Penguins (feeding close inshore), Northern Giant Petrels (feeding inshore when Antarctic fur seal carrion is unavailable ashore), and Blue Petrels (feeding well distant, probably in the general direction of the South Sandwich Islands, judging from the pumice stones that they ingest at this time (Prince, 1980b)) have just started and Wandering Albatrosses have nearly finished. The bulk of the food consumption at this time of year is probably by pre-breeding birds of both resident and migratory species. By mid-December a greater variety of species is breeding but food consumption is dominated by the activities of Macaroni Penguins, mainly in the 80-100 km zone and probably confined to shelf and shelf-slope waters. Field observations suggest that their foraging grounds lie to the north of Bird Island, perhaps coextensive with those suggested for the krill-eating Antarctic Fur Seals (Croxall and Pilcher, 1984). The other species are distributed close inshore (mainly Gentoo Penguins and Blue-eyed Shags), far offshore (Light-mantled Sooty Albatrosses and White-chinned Petrels) and at different intermediate positions between these as exemplified by the two albatrosses and Common Diving Petrels. The mid-January pattern is broadly similar but the consumption inshore has been greatly augmented by the activities of South Georgia Diving Petrels, replacing their congeners, Wilson's Storm Petrels and especially Dove Prions, while Blue Petrels have disappeared offshore.

Macaroni Penguin chicks have fledged by mid-February (though the adults are at sea somewhere building up reserves for the molt fast). With the fledging of Blue-eyed Shags and Gentoo Penguins, the Dove Prions exert even greater influence in inshore waters, the offshore pattern remaining similar to previous months. By mid-March most species have finished rearing chicks, only Southern Giant Petrels remaining inshore, and four albatrosses (Wandering Albatross chicks having just hatched) and White-chinned Petrels offshore. The main consumers at this time will undoubtedly be post-breeding adults and recently fledged juveniles.

Consumption in Relation to Prey Stocks

To examine this effectively requires detailed knowledge of the age (and/or size) and status of the prey taken in order to assess the significance of the predation in terms of the structure, size and dynamics of the natural prey populations. This has not yet been satisfactorily done for any marine predator-prey interaction and the current limited state of knowledge with respect to even the best-known part of the Southern Ocean was reviewed by Croxall et al. (in press b). The main features of the situation, together with some additional information on feeding performance, are summarised here.

Krill. Preliminary and provisional acoustic surveys for krill in the shelf waters around South Georgia produced estimates of c 0.5 million tonnes (B.A.S. Offshore Biological Programme). This seems very low in view of the many other krill predators present and our estimates of an annual avian consumption of about 6 million tonnes, although this would be reduced appreciably (to perhaps 2 - 3 million tonnes) if only the breeding season and species restricted then to shelf waters were considered. The krill stock, however, is underestimated for a number of reasons: a) it was surveyed in December, when stocks will have already been reduced by consumption, b) the survey is effectively instantaneous, and we do not know the replenishment rate of krill via the major currents from the Bellingshausen and Weddell Sea, but this must be important, c) krill in the upper 10 m of the water column (of most relevance to predators) could not be surveyed with the equipment used.

It is also very difficult to evaluate the significance of the size and status of krill taken by bird predators (Fig. 7a-f). By numbers alone, most predators mainly take mature individuals (krill of over 35 mm length). The contribution by weight of these large krill to the diet is overwhelming as a 45 mm krill weighs seven times as much as a 25 mm one. The only birds that take appreciable quantities of juvenile krill are Macaroni Penguins (18% by weight), Dove Prions (10% by weight; presumably catching them like copepods when filter feeding) and especially diving petrels (100% of their krill; they also take many copepods in the diet). However, while in the southern Scotia Sea net haul and predator samples match quite closely (Croxall et al., in press b; Lishman, in press a), the South Georgia situation is more complex (Fig. 7 g, h). Net mesh is usually too large to retain small krill (<25 mm) but the absence of large ones cannot be explained by net avoidance (as the same nets catch 50 - 60 mm krill around the South Orkney Islands). Clearly the main bird predators readily locate large krill and it is not clear why net haul samples, including those from the 'Discovery' investigations (Marr,1962), should contain so few. It may be that the birds (especially penguins that must be restricted to shelf and shelf-slope waters during the breeding season) feed in restricted local areas, hitherto largely unsampled by nets, but the relatively few samples from around north-west South Georgia provide no better match at present (Croxall and Pilcher, 1984). The implications of this discrepancy, in terms of calibrating the acoustic echo-integration surveys of krill abundance, have been discussed elsewhere (Croxall et al., in press b).

To complete the picture of predation on krill one needs to know the sex and status of the prey. Croxall et al. (in press b) noted that the greatly increased energy content of pre-spawning female krill in February (Clarke, in press) make them a particularly attractive prey and that at this time penguin diets contain on average 10% more mature female krill, and Fur Seals about four

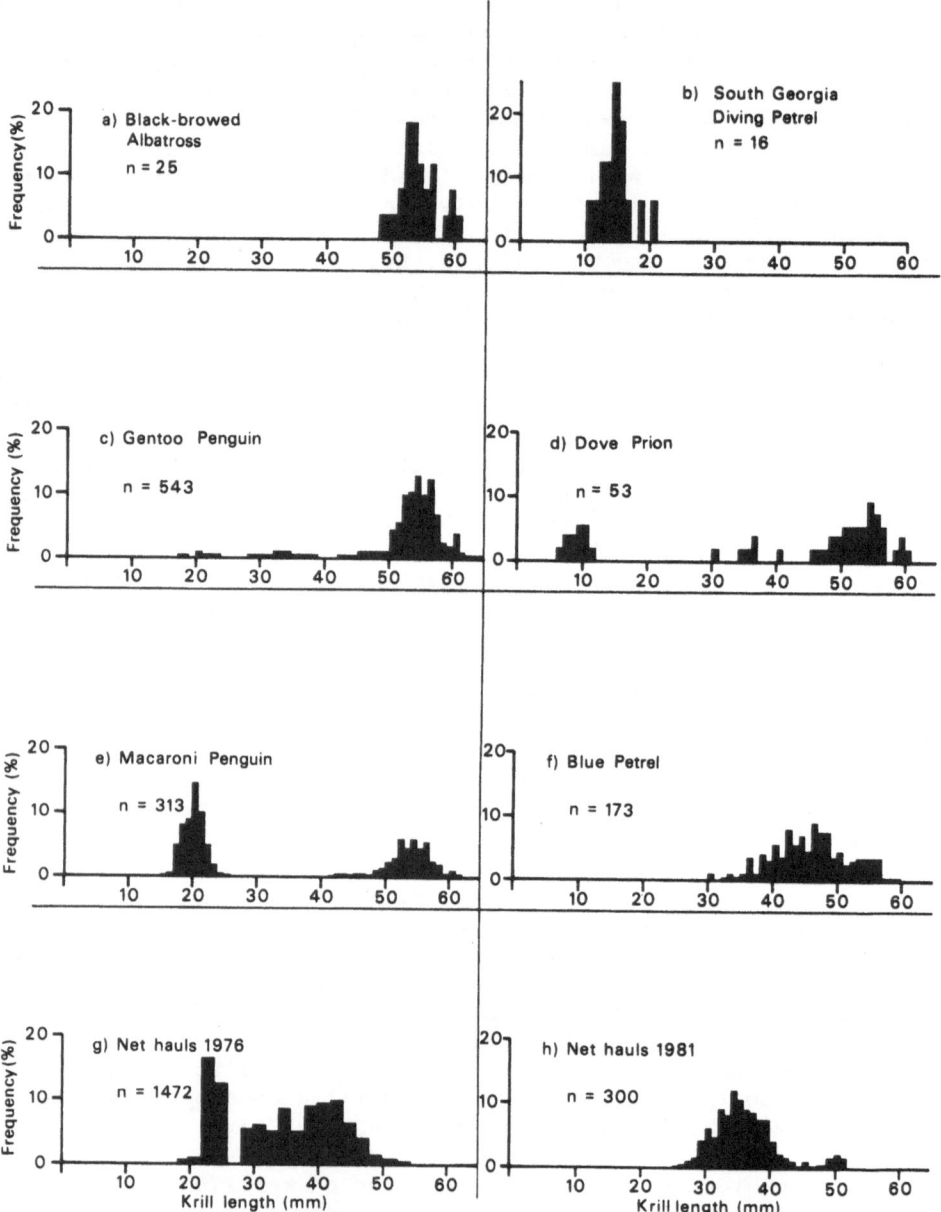

Fig. 7. Length-frequency distribution of krill at South Georgia taken by seabirds (a-f) and net hauls (g: Jazdzewski et al., 1978; h: BIOMASS, 1982).

times more than males. Thus, from the standpoint of krill
population dynamics, seabirds are largely avoiding the least
important segment, the juveniles, and concentrate on sexually
mature adults. The impact of predation would be least if directed
towards pre-spawning individuals in their final year of life and
greatest if on mature females just prior to their first spawning.
Unfortunately, present uncertainties over the ageing of krill
(Ettershank, in press) mean that the longevity and age structure of
natural krill populations are unknown and so the necessary detailed
comparisons could not be made, even if the data from the seabird
prey were fully adequate.

Apart from penguins (and to a lesser extent diving petrels)
krill eating seabirds are virtually restricted to feeding at the
sea surface, employing several techniques to do so (Ainley, 1977;
Croxall and Prince, 1980a). Krill, like many marine zooplankton,
migrate towards the surface at night and are presumably most easily
caught by predators then. Penguins are able to exploit a considerable portion of the water column and, in theory, have fewer constraints on when they feed. However, data for Chinstrap Penguins
(Lishman and Croxall, 1983) indicate that 40% of dives are
shallower than 10 m (and 90% less than 40 m) and activity budgets
and times of attendance at the breeding colony suggest that they
may feed mainly at night (Lishman, in press a). This would be
similar to the feeding patterns of Antarctic Fur Seals where 75% of
dives are at night and most of these are shallower than 30 m, in
contrast to the consistently deeper daytime dives, which, however,
do not exceed 105 m (Croxall et al., in press c; Kooyman et al., in
press). Although both Fur Seals and penguins on foraging trips
while rearing offspring make many dives, they need to average a
catching rate of one krill every 2 s per dive (Fur Seals) and one
krill per 6 s per dive (Chinstrap Penguins) to meet their energy
requirements. In the absence of information on their methods of
catching krill (and especially when in swarms) it is difficult to
interpret these results. However, Fur Seals spend lengthy periods
resting between feeding bouts (Kooyman et al., in press) and we
have no indication that they are usually limited by their ability
to find and catch krill in summer (Doidge et al., 1984).

Fish. The composition of the fish diet of South Georgia seabirds
is poorly known. White-chinned Petrels, the main consumers,
probably usually take lantern fish Myctophidae and Grey-headed
Albatrosses are calculated to take about 1 million sub-adult
Southern Lampreys Geotria australis (Prince, 1980a); in neither
case is there any estimate of the stocks nor are they commercially
fished. Species of Antarctic cod Nototheniidae are the main fish
prey of Gentoo Penguins, Blue-eyed Shags and albatrosses. The
benthic-demersal Notothenia rossii was for many years the main
commercially fished species but the annual catch has fallen from
c. 400,000 tonnes in 1969-70 to 45,000 in 1979-80, when the total

stock (after natural predation) was estimated at 117,000 tonnes (Burchett and Ricketts, 1984). Seabirds probably take c. 11,000 tonnes but as Gentoo Penguins and Blue-eyed Shags almost exclusively catch immature fish (which mature in the inshore kelp beds) the impact on the population and competition with fisheries are probably insignificant. The pelagic ice fish Champsocephalus gunnari is now the principal fishery target and few seabirds are recorded to take it, though adult fish are important in the diet of male Fur Seals (North et al., 1983). Only for the Fur Seal and the Blue-eyed Shag are detailed quantitative data, based on analysis of otoliths to determine species, size and age of prey, becoming available for South Georgia. In general, however, we cannot properly evaluate the quantitative relationship between the seabirds and their fish prey - a situation not dissimilar to that in much better studied systems elsewhere (Straty and Haight, 1979).

Squid. Quantitative data on the squid taken by all South Georgia albatrosses and several petrels are available from studies of the keratinous beaks which accumulate in predator stomachs and from whole squid in regurgitations by adults (Clarke and Prince, 1981; Clarke et al., 1981; Thomas, 1982). These data were reviewed, in comparison with the squid diets of the seals and Sperm Whales in the region, by Croxall et al. (in press b) and on a broader scale by Clarke (1983). Although several genera of squid are taken by many predators, the main squid prey of each tend to be rather different. Given the differences in size of the seabirds (and also in the timing and likely provenance of the prey) this is perhaps not surprising and there is much scope for detailed studies. Although over 40 species of squid (and an octopus in shag samples) have occurred in seabird diets at South Georgia, only three of the rather less important species are at all common in net haul samples. This is because albatrosses in particular take relatively large squid (200-1000 g) and these are adept at avoiding nets (Clarke, 1977). Consequently it is impossible to compare predator consumption with any realistic estimate of stocks; indeed, although biased to large squid, the predator samples are presently the best index of the relative abundance of Southern Ocean squid!

How surface-feeding seabirds catch large and potentially swift-moving prey such as fish and squid is still a puzzle. Much circumstantial evidence suggests that they do so at night when squid and fish follow to the surface the upward movement of krill swarms; this is strongly supported by the at-sea activity budget data showing that albatrosses spend much more time on the water during the (short) nights than in the daytime (Prince and Francis, in press). Pursuit-diving species may reach considerable depths e.g half the dives of King Penguins were deeper than 50 m (and one exceeded 240 m), a very different pattern from that of the krill-eating penguins. Also, only about 10% of these dives needed to result in prey capture to meet the energy requirements of the adult and its chick (Kooyman et al., 1982).

CONCLUSIONS

The few quantitative studies of the impact of seabird communities on marine resources have each approached the problem with rather different data bases and methods, so instructive detailed comparison is difficult. However, because all depend to a greater or lesser extent on similar generalised equations, broad-scale comparisons are likely to be fairly realistic.

By any criteria the South Georgia seabird community is both diverse in species and exceptionally rich in individuals and biomass. Thus the annual consumption by 53 million seabirds in the shelf waters of the eastern Bering Sea (including some of the most important Pribilof colonies) was estimated as 55-109,000 tonnes (Hunt et al., 1981) and that of the Peruvian guano colonies in the 1960's as c. 2.8 million tonnes (Schaefer, 1970). Consumption by 36 million seabirds in South Georgia shelf waters alone might reach c. 5 million tonnes annually and is unlikely to be less than 3 million tonnes, over 2 million of which will be krill.

Schneider and Hunt (1982) compared various estimates of trophic transfer by seabirds. They reported values of 0.09 g (of food) m^{-2} yr^{-1} for the eastern North Pacific (Sanger, 1972), 0.7 - 2.5 g m^{-2} yr^{-1} for the southeast Bering Sea, 1.9 g m^{-2} yr^{-1} for a small area of the North Sea (Furness, 1978), 8 g m^{-2} yr^{-1} for the upwelling system of the Oregon Coast (Wiens and Scott, 1975) and 11 - 45 g m^{-2} yr^{-1} for Peru (Schaefer, 1970). Obviously such values are very sensitive to the area being considered. Thus, if the South Georgia consumption is averaged over ranges of 1000 km and 100 km the results are 2.5 g m^{-2} yr^{-1} and 250 g m^{-2} yr^{-1}, respectively. Possibly more realistic figures, using the chick-rearing foraging ranges, are given for the principal species of each foraging type in Table 2. This reinforces the importance of Macaroni Penguins, especially compared with any petrel, but also shows the potential local importance of species with restricted foraging ranges like Gentoo Penguins (and, by analogy, Blue-eyed Shag). However, it is more appropriate to view penguins in terms of the volume of sea they can exploit. Thus a krill-eating penguin with a mean diving depth of 10 m (like our data for Chinstrap Penguins) has access to as much water within a range of 100 km as an albatross, restricted to the surface 1 m, over 316 km. Some idea of the density of krill required to support these seabirds can be gauged from the daily requirements at the peak period in what we estimate to be the zone of main feeding activity (Table 2). To satisfy Macaroni Penguin demands 2.5 g (i.e. about three individuals) of krill must be available per cubic metre of sea in its main foraging area at this time.

Table 2. Rates of transfer of krill to some seabirds breeding at South Georgia.

Species	Macaroni Penguin	Black-browed Albatross	Dove Prion	Gentoo Penguin
Foraging type	Offshore	Offshore	Uniform	Inshore
Foraging range (km)	100	430	240	30
Krill consumption:				
Total (t x 10^3 yr^{-1})	3870	8	1245	61
Annual (g m^{-2} yr^{-1})	123	0.04	7.4	19.4
Daily: outer 10 km (g m^{-2} d^{-1})	2.5	0.0003	0.0005	–
: inner 10 km	–	–	0.02	0.13

The role of seabirds in terms of energy transfer should also be evaluated with respect to both direct and indirect consumption. There is not space to do this here, but it should be noted that although South Georgia seabirds feeding on crustaceans function as top predators they are also mainly only primary (sometimes secondary) carnivores, as krill and other crustaceans feed directly on phytoplankton (krill also sometimes on microzooplankton). The main prey of many Southern Ocean squid and fish is also crustaceans (including krill) but small fish and squid are not infrequently eaten by large ones so seabirds function as both secondary and tertiary (but very rarely as higher order) carnivores.

Improving assessments of the energy requirements and predatory role of seabirds is not difficult as every part of any model would benefit from better empirical data. The most pressing needs, as almost every writer has emphasised, are for accurate information on a) population size and structure (in particular to define the size and habits of the non-breeding population), b) diet, distribution and behavior in winter and c) the energy costs of flight and swimming especially in relation to known activity budgets. The use of radio-isotopes (see Ellis, this volume), especially in conjunction with simple activity recorders (e.g. Prince and Francis, in press), offers considerable prospects in addressing c), but a) and b) require major long term investments of time and effort. Without this, however, our knowledge of seabird energetics will quickly outstrip our ability to use it constructively in addressing both the problems of predator-prey interactions and also the larger questions concerning the management and conservation of marine ecosystems.

ACKNOWLEDGEMENTS

We are particularly grateful to the many colleagues, past and present, who have worked with us at Bird Island. We thank M. O'Connell for assistance in data compilation, A. Sylvester for preparing the illustrations, E. Bailey for typing the manuscripts and especially Dr J. Mitton for much help in producing the final version. I. Everson kindly commented on the text.

REFERENCES

Adams, N.J., and Brown, C.R., 1984, Metabolic rates of sub-Antarctic Procellariiformes: a comparative study, Comp. Biochem. Physiol., 77A: 169.

Ainley, D.G., 1977, Feeding methods of seabirds: a comparison of polar and tropical communities, pp 669-685, in "Adaptations within Antarctic Ecosystems", G.A. Llano, ed., Smithsonian Institution, Washington D.C.

Ainley, D.G., and Sanger, G.A., 1979, Trophic relations of seabirds in the northeastern Pacific Ocean and Bering Sea, pp 95-122, in "Conservation of Marine Birds of Northern North America", J.C. Bartonek and D.N. Nettleship, eds, U.S. Dept. Interior, Wildlife Res. Report 11, Washington D.C.

Baudinette, R.V., and Schmidt-Nielsen, K., 1974, Energy cost of gliding in Herring Gulls, Nature (Lond.), 248: 83.

Beck, J.R., 1969, Food, moult and age of first breeding in the Cape Pigeon Daption capensis Linnaeus. Bull. Br. Antarct. Surv., 21: 33.

Beck, J.R., 1970, Breeding seasons and moult in some smaller Antarctic petrels, pp. 542-550, in "Antarctic Ecology", Vol. 1, M.W. Holdgate, ed., Academic Press, New York and London.

Beck, J.R., and Brown, D.W., 1971, The breeding biology of the Black-bellied Storm Petrel Fregetta tropica, Ibis, 113: 73.

Beck, J.R., and Brown, D.W., 1972, The biology of Wilson's Storm Petrel, Oceanites oceanicus (Kuhl), at Signy Island, South Orkney Islands. Sci. Rep. Br. Antarct. Surv., 69: 54 pp.

Bernstein, N.P., 1982, Activity patterns, energetics and parental investment of the Antarctic Blue-eyed Shag (Phalacrocorax atriceps bransfieldensis), PhD thesis, Univ. of Minnesota.

BIOMASS, 1982, Post-Fibex data interpretation workshop, Biomass Rep. Ser., 20: 38pp.

Brooke, M. de L., 1978, Weights and measurements of the Manx Shearwater, Puffinus puffinus, J. Zool., Lond., 186: 359.

Burchett, M.S., and Ricketts, C., 1984, The population dynamics of Notothenia rossii Fischer (1885) from South Georgia (Antarctica), Polar Biol., 3: 35.

Clark, B.D., and Bemis, W., 1979, Kinematics of swimming of penguins at the Detroit Zoo, J. Zool. Lond., 188: 411.

Clarke, A., in press, Lipid content and composition of Antarctic

Krill <u>Euphausia superba</u> Dana, <u>J. Crustacean Biol.</u>

Clarke, A., and Prince, P.A., 1980, Chemical composition and calorific value of food fed to mollymawk chicks at Bird Island, South Georgia, <u>Ibis</u>, 122: 488.

Clarke, M.R., 1977, Beaks, nets and numbers. <u>Symp. Zool. Soc. Lond.</u>, 38: 89.

Clarke, M.R., 1983, Cephalopod biomass - estimtion from predation, <u>Mem. Nat. Mus. Vict.</u>, 44: 95.

Clarke, M.R., and Prince, P.A., 1981, Cephalopod remains in regurgitations of Black-browed Albatross <u>Diomedea melanophris</u> and Grey-headed Albatross <u>D. chrysostoma</u> at South Georgia, <u>Bull. Br. Antarct. Surv.</u>, 54: 1.

Clarke, M.R., Croxall, J.P., and Prince, P.A., 1981, Cephalopod remains in regurgitations of the Wandering Albatross at South Georgia, <u>Bull. Br. Antarct. Surv.</u>, 54: 9.

Croxall, J.P., 1979, Distribution and population changes in the Wandering Albatross <u>Diomedea exulans</u> L. at South Georgia, <u>Ardea</u>, 67: 15.

Croxall, J.P., 1982, Energy costs of incubation and moult in petrels and penguins, <u>J. Anim. Ecol.</u>, 51: 177.

Croxall, J.P., 1984, Seabirds, pp. 533-619, <u>in</u> "Antarctic Ecology", Vol. 2, R.M. Laws, ed., Academic Press, London and New York.

Croxall, J.P., and Hunter, I., 1982, The distribution and abundance of burrowing seabirds (Procellariiformes) at Bird Island, South Georgia. II. South Georgia Diving Petrel <u>Pelecanoides georgicus</u>, <u>Bull. Br. Antarct. Surv.</u>, 56: 69.

Croxall, J.P., and Pilcher, M.N., 1984, Characteristics of krill <u>Euphausia superba</u> eaten by Antarctic Fur Seals <u>Arctocephalus gazella</u> at South Georgia, <u>Bull. Br. Antarct. Surv.</u>, 63: 117.

Croxall, J.P., and Prince, P.A., 1979, Antarctic seabird and seal monitoring studies, <u>Polar Rec.</u>, 19: 573.

Croxall, J.P., and Prince, P.A., 1980a, Food, feeding ecology and ecological segregation of seabirds at South Georgia, <u>Biol. J. Linn. Soc.</u>, 14: 103.

Croxall, J.P., and Prince, P.A., 1980b, The food of Gentoo Penguins <u>Pygoscelis papua</u> and Macaroni Penguins <u>Eudyptes chrysolophus</u> at South Georgia, <u>Ibis</u>, 122: 245.

Croxall, J.P., and Prince, P.A., 1982a, A preliminary assessment of the impact of seabirds on marine resources at South Georgia, <u>Com. Nat. Fr. Recherch. Antarct.</u>, 51: 501.

Croxall, J.P., and Prince, P.A., 1982b, Calorific contents of squid (Mollusca: Cephalopoda), <u>Bull. Br. Antarct. Surv.</u>, 55: 27.

Croxall, J.P., and Ricketts, C., 1983, Energy costs of incubation in the Wandering Albatross <u>Diomedea exulans</u>, <u>Ibis</u>, 125: 33.

Croxall, J.P., Prince, P.A., Hunter, I., McInnes, S., and Copestake, P.G., in press a, The seabirds of the Antarctic Peninsula, islands of the Scotia Sea and Antarctic Continent between 80°W and 20°W: their status and conseravation, <u>in</u> "The Status and Conservation of Seabirds", J.P. Croxall, P.G.H. Evans and R.W. Schreiber, eds, ICBP, Cambridge.

Croxall, J.P., Prince, P.A., and Ricketts, C., in press b,
Relationships between prey life-cycles and the timing nature
and extent of seal and seabird predation in the Scotia Sea, in
"Nutrient Cycling and Food Webs in the Antarctic: the
Proceedings of the Fourth SCAR Symposium on Antarctic Biology",
W.R. Siegfried, P. Condy and R.M. Laws, eds., Springer-Verlag,
Berlin.
Croxall, J.P., Everson, I., Kooyman, G.L., Ricketts, C., and Davis,
R.W., in press c, Fur seal diving behaviour in relation to
vertical distribution of krill, Polar Biol.
Davis, R.W., Kooyman, G.L., and Croxall, J.P., 1983, Water flux and
estimated metabolism of free-ranging Gentoo and Macaroni
Penguins at South Georgia, Polar Biol., 2: 41.
Doidge, D.W., Croxall, J.P., and Ricketts, C., 1984, Growth rates
of Antarctic Fur Seal pups at South Georgia, J. Zool. Lond.
Ettershank, G., in press, A new approach to the problem of
longevity in the Antarctic Krill Euphausia superba Dana
(Euphausiacea: Crustacea), Aust. J. Mar. Freshwater Biol.
Ford, R.G., Wiens, J.A., Heinemann, D., and Hunt, G.L., 1982,
Modelling the sensitivity of colonially breeding marine birds
to oil spills: guillemot and kittiwake populations on the
Pribilof Islands, Bering Sea, J. Appl. Ecol., 19: 1.
Furness, R.W., 1978, Energy requirements of seabird communities: a
bioenergetics model, J. Anim. Ecol., 47: 39.
Furness, R.W., and Cooper, J.C., 1982, Interactions between
breeding seabird and pelagic fish populations in the Southern
Benguela region, Mar. Ecol. Progr. Ser., 8: 243.
Hunt, G.L., Burgeson, B., and Sanger, G.A., 1981, Feeding ecology
of seabirds of the eastern Bering Sea, pp 629-247 in "The
Eastern Bering Sea Shelf: Oceanography and Resources, Vol 2",
D.W. Hood and J.A. Calder, eds, Univ. Washington Press,
Seattle.
Hunter, S., 1983, The food and feeding ecology of the giant petrels
Macronectes halli and M. giganteus at South Georgia, J. Zool.
Lond., 200: 521.
Hunter, S., 1984, Breeding biology and population dynamics of giant
petrels Macronectes spp. at South Georgia, J. Zool. Lond.
Hunter, I., Croxall, J.P., and Prince, P.A., 1982, The distribution
and abundance of burrowing seabirds (Procellariiformes) at Bird
Island, South Georgia. I. Introduction and methods, Bull. Br.
Antarct. Surv., 56: 49.
Jazdzewski, K., Dzik, J., Porebski, J., Rakusa-Suszczewski, S.,
Witek, Z, and Wolnomiejski, N., 1978, Biological and population
studies on krill near South Shetland Islands, Scotia Sea and
South Georgia in the summer 1976, Pol. Arch. Hydrobiol., 25:
607.
Kendeigh, S.C., Dol'nik, V.R., and Gavrilov, V.M., 1977, Avian
energetics, pp. 127-204, in "Granivorous Birds in Ecosystems",
J. Pinowski and S.C. Kendeigh, eds, Cambridge University Press,
Cambridge.

Kooyman, G.L., 1975, The physiology of diving in penguins, pp.
115-137, in "The Biology of Penguins", B. Stonehouse, ed.,
Macmillan, London.

Kooyman, G.L., Davis, R.W., Croxall, J.P., and Costa, D.P., 1982,
Diving depths and energy requirements of King Penguins,
Science, 217: 726.

Kooyman, G.L., Davis, R.W., and Croxall, J.P., in press, Diving
behaviour of the Antarctic Fur Seal Arctocephalus gazella, in
"Fur Seals: Maternal Strategies on Land and at Sea", R.L. Gentry
and G.L. Kooyman, eds, Princeton University Press, Princeton.

Lishman, G.S., in press a, The food and feeding ecology of Adelie
and Chinstrap Penguins at Signy Island, South Orkney Islands,
J. Zool. Lond.

Lishman, G.S., in press b, The comparative breeding biology of
Adelie and Chinstrap Penguins at Signy Island, South Orkney
Islands, Ibis.

Lishman, G.S., and Croxall, J.P., 1983, Diving depths of the
Chinstrap Penguin Pygoscelis antarctica, Bull. Br. Antarct.
Surv., 61: 21.

Marr, J.W.S., 1962, The natural history and geography of the
Antarctic Krill (Euphausia superba Dana), Discovery Rep., 32:
33.

North, A.W., Croxall, J.P., and Doidge, D.W., 1983, Fish prey of
the Antarctic Fur Seal Arctocephalus gazella at South Georgia:
methods and results of otolith examination, Bull. Br. Antarct.
Surv., 61: 27.

Payne, M.R., and Prince, P.A., 1979, Identification and breeding
biology of the diving petrels Pelecanoides georgicus and P.
urinatrix exsul at South Georgia, N.Z. J. Zool., 6: 299.

Pennycuick, C.J., 1982, The flight of petrels and albatrosses
(Procellariiformes), observed in South Georgia and its
vicinity, Phil. Trans. Roy. Soc. B., 300: 75.

Pennycuick, C.J., Croxall, J.P., and Prince, P.A., 1984, Scaling of
foraging radius and growth rate in petrels and albatrosses,
Ornis Scand.

Pinder, R., 1966, The Cape Pigeon Daption capensis Linnaeus at
Signy Island, South Orkney Islands, Bull. Br. Antarct. Surv.,
8: 19.

Prince, P.A., 1980a, The food and feeding ecology of Grey-headed
Albatross Diomedea chrysostoma and Black-browed Albatross D.
melanophris, Ibis, 122: 476.

Prince, P.A., 1980b, The food and feeding ecology of Blue Petrel
Halobaena caerulea and Dove Prion Pachyptila desolata, J.
Zool. Lond., 190: 59.

Prince, P.A., in press, Population and energetic aspects of the
relationships between Black-browed and Grey-headed Albatrosses
and the Southern Ocean marine environment, in "Nutrient Cycling
and Food Webs in the Antarctic, the Proceedings of the Fourth
SCAR Symposium on Antarctic Biology", W.R. Siegfried, P. Condy
and R.M. Laws, eds, Springer-Verlag, Berlin.

Prince, P.A., and Croxall, J.P., 1983, Birds of South Georgia: new records and re-evaluations of status, Bull. Br. Antarct. Surv., 61: 27.

Prince, P.A., and Francis, M.R., in press, Activity budgets of foraging Grey-headed Albatrosses, Condor.

Prince, P.A., and Payne, M.R., 1979, Current status of birds at South Georgia, Bull. Br. Antarct. Surv., 48: 103.

Prince, P.A., and Ricketts, C., 1981, Relationships between food supply and growth in albatrosses: an interspecies fostering experiment, Ornis Scand. 12: 207.

Prince, P.A., Ricketts, C. and Thomas, G., 1981, Weight loss in incubating albatrosses and its implications for their energy and food requirements, Condor, 83: 238.

Roby, D.D., and Ricklefs, R.E., 1983, Some aspects of the breeding biology of the diving petrels Pelecanoides georgicus and P. urinatrix exsul at Bird Island, South Georgia. Bull. Br. Antarct. Surv., 59: 29.

Sanger, G.A., 1972, Preliminary standing stock and biomass estimates of seabirds in the subarctic Pacific Region, pp. 589-611, in "Biological Oceanography of the northern North Pacific", A.Y. Takenouti, ed., Idemitsu Shoten, Tokyo.

Sanger, G.A., 1983, Diets and food web relationships of seabirds in the Gulf of Alaska and adjacent marine regions. Final Report to Outer Continental Shelf Environmental Assessment Programme, 91 pp.

Schaefer, M.B., 1970, Birds and anchovies in the Peru current - dynamic interactions, Trans. Am. Fish. Soc., 99: 461.

Schneider, D., and Hunt, G.L., 1982, Carbon flux to seabirds in waters with different mixing regimes in the southeastern Bering Sea, Mar. Biol., 67: 337.

Shaw, P., 1984, Factors affecting the breeding performance of the Antarctic Blue-eyed Shag Phalacrocorax atriceps bransfieldensis, PhD thesis, Univ. of Durham.

Straty, R.R., and Haight, R.E., 1979, Interactions among marine birds and commercial fisheries in the eastern Bering Sea, pp. 201-219, in "Conservation of Marine Birds of Northern North America", J.C. Bartonek and D.N. Nettleship, eds, U.S. Dept. Interior, Wildlife Res. Report 11, Washington DC.

Thomas, G., 1982, The food and feeding ecology of the Light-mantled Sooty Albatross at South Georgia, Emu, 82: 92.

Thomas, G., Croxall, J.P., and Prince, P.A., 1983, Breeding biology of the Light-mantled Sooty Albatross at South Georgia, J. Zool. Lond., 199: 123.

Tickell, W.L.N., 1967, Movements of Black-browed and Grey-headed Albatrosses in the South Atlantic, Emu, 66: 357.

Wiens, J.A., and Scott, J.M., 1975, Model estimation of energy flow in Oregon coastal seabird populations, Condor, 77: 439.

Appendix 1. Body weight, breeding success and duration of breeding cycle event in South Georgia seabirds.

Species	Weight (g)		Breeding population (pairs x 10³)		Mortality (fraction lost d⁻¹)		Pre-lay attendance[a] (d)	Incubation period (d)	o share (%)	Brooding (d)	Total chick rearing (d)
	♂	♀	SG	NWSG	Eggs	Chicks					
King Penguin	13450	13760	343		0.0083	0.00048	17	54	58[b]	40	313
Gentoo Penguin	5890	5890	100	10	0.0029	0.0048	R	36	50	31	81
Chinstrap Penguin	4000	3600	4	0.01	0.0096	0.00	25	37	59[b]	18	54
Macaroni Penguin	5000	4600	5000	3000	0.018	0.0035	20	34	49[b]	18	59
Wandering Albatross	10580	9020	4.7	1.6	0.0043	0.00064	27	78	55	32	278
Black-browed Albatross	3922	3694	60	30	0.0057	0.0040	16	68	51	23	116
Grey-headed Albatross	3751	3624	60	38	0.0046	0.0020	26	72	48	23	141
Light-mantled Sooty Albatross	2840	2840	8	0.25	0.0064	0.0034	21	67	53	21	141
Southern Giant Petrel	5035	3798	5	1	0.0025	0.0015	R	61	54	18	118
Northern Giant Petrel	4902	3724	3.5	1.3	0.0023	0.00064	60	62	55	22	112
Cape Pigeon	442	407	20	0.5	0.011	0.0073	40	45	53	12	48
Snow Petrel	340	286	3	0.25	0.013	0.0033	55	43	54	12	51
Dove Prion	168	168	22000	1000	0.013	0.0062	52	45	49	4	51
Blue Petrel	193	193	70	15	0.013	0.0069	52	46	50	2	49
White-chinned Petrel	1368	1368	2000	35	0.013	0.00365	51	54	51	7	92
Wilson's Storm Petrel	34	34	600	30	0.014	0.0049	45	41	(50)	3	55
Black-bellied Storm Petrel	53	53	10	0.5	0.015	0.0040	45	40	(50)	3	67
Common Diving Petrel	133	135	3800	100	0.011	0.0050	21	54	(50)	10	54
South Georgia Diving Petrel	107	107	2000	5	0.013	0.0060	21	46	(50)	10	45
Blue-eyed Shag	2867	2473	7.5	0.2	0.008	0.0052	R	27	(50)	14	65
Antarctic Tern	151	151	2.6	0.05	0.013	0.0090	R	40	(50)	8	45

a R: resident; b Actual shifts entered; Parentheses indicate estimated values

Species	Diet[a] (% by weight)						Foraging type[b]	Chick meal size (g)	Flight speed (ms⁻¹)	Zig-zag factor	% trip travelling	Meals chick⁻¹ adult⁻¹ d⁻¹	Foraging range (km)
	Krill	Copepod	Amphipod	Other crustacea	Fish	Squid							
King Penguin	68				(30)	(70)	O	(2500)	1.9	1.0	93	0.18	424
Gentoo Penguin	100				32		I	860	1.9	1.0	42	0.50	35
Chinstrap Penguin	98						U	500	1.9	1.0	83	0.50	68
Macaroni Penguin	98				2		O	690	1.9	1.0	75	0.60	123
Wandering Albatross	(10)				(10)	(80)	O	750	14.0	1.50	66	0.18	1478
Black-browed Albatross	38		1		39	21	O	570	12.4	1.54	51	0.42	428
Grey-headed Albatross	15		1		35	49	O	600	12.4	1.47	76	0.45	615
Light-mantled Sooty Albatross	37			4	12	47	I	700	(11.0)	(1.5)	(75)	0.34	709
Southern Giant Petrel	12				1	2	I	310	13.1	1.55	(50)	0.97	189
Northern Giant Petrel	15				2	6	I	250	13.1	1.55	(50)	1.01	181
Cape Pigeon	85				15		O	75	11.0	1.33	(50)	0.50	357
Snow Petrel	80				10	10	U	(40)	(11.0)	(1.33)	(50)	0.50	357
Dove Prion	58	31	8		2	1	U	25	10.0	1.33	(50)	0.66	244
Blue Petrel	82	4	5		8	1	O	35	(11.0)	(1.33)	(50)	0.40	670
White-chinned Petrel	27		1		24	47	O	150	14.0	1.49	(75)	0.25	1218
Wilson's Storm Petrel	(45)	(40)	(10)		(5)		U	(8)	8.8	1.34	(50)	0.75	189
Black-bellied Storm Petrel	(45)	(40)	(10)		(5)		U	(8)	(8.8)	(1.34)	(50)	0.75	189
Common Diving Petrel	15	68	17				U	40	(9.0)	(1.0)	(50)	0.80	243
South Georgia Diving Petrel	76	20	4				U	40	(9.0)	(1.0)	(50)	0.90	216
Blue-eyed Shag				10	70	20	I	80	13.3	1.0	20	10	12
Antarctic Tern	(15)	(15)		(20)	(50)		I	(2.5)	(6.0)	(1.0)	(33)	40	2

a Excluding carrion; b I: inshore; O: offshore; U: uniform; Parentheses indicate estimated values

CONTRIBUTORS

R.A. Ackerman, Department of Zoology, Iowa State University, Ames, Iowa.

J.P. Croxall, British Antarctic Survey, Natural Environment Research Council, Madingley Road, Cambridge, U.K.

R.W. Davis, Physiological Research Laboratory, Scripps Institution of Oceanography, La Jolla, California.

W.R. Dawson, Museum of Zoology and Division of Biological Sciences, The University of Michigan, Ann Arbor, Michigan.

G. Dewasmes, Laboratoire de Thermorégulation, CNRS, Université Claude Bernard, Lyon, France.

H.I. Ellis, Department of Biology, University of San Diego, Alcala Park, San Diego.

G.S. Grant, North Carolina State Museum of Natural History, Raleigh, North Carolina.

C.R. Grau, Department of Avian Sciences, University of California, Davis, California.

G.L. Kooyman, Physiological Research Laboratory, Scripps Institution of Oceanography, La Jolla, California.

Y. Le Maho, Laboratoire de Physiologie Respiratoire, CNRS, Associé Université Louis Pasteur, Strasbourg, France.

S. Lustick, Department of Zoology, The Ohio State University, Columbus, Ohio.

C.V. Paganelli, Department of Physiology, Schools of Medicine and Dentistry, State University of New York, Buffalo, New York.

T.N. Pettit, Department of Physiology, John A. Burns School of Medicine, and P.B.R.C. Kewalo Marine Laboratory, University of Hawaii, Honolulu, Hawaii.

P.A. Prince, British Antarctic Survey, Natural Environment Research Council, Madingley Road, Cambridge, U.K.

H. Rahn, Department of Physiology, Schools of Medicine and Dentistry, State University of New York, Buffalo, New York.

C. Ricketts, British Antarctic Survey, Natural Environment Research Council, Madingley Road, Cambridge, U.K.

R.C. Seagrave, Department of Chemical Engineering, Iowa State University, Ames, Iowa.

T.R. Simons, Wildlife Science Group, College of Forest Resources, University of Washington, Seattle, Washington.

G.C. Whittow, Department of Physiology, John A. Burns School of Medicine, and P.B.R.C. Kewalo Marine Laboratory, University of Hawaii, Honolulu, Hawaii.

J.A. Wiens, Department of Biology, University of New Mexico, Albuquerque, New Mexico.